# Select Topics in Signal Analysis

# Select Topics in Signal Analysis

**Harish Parthasarathy**
Professor
Electronics & Communication Engineering
Netaji Subhas Institute of Technology (NSIT)
New Delhi, Delhi-110078

CRC Press
Taylor & Francis Group
Boca Raton London New York

CRC Press is an imprint of the
Taylor & Francis Group, an **informa** business

Manakin
PRESS

First published 2023
by CRC Press
4 Park Square, Milton Park, Abingdon, Oxon, OX14 4RN

and by CRC Press
6000 Broken Sound Parkway NW, Suite 300, Boca Raton, FL 33487-2742



*CRC Press is an imprint of Informa UK Limited*



*British Library Cataloguing-in-Publication Data*
A catalogue record for this book is available from the British Library

*Library of Congress Cataloging-in-Publication Data*
A catalog record has been requested

ISBN: 9781032384153 (hbk)
ISBN: 9781032384177 (pbk)
ISBN: 9781003344957 (ebk)

DOI: 10.1201/9781003344957

Typeset in Arial, MinionPro, Symbol, CalisMTBol, TimesNewRoman, RupeeForadian, Wingdings
by Manakin Press, Delhi

Manakin
PRESS

# Preface

This book developed from a course given by the author to undergraduate and post-graduate students at the NSUT over a period of two semesters. The first chapter on matrix theory discusses in reasonable depth the theory of Lie algebras leading upto Cartan's classification theory. It also discusses some basic elements of functional analysis and operator theory in infinite dimensional Banach and Hilbert spaces. The theory of Lie groups and Lie algebras play a fundamental role in robotic dynamics with 3-D links as well as in image processing. We discuss here in full details the construction of Cartan subalgebras of a semisimple Lie algebra leading to the root space decomposition and consequent classification of the simple Lie algebras. In functional analysis, we discuss some issues related to graphs of operators, the adjoint map, the basic theorems of Banach space theory and some aspects of operators in Hilbert spaces. These topics are important mainly because of the applications they find in quantum mechanics and modern infinite dimensional control theory. The second chapter is about basic probability theory and the topics discussed in this book find applications to stochastic filtering theory for differential equations driven by white Gaussian noise. The third chapter is on antenna theory with a focus on modern quantum antenna theory in which we discuss quantum electrodynamics within a cavity resonator in which the Maxwell photon field interacts with the Dirac field of electrons and positrons to radiate out a quantum electromagnetic field into space whose statistics in any state of the cavity like say a joint coherent state of the electrons, positrons and photons can be calculated and controlled by insertion of classical photon probes using a laser and classical current probes using a wire. The last chapter of Part I is based upon many interesting and useful discussion I have had over the past few months with Dr.Steven A.Langford, a geophysicist who has made a variety of experiments on the refractive index of materials and liquids. Dr.Langford has suggested to me many ways of analysing and modeling the refractive index of materials using classical and quantum statistics always stressing on the fact that the problem of modeling the refractive index is intimately connected with resonance with dark matter, the cosmic expansion of the universe, the cosmic microwave background radiation and the phenomenon of reflection and refraction of De-Broglie matter waves at a boundary between two matrix valued potentials which take into account the spin of the quantum mechanical particle. My debt to Dr.Langford is too heavy to be repaid in words. The material in Part II deals primarily with applications of large deviation theory to problems in engineering and physics. The main idea is that when weak noise enters into a system, then we can use large deviation theory to approximately compute the probability of the system state deviating from the noiseless system state by an amount greater than a prescribed threshold in terms of the system control parameters. Then, we adpat the control parameters so that this probability is minimized. We have adapted this principle to a variety of problems including fluid dynamics, plasma physics, hydrodynamics in curved space-time and even to quantum filtering theory. It should be noted that large deviations cannot be directly applied to quantum mechanics because the evolving observables at different times do not commute and hence by the Heisenberg uncertainty principle they do not have a joint probability distribution. However, if we filter the state using Belavkin's non-demoltion measurement based quantum filtering theory, then these filtered states form an Abelian family and hence have a joint probability density to which large deviation theory can be applied. The final chapter in this book deals with quantum field theoretic aspects of the human eye modeled as an antenna. I take this opportunity to thank my colleague Prof.Rajveer Yaduvanshi for posing the problem of performing a quantum mechanical analysis of the eye cavity as a quantum antenna and for some nice suggestions. We can formulate large deviations problems even this context, for example, by including classical external random gauge fields and currents interacting with the quantum Dirac and Maxwell fields and then asking the question, what is the rate function for the quantum average of a field observable like the quantum current field or quantum electromagnetic field (or even the space-time moments for the far field radiation pattern generated by the quantum currents within the cavity body and on its surface) in a given state of electrons, positrons and photons and hence deduce the probability for these quantum averages to deviate from the corresponding noiseless (ie, zero classical noise) values by an amount greater than a given threshold. This would enable us to design control algorithms by introducing non-random classical control fields and current sources so that this deviation probability is a minimum.

Author

# Table of Contents

# Chapter 1

# Matrix Theory

## 1.1 Course Outline

[0] Prerequisites of linear algebra.

Fields, rings, vector spaces over a field, modules over a ring, algebras, ideals in a ring and an algebra, bases for vector spaces, linear transformations in a vector space, basis for a vector space, matrix of a linear transformation relative to a basis, inner product spaces, unitary, Hermitian and normal operators in an inner product space, spectral theorem for normal operators.

[1] Quotient of a vector space by another space.

[2] Simultaneous triangulability of commuting matrices relative to an onb.

[3] Simultaneous diagonability of commuting normal matrices relative to an onb.

[4] Tensor products of vector spaces.

[5] Variational principles for calculating the eigenvalues of a Hermitian matrix.

[6] Positive definite matrices.

[7] The basic decomposition theorems of matrix theory.

[a] Row reduced Echelon form.

[b] Spectral theorem for normal matrices.

[c] Polar decomposition.

[d] Singular value decomposition.

[e] QR decomposition based on the Gram-Schmidt orthonormalization process.

[f] LDU decomposition of positive definite matrices.

[8] Applications of matrix theory to finite state quantum systems.

[a] Schrodinger and Heisenberg evolution in finite dimensional Hilbert spaces.

[b] Different kinds of unitary gates for finite state quantum computation: CNOT, Swap, Fredkin, Toffoli, phase gate, Quantum Fourier transform gate.

[c] Perturbation theory for quantum systems in finite dimensional state space.

[8] Lie groups and Lie algebras.

[a] Group action on a differentiable manifold.

[b] Differential of a group action on a manifold.

[c] Differentiable and analytic structures on a group: The notion of a Lie group based on differentiability of the composition and inversion operation.

[d] The Lie algebra of vector fields on a differentiable manifold

[e] The notion of Lie algebra of a Lie group determined by left invariance of vector fields.

[f1] Identification of the Lie algebra of a Lie group with its tangent space at the identity.

[f2] The exponential map from the Lie algebra into a Lie group.

[f3] Examples when the exponential map is not surjective: $O(3)$

[f4] The one-one correspondence between Lie algebra elements and one parameter subgroups of a Lie group.

[g] Structure constants associated with a basis for a Lie algebra.

[h] Examples of linear Lie groups and their Lie algebras: The classical Lie groups

[i] Examples of nonlinear Lie groups taken from dynamical systems.

[j] Homotopy and covering groups of a Lie group.

[k] The universal enveloping algebra of a Lie algebra.

[l] The invariant bilinear form on a Lie algebra.

[m] Solvable and semisimple Lie algebras.

[n] Simple Lie algebras, root space decomposition and Cartan's classification theory based on the theory of roots.

[9] Representation theory for Lie groups and Lie algebras.

[a] Completely reducible representations.

[b] irreducible representations.

[c] Classification of finite dimensional irreducible representations of a semisimple Lie algebra based on dominant integral weights: The Cartan-Weyl-HarishChandra theory.

[d] The character of a representation.

[e] Representations of compact groups:Schur Lemmas and the Peter-Weyl theory.

[f] Characters of a compact semisimple group: Weyl's character formula.

[g] Applications of representation theory to image processing problems.

[10] Algebraic varieties in a multivariable polynomial ring.

[11] Prime ideals and maximal ideals of a ring.

[12] Grothendieck's generalization of an algebraic variety to schemes on a ring consisting of prime ideals.

[13] Some remarks on algebraic groups.

[a] Examples.

[b] Flag variety.

[c] The Grassmannian variety and the Schubert variety.

[d1] The relationship between irreducible representations of an algebraic group and Flag varieties.

[d2] The relationship between Schubert cells in a Grassmannian variety and the Schubert variety constructed from the Bruhat decomposition of a semisimple algebraic group associated with a parabolic subgroup.

[e] Plucker's coordinates on a Grassmannian variety.

[f] Quadratic relations satisfied by Plucker's coordinates on the Grassmannian variety.

## 1.2 Prerequisites of linear algebra

### 1.2.1 Fields

[1] A **field** is a set of elements $\mathbb{F}$ that has two binary operations called addition and multiplication, a zero element 0 and a unit element 1 such that under addition, $\mathbb{F}$ is an Abelian group with identitity 0, under multiplication, $\mathbb{F} - \{0\}$ (ie, $\mathbb{F}$ without the zero element) is an Abelian group with identity 1 and multiplication distributes over addition. Examples of fields are

[a] $\mathbb{R}, \mathbb{C}, \mathbb{Q}$, the set of real, complex and rational numbers.

[b] $\{0, 1, ..., p-1\}$ where $p$ is any prime, with addition and multiplication defined as for real numbers but modulo $p$.

[c] If $x$ is any indeterminate, and $\mathbb{F}$ is any field, then $\mathbb{Q}_F[x]$, the set of all rational functions in $x$ over $\mathbb{F}$, ie, the set of all ratios of polynomials in $x$ with the coefficients of the polynomials coming from $\mathbb{F}$ and with the denominator polynomial never being the zero polynomial is a field.

### 1.2.2 Vector spaces

[2] A **vectorspace** over a field $\mathbb{F}$ is a set of elements endowed with a binary operation called vector addition and denoted by $+$, a scalar multiplication, ie, a map $\mathbb{F} \times V \rightarrow V$ denoted by a dot and a zero element called the zero vector so that under vector addition, $V$ forms an Abelian group with identity being the zero vector $\mathbf{0}$, scalar multiplication distributes over vector addition, if $c_1, c_2 \in \mathbb{F}, \mathbf{x} \in V$, then

$$c_1.(c_2.\mathbf{x}) = (c_1 c_2).\mathbf{x},$$

$$(c_1 + c_2).\mathbf{x} = c_1.\mathbf{x} + c_2.\mathbf{x}$$

if 0 denotes the zero element of $\mathbb{F}$, then

$$0.\mathbf{x} = \mathbf{0},$$

and if 1 denotes the unit element in $\mathbb{F}$, then

$$1.\mathbf{x} = \mathbf{x}$$

if $-\mathbf{x}$ denotes the additive inverse of $\mathbf{x}$ in the Abelian group $V$ (under vector addition), then,

$$-1.\mathbf{x} = -\mathbf{x}$$

Examples of vector spaces:

[a] $V = \mathbb{F}^n$, the set of all column vectors (or equivalently row vectors), ie, ordered $n$-tuples with the $n$ entries in $\mathbb{F}$ and vector addition and scalar multiplication being defined component wise, ie,

$$[c_1, ..., c_n]^T + [d_1, ..., d_n]^T = [c_1 + d_1, ..., c_n + d_n]^T, c_i, d_i \in \mathbb{F}, i = 1, 2, ..., n$$

$$c.[c_1, ..., c_n]^T = [cc_1, ..., cc_n]^T, c, c_1, ..., c_n \in \mathbb{F}$$

In other words, vector addition and scalar multiplication in $\mathbb{F}^n$ are induced by addition and multiplication in $\mathbb{F}$.

### 1.2.3 Rings

3. A **ring** is a set $R$ with two binary operations $+, .$ and a zero element 0 such that under $+$, $R$ is an Abelian group and under ., $R$ is a semigroup, (A semigroup has all the properties of the group like closure under composition, associativity under composition except that there may not be any identity and inverse of an element) and such that . distributes over $+$. Clearly, a field is a special case of a ring. Some examples of a rings that is not fields are

[a] $\mathbb{F}[x]$, the set of all polynomials in an indeterminate $x$ with coefficients taken from the field $\mathbb{F}$. This is called the ring of polynomials over the field $\mathbb{F}$. Note that a polynomial will not have a multiplicative inverse unless it is a constant polynomial. $\mathbb{F}[x]$ is an example of a commutative/Abelian ring.

[b] $R = M_n(\mathbb{F})$, the space of $n \times n$ matrices with elements from $\mathbb{F}$. The . and $+$ operations in this ring are respectively matrix multiplication and matrix addition. Note that this ring is non-Abelian.

[c] As a generalization of $[a]$, we define $R$ to be the commutative ring of all functions $f$ from a set $X$ to a field $\mathbb{F}$. The . operation is simply the ordinary multiplication of functions with the multiplication operation induced by that in $\mathbb{F}$ and the addition operation is likewise the ordinary addition operation of functions induced by addition in $\mathbb{F}$.

[d] As a generalization of $[a]$ and $[c]$ we consider the ring $R$ of all polynomials in an indeterminate $x$ with coefficients coming from $M_n(\mathbb{F})$, ie, all $n \times n$ matrix polynomials.

### 1.2.4 Simultaneous triangulability and diagonability

1.Quotient of a vector space.

2. Simultaneous triangulability of a family of commuting matrices relative to an o.n.b.

3. Simultaneous diagonability of a family of commuting normal matrices relative to an o.n.b.

2[a]. First we prove the triangulability of a matrix over the complex field relative to an o.n.b.

Let $V$ be a complex inner product space of dimension $n$ and let $T : V \to V$ be a linear operator. $T$ has at least on eigenvalue: Say

$$Te_1 = c_1 e_1, c_1 \in \mathbb{C}, \| e_1 \| = 1$$

Define

$$W_1 = N(T - c_1 I)$$

Then

$$W_1 \neq \{0\}$$

Consider the quotient vector space

$$V/W_1`, dim(V/W_1) < n$$

$W_1$ is $T$-invariant and hence

$$T_1 : V/W_1 \to V/W_1, T(x + W_1) = T(x) + W_1$$

is a well defined linear operator. We define an inner product on $V/W_1$ as follows: For $x, y \in V$,

$$< x + W_1, y + W_2 > = < P^{\perp}x, P^{\perp}y >$$

where $P$ is the orthogonal projection of $V$ onto $W$ and $P^{\perp} = I - P$. It is easy to see that this is a well defined inner product on $V/W_1$. Now by the induction hypothesis on the vector space dimension, $T_1$ can be brought into upper triangular form relative to an onb say $\{f_k + W_1 : k = 1, 2, ..., m\}$ of $V/W_1$. We note that

$$P^{\perp}f_k + W_1 = f_k + W_1$$

and hence we can replace the above onb for $V/W_1$ by $\{P^{\perp}f_k + W_1 : k = 1, 2, ..., m\}$. Then by construction, $P^{\perp}f_k, k = 1, 2, ..., m$ is an orthonormal set in $V$ and further the span of this set is orthogonal to $W_1$. Thus, if we choose an orthogonormal basis $\{e_1, ..., e_{n-m}\}$ for $W_1$, then it is immediate that $T$ is upper-triangular w.r.t the onb

$$\mathcal{B} = \{P^{\perp}f_1, ..., P^{\perp}f_m, e_1, ..., e_{n-m}\}$$

for $V$.

2[b]. Now we prove simultaneous triangulability of a family of commuting operators on a finite dimensional complex inner product space w.r.t an onb. Let $dim_{\mathbb{C}}V = n < \infty$ and let $\mathcal{F}$ be a family of commuting operators in this space. Choose one operator say $T$ from this family and choose an eigenvalue $c_1$ with a corresponding normalized eigenvector $e_1$ for this operator:

$$Te_1 = c_1 e_1, \| e_1 \| = 1$$

Then

$$W_1 = N(T - c_1)$$

is a non-zero subspace of $V$ and is invariant under every element of $\mathcal{F}$ since all these elements commute with $T$. Thus for each $S \in \mathcal{F}$, we can define a linear operator

$$S_1 : V/W_1 \rightarrow V/W_1$$

by

$$S_1(x + W_1) = S(x) + W_1, x \in V$$

and it is immediate that $\{S_1 : S \in \mathcal{F}\}$ is a commuting family of operators in $V/W_1$ and hence by the induction hypothesis on the vector space dimension, this family can be simultaneously triangulated w.r.t an onb for $V/W_1$. Let $\{f_k + W_1 : k = 1, 2, ..., m\}$ be such an onb. Then again by the induction hypothesis, $\mathcal{F}$ restricted to $W_1$ can also be simultaneously triangulated w.r.t an onb for it. Denote this onb by $\{e_1, ..., e_{n-m}\}$. It is immediate then that $\mathcal{F}$ is simultaneously triangulable relative to the onb $\{P^{\perp}f_1, ..., P^{\perp}f_m, e_1, ..., e_{n-m}\}$ for $V$. The proof is complete.

Reference: This problem was taken from

Rajendra Bhatia, "Matrix Analysis", Springer.

[3] Simultaneous diagonability of a family of commuting normal operators w.r.t an onb in a finite dimensional complex inner product space.

Let $\mathcal{F}$ be a commuting family of normal operators in $V$ with $dim_{\mathbb{C}} V = n < \infty$. We first choose an element $T \in \mathcal{F}$ and a vector $e_1$ of unit norm such that

$$Te_1 = c_1 e_1$$

for some $c_1 \in \mathbb{C}$. Again define

$$W_1 = N(T - c_1)$$

Choose any $S \in \mathcal{F}$. Then, $W_1$ is $S$-invariant since $[T, S] = 0$. Now define $S_1 : V/W_1 \rightarrow V/W_1$ in the usual way. We claim that $S_1$ is normal. Indeed, let $u, v \in V$. Then,

$$< u + W_1 | S_1^* S_1 | v + W_1 >=< S(u) + W_1 | S(v) + W_1 >=< P^{\perp} Su | P^{\perp} Sv >$$

where $P$ is the orthogonal projection onto $W_1$. Then, $P$ is expressible as a function of $T$ using the spectral theorem for normal operators, ie, $P = f(T)$. From that it is immediate that $S$ commutes with $P$ and hence also with $P^{\perp} = I - P$. Since $P^{\perp}$ is Hermitian, $S^*$ also commutes with $P^{\perp}$. Then, using this and the normality of $S$,

$$< P^{\perp} Su | P^{\perp} Sv >=< u | S^* P^{\{perp} Sv >=$$

$$< u | P^{\perp} S^* S | v >=< u | P^{\perp} SS^* | v >$$

$$=< [S^*]_1 (u + W_1) | [S^*]_1 (v + W_1 >$$

On the other hand, we claim that

$$[S^*]_1 = S_1^*$$

for the following reason:

$$< u + W_1|[S^*]_1|v + W_1 >=< u + W_1|S^*(v) + W_1 >=< P^{\perp}u|P^{\perp}S^*v >$$
$$=< u|P^{\perp}S^*v >$$

while on the other hand,

$$< u + W_1|S_1^*|v + W_1 >=< S_1(u + W_1)|v + W_1 >=< S(u) + W_1|v + W_1 >$$
$$=< P^{\perp}S(u)|P^{\perp}v >=< Su|P^{\perp}v >=< u|S^*P^{\perp}v >$$
$$=< u|P^{\perp}S^*v >$$

proving the claim. Thus, we get

$$< u+W_1|S_1^*S_1|v+W_1 >=< S_1^*(u+W_1)|S_1^*(v+W_1) >=< u+W_1|S_1S_1^*|v+W_1 >$$

and therefore

$$S_1^*S_1 = S_1S_1^*$$

ie, $S_1$ is normal for all $S \in \mathcal{F}$. Moreover, it is clear that $\{S_1 : S \in \mathcal{F}\}$ is a commuting family since if $S, L \in \mathcal{F}$, then

$$S_1L_1(u+W_1) = S_1(L(u)+W_1) = SL(u)+W_1 = LS(u)+W_1 = L_1S_1(u+W_1), u \in V$$

Thus, $\{S_1 : S \in \mathcal{F}\}$ is a commuting family of normal operators in $V/W_1$ and by the induction hypothesis (on the dimension of the vector space), it follows that this family is simultaneously diagonable w.r.t an onb say $\{f_k + W_1 : k = 1, 2, ..., m\}$ for $V/W_1$. Likewise by the same induction hypothesis, $\mathcal{F}$ is simultaneously diagonable on $W_1$ w.r.t. an onb say $\{e_1, .., e_{n-m}\}$ for $W_1$. In other words, for any $S \in \mathcal{F}$, we have

$$S(f_k) - c_k(S)f_k \in W_1, k = 1, 2, ..., m$$

Thus,

$$S(P^{\perp}f_k) - c_k(S)P^{\perp}f_k = P^{\perp}(Sf_k - c_k(S)f_k) = 0$$

This proves that $\{P^{\perp}f_k, k = 1, 2, ..., m, e_1, ..., e_{n-m}\}$ is an onb for $V$ that simultaneously diagonalizes $\mathcal{F}$.

### 1.2.5 Tensor products of vectors and matrices

[a] Tensor products of vector spaces.

[b] The symmetric tensor product and permanents.

[c] The antisymmetric tensor product and determinants.

[d] Tensor product of matrices.

[e] Eigenvalues and eigenvectors of the tensor product, antisymmetric tensor product and symmetric tensor product of matrices.

### 1.2.6 The minimax variational principle for calculating all the eigenvalues of a Hermitian matrix

Let $X$ be an $n \times n$ Hermitian matrix and let $\{v_1, ..., v_n\}$ be an orthonormal eigenbasis for $X$ with corresponding eigenvalues $\{c_1, ..., c_n\}$ so that $c_1 \geq c_2 \geq ... \geq c_n$. Let $M$ be any $k$ dimensional subspace of $\mathbb{C}^n$. Consider the $n-k+1$ dimensional subspace $W_k = span\{v_k, ..., v_n\}$ of $\mathbb{C}^n$. It is clear that

$$dim(M \cap W_k) = dimM + dimW_k - dim(M + W_k) \geq k + (n-k+1) - n = 1$$

and hence we can choose

$$v \in M \cap W_k, \| v \| = 1$$

and therefore,

$$< v|X|v > \leq c_k$$

It follows that

$$inf_{x \in M} < x|X|x > \leq c_k$$

Thus,

$$sup_{dimN=k} inf_{x \in N} < x|X|x > \leq c_k$$

Now choosing

$$N = span\{v_1, ..., v_k\}$$

it follows that $dimN = k$ and $v_k \in N$,

$$< v_k|X|v_k >= c_k$$

Thus we have proved

$$sup_{dimN=k} inf\{x \in N\} < x|X|x >= c_k$$

This is the first minimax theorem for computing the eigenvalues of a Hermitian matrix. Now we prove the second minimax theorem.

### 1.2.7 The basic decomposition theorems of matrix theory

[a] LDU decomposition of a positive definite matrix. Let $R$ be a positive definite matrix. It can be looked upon as the correlation matrix of a random vector $\mathbf{X}$. We write

$$\mathbf{X} = [X_1, ..., X_n]^T$$

and then Gram-Schmidt orthonormalize this vector relative to the standard inner product on $L^2(\Omega, \mathcal{F}, P)$: $< u, v >= \mathbb{E}(uv)$. Then we get an orthonormal set $\{e_1, e_2, ..., e_n\}$ of random variables in $L^2(\Omega, \mathcal{F}, P)$ where

$$X_1 = a(1,1)e_1, X_2 = a(2,1)e_1 + a(2,2)e_2, ..., X_k = a(k,1)e_1 + ... + a(k,k)e_k,$$
$$k = 1, 2, ..., n$$

writing

$$\mathbf{e} = [e_1, ..., e_n]^T$$

we can express this as

$$\mathbf{X} = \mathbf{Le}$$

where $\mathbf{L}$ is the lower triangular matrix $((a(i,j)))$. Then taking correlations on both sides gives us

$$\mathbf{R} = \mathbb{E}(\mathbf{XX}^T) = \mathbf{L}\mathbb{E}(\mathbf{ee}^T)\mathbf{L}^T = \mathbf{LL}^T$$

which is known as the *LU* decomposition of $\mathbf{R}$. Defining $\mathbf{B}$ as $\mathbf{L}$ except that its diagonal entries are all 1 and defining the diagonal matrix

$$\mathbf{D} = diag[a(i,i)^2, i = 1, 2, ..., n]$$

we get

$$\mathbf{R} = \mathbf{BDB}^T$$

with $B$ lower triangular and having ones on its diagonal. This is called the *LDU* decomposition of the matrix $\mathbf{R}$. This decomposition is important in linear prediction theory of stochastic processes as its derivation suggests.

There is another way to derive the LDU decomposition of a positive definite matrix $R$. Using the spectral decomposition theorem, we can write $R = XX^T$ where $X$ is a square matrix. Then using the QR decomposition, write

$$X^T = QY$$

where $Y$ is upper triangular and $Q$ is orthogonal. This can be achieved for example, by applying the Gram-Schmidt decomposition to the columns of $X^T$ or equivalently to the rows of $X$. In other words

$$Q = [e_1, e_2, ..., e_n], e_a^T e_b = \delta_{ab}$$

$$X^T = [x_1, x_2, ..., x_n]$$

Then,

$$x_1 = y(1,1)e_1, x_2 = y(1,2)e_1 + y(2,2)e_2, ..., x_k = y(1,k)e_1 + ...y(k,k)e_k,$$
$$k = 1, 2, ..., n$$

Then,

$$R = XX^T = Y^T Q^T QY = Y^T Y$$

and $Y^T$ is lower triangular.

[b] The $QR$ decomposition of any rectangular matrix. First assume that $A \in \mathbb{C}^{m \times n}$. Let

$$A = [a_1, ..., a_n], a_j \in \mathbb{C}^{m \times 1}, j = 1, 2, ..., n$$

Choose a permutation matrix $P$ so that if

$$B = AP$$

so that the columns of $B$ are obtained by permuting the columns of $A$, then

$$B = [b_1, ..., b_n]$$

and $b_1, ..., b_r$ are linearly independent while $b_{r+1}, ..., b_n$ can be expressed as linear combinations of $b_1, ..., b_r$. We can thus write

$$[b_{r+1}, ..., b_n] = [b_1, .., b_r]C$$

for some appropriate $r \times n$ matrix $C$. Then

$$B = [B_1|B_1C] = B_1[I_r|C], B_1 = [b_1, ..., b_r] \in \mathbb{C}^{m\times r}$$

Apply the Gram-Schmidt process to the columns of $B_1$ which are linearly independent and then obtain

$$B_1 = [e_1, ..., e_r]R_1$$

where $e_1, ..., e_r$ are orthonormal and $R_1$ is an $r \times r$ upper triangular matrix. Extend the set $\{e_1, ..., e_r\}$ to an orthonormal basis $\{e_1, ..., e_m\}$ for $\mathbb{C}^m$ and thus define the unitary matrix

$$Q = [e_1, ..., e_m] \in \mathbb{C}^{m\times m}$$

Define

$$R_2 = \begin{pmatrix} R_1 \\ 0 \end{pmatrix} \in \mathbb{C}^{m\times r}$$

Then, we can write

$$B_1 = QR_2$$

and

$$B = [QR_2|QR_2C] = QR_2[I|C]$$

and hence

$$A = BP^{-1} = QR_2[I|C]P^{-1}$$

In the special case when the columns of $A$ are all linearly independent, this formula reduces to

$$A = QR_2$$

An application of the QR decomposition to linear least squares problems.

Let $A$ be an $m \times n$ matrix of full column rank $n$. Then, by Gram-Schmidt orthornormalization of its columns, we get

$$A = Q_1R_1$$

where $Q_1$ is $m \times n$ with orthonormal columns and $R_1$ is $n \times n$ upper triangular and non-singular, ie, with nonzero columns on its diagonals. Extend $Q_1$ by appending more orthonormal columns to $Q$ so that $Q$ becomes a $m \times m$ orthogonal matrix and define

$$R = \begin{pmatrix} R_1 \\ 0 \end{pmatrix} \in \mathbb{C}^{m \times n}$$

Then

$$A = QR$$

Now consider the problem of minimizing

$$E(\theta) = \| x - A\theta \|^2, \theta \in \mathbb{R}^n$$

We have

$$\| x - QR\theta \|^2 = \| y - R\theta \|^2, y = Q^T x$$

Write

$$\theta = \begin{pmatrix} \theta_1 \\ \theta_2 \end{pmatrix}$$

where

$$\theta_1 \in \mathbb{R}^n, \theta \in \mathbb{R}^{m-n}$$

Then, clearly,

$$E(\theta) = \| y_1 - R_1\theta_1 \|^2 + \| y_2 \|^2$$

where

$$y = \begin{pmatrix} y_1 \\ y_2 \end{pmatrix}, y_1 \in \mathbb{R}^n, y_2 \in \mathbb{R}^{m-n}$$

Thus, $E(\theta)$ is minimized when and only when

$$\theta_1 = R_1^{-1} y_1$$

In other words, the set of all $\theta'$ that minimize $E(\theta)$ are of the form

$$\theta = \begin{pmatrix} R_1^{-1} y_1 \\ \theta_2 \end{pmatrix}, \theta_2 \in \mathbb{R}^{m-n}$$

### 1.2.8 A computational problem in Lie group theory

Let $G$ be a Lie group and $\mathfrak{g}$ is Lie algebra. For $X \in \mathfrak{g}$, define the left invariant vector field on $G$ by the formula

$$v_X f(g) = \frac{d}{dt} f(g.exp(tX))|_{t=0}$$

where $f : G \to \mathbb{C}$ is a differentiable function.

Now do the following:

[a] Take $G = SL(2, \mathbb{R})$ so that $\mathfrak{g}$ has a basis $H, X, Y$ where

$$H = \begin{pmatrix} 1 & 0 \\ 0 & -1 \end{pmatrix},$$

$$X = \begin{pmatrix} 0 & 1 \\ 0 & 0 \end{pmatrix}, Y = \begin{pmatrix} 0 & 0 \\ 1 & 0 \end{pmatrix}$$

Express $v_H, v_X, v_Y$ as linear first order differential operators in the coordinates $(t, x, y)$ where $g \in G$ is parametrized as

$$g = g(t, x, y) = exp(tH + xX + yY)$$

[b] Consider the Lie group $G = SO(3)$. Its Lie algebra is the real vector space spanned by all $3 \times 3$ skew symmetric real matrices. We choose the standard basis for this Lie algebra, namely

$$X_1 = \begin{pmatrix} 0 & 0 & 0 \\ 0 & 0 & -1 \\ 0 & 1 & 0 \end{pmatrix}$$

$$X_2 = \begin{pmatrix} 0 & 0 & 1 \\ 0 & 0 & 0 \\ -1 & 0 & 0 \end{pmatrix}$$

$$X_3 = \begin{pmatrix} 0 & -1 & 0 \\ 1 & 0 & 0 \\ 0 & 0 & 0 \end{pmatrix}$$

Prove that the element

$$R(n) = exp(n_1X_1 + n_2X_2 + n_3X_3), n_1, n_2, n_3 \in \mathbb{R}$$

defines a rotation around the axis

$$\hat{n} = (n_1, n_2, n_3)/\sqrt{n_1^2 + n_2^2 + n_3^2}$$

by an angle

$$\phi = \sqrt{n_1^2 + n_2^2 + n_3^2}$$

in the counterclockwise sense. Evaluate the vector fields $v_{X_k}, k = 1, 2, 3$ in terms of the Euler angles which parametrize $R \in SO(3)$ in terms of the Euler angles:

$$R = R_z(\phi)R_x(\theta)R_z(\psi) = R(\phi, \theta, \psi)$$

by expressing

$$\frac{d}{dt} f(R(\phi, \theta, \psi)exp(tX_k))|_{t=0}, k = 1, 2, 3$$

as a linear partial differential operators of the first order in $(\phi, \theta, \psi)$. Now writing

$$v_{X_k} = v_{k1}(\phi, \theta, \psi)\partial/\partial\phi + v_{k2}(\phi, \theta, \psi)\partial/\partial\theta + v_{k3}\partial/\partial\psi, k = 1, 2, 3$$

compute

$$f(\phi, \theta, \psi) = det((v_{ij}(\phi, \theta, \psi)))^{-1}$$

and prove using general arguments that

$$f(\phi, \theta, \psi)d\phi.d\theta, d\psi$$

is the Haar measure on $SO(3)$, ie, invariant under left and right translations.

### 1.2.9 The Primary Decomposition Theorem

Let $T$ be a linear operator in a finite dimensional complex vector space $V$. Prove the primary decomposition theorem:

$$V = \bigoplus_{k=1}^{r} W_k$$

where

$$W_k = N((T - c_k)^{m_k}), k = 1, 2, ..., r$$

with $c_1, ..., c_r$ being the distinct eigenvalues of $T$ and

$$p(t) = \Pi_{k=1}^{r}(t - c_k)^{m_k}$$

being the minimal polynomial of $T$. Prove that

$$W_k = \bigcup_{m \geq 1} N((T - c_k)^m)$$

Let $E_k$ denote the projection onto $W_k$ corresponding to this decomposition, ie,

$$R(E_k) = W_k, E_k E_j = 0, k \neq j, \sum_{k=1}^{r} E_k = I$$

Show that an operator $S$ in $V$ commutes with $T$ iff

$$E_k S E_m = 0 \forall k \neq m$$

iff

$$S = \sum_{k=1}^{m} E_k S E_k$$

Now suppose $ad(T)(S) = [T, S]$ leaves all the subspaces $W_k, k = 1, 2, ..., r$ invariant. Then, we claim that $S$ also shares this same property. Indeed, consider for some $k \neq l$ the operator $S_{kl} = E_k S E_l$. Then since the $E_j's$ commute with $T$ (in fact, they are polynomials in $T$), it follows that

$$E_k ad(T)(S) E_l = E_k [T, S] E_l = [T, S_{kl}]$$

and the lhs is zero by hypothesis. Thus, $S_{kl}$ commutes with $T$ and hence leaves all the subspaces $W_j, j = 1, 2, ..., r$ invariant. This means that $S_{kl} = 0$ and since $k \neq l$ are arbitary, it follows that $S = \sum_{j=1}^{r} E_j S E_j$ leaves all the subspaces $W_j, j = 1, 2, ..., r$ invariant. Now we can prove the following theorem: If for some positive integer $n$, $ad(T)^n(S) = 0$, then $S$ leaves all the subspaces $W_j, j = 1, 2, ..., r$ invariant. In fact, $0 = ad(T)^n(S) = [T, ad(T)^{n-1}(S)]$ implies that $ad(T)^{n-1}(S)$ leaves all the subspaces $\{W_j\}$ invariant and by induction, it follows that $ad(T)^j(S), j = n - 2, n - 3, ..., 1, 0$ also leave all these subspaces invariant. The proof of the theorem is complete.

Now we wish to use the primary decomposition theorem to prove that

$$V = [\bigcup_{m\geq 1} N(T^m)] \cap [\bigcap_{m\geq 1} R(T^m)]$$

Indeed, suppose $T$ has no zero eigenvalue. Then, the primary decomposition theorem for $T$ reads

$$V = \bigoplus_{k=1}^{r} N((T - c_k)_k^m)$$

where none of the $c_k's$ are zero. Now, if for some $k$

$$v \in N((T - c_k)^{m_k})$$

then

$$(T - c_k)^{m_k} v = 0$$

and this equation can be expressed as

$$(-1)^{m_k} c_k v + T f_k(T) v = 0$$

where $f(T)$ is a polynomial in $T$. Since $c_k \neq 0$, it follows that

$$v = T g_k(T) v$$

where $g_k$ is also a polynomial. By induction, it follows that

$$v \in R(T^m), m = 1, 2, ...$$

and hence

$$v \in \bigcap_{m\geq 1} R(T^m)$$

Since $T$ has no zero eigenvalue, we also have

$$N(T^m) = 0, m = 1, 2, ...$$

and the proof of the theorem for this case is complete. Now suppose that zero is an eigenvalue of $T$. Then we can take $c_1 = 0$ and since $c_k \neq 0, k > 1$, the result follows in the same way from the primary decomposition theorem by noting that

$$W_1 = N(T^{m_1}) = \bigcup_{m\geq 1} N(T^m),$$

and for $k > 1$,

$$W_k = N((T - c_k)^{m_k}) \subset \bigcap_{m\geq 1} R(T^m)$$

Remark: If $T^m v = 0$ for some positive integer $m$ and simultaneously $v \in R(T^r)$ for all $r \geq 1$, then we can write $v = T^r v_r, r \geq 1$ and hence $T^{m+r} v_r = 0$ for all $r \geq 1$ and hence $v_r \in N(T^{m_1})$ for all $r \geq 1$. Therefore, $v = T^{m_1+r} v_{m_1+r} = 0$. This proves that

$$[\bigcup_{m\geq 1} N(T^m)] \bigcap [\bigcap_{m\geq 1} R(T^m)] = \{0\}$$

## 1.3 Cartan Subalgebras of a Semisimple Lie algebra

Now we apply this result to the proof of the existence of a Cartan subalgebra of a semisimple Lie algebra leading thereby to the root space decomposition and hence Cartan's classification of all the simple Lie algebras.

Let $\mathfrak{g}$ be a any finite dimensional Lie algebra over the complex field. A Lie subalgebra $\mathfrak{h}$ of $\mathfrak{g}$ is said to be a Cartan algebra (or a Cartan subalgebra of $\mathfrak{g}$), if (a) $\mathfrak{h}$ is nilpotent and (b) $\mathfrak{h}$ is its own normalizer in $\mathfrak{g}$, ie $X \in \mathfrak{g}$ and $[X, \mathfrak{h}] \subset \mathfrak{h}$ together imply $X \in \mathfrak{h}$. We first show that a Cartan algebra is maximal nilpotent. Indeed, suppose that $\mathfrak{n}$ is a nilpotent Lie algebra that properly contains $\mathfrak{h}$. Consider the Lie algebra $\mathfrak{n}/\mathfrak{h}$ with its canonical Lie bracket. This is a non-trivial Lie algebra and it is nilpotent. Thus, its adjoint representation is a nil representation. By a basic theorem in nil-representations, it follows therefore that there exists a non-zero element $\xi = X + \mathfrak{h} \in \mathfrak{n}/\mathfrak{h}$ (ie $X \notin \mathfrak{h}$) such that $ad(\mathfrak{n}/\mathfrak{h})(\xi) = \mathfrak{h}$ (Note that $\mathfrak{h}$ is the zero element in $\mathfrak{n}/\mathfrak{h}$. It follows therefore that $[\mathfrak{n}, X] \subset \mathfrak{h}$ and this obviously implies $[X, \mathfrak{h}] \subset \mathfrak{h}$. This contradicts the fact that a Cartan algebra is its own normalizer.

Regular element: Now given a $X \in \mathfrak{g}$ define the characteristic polynomial of $ad(X)$:

$$p_X(t) = det(tI - ad(X)) = t^n + t^{n-1}c_1(X) + ... + tc_{n-1}(X) + c_n(X)$$

and set

$$l(X) = min(m : c_{n-m}(X) \neq 0)$$

Then $0 \leq l(X) \leq n$. Define

$$rk(\mathfrak{g}) = min(l(X) :\in \mathfrak{g})$$

Further, if $\mathfrak{h}$ is any subspace of $\mathfrak{g}$, define

$$\zeta_{\mathfrak{h}}(X) = det(ad(X)|_{\mathfrak{g}/\mathfrak{h}})$$

If $\mathfrak{h}$ is a subspace of $\mathfrak{g}$ such that $rk(\mathfrak{g}) = dim\mathfrak{h}$ and if $X$ a regular element of $\mathfrak{h}$ in $\mathfrak{g}$, then it is clear that $\zeta_{\mathfrak{h}}(X) \neq 0$ since in this case, $l(X) = rk(\mathfrak{g}) = dim\mathfrak{h}$.

Construction of Cartan algebras: Let $X$ be any regular element in $\mathfrak{g}$. Define

$$\mathfrak{h} = \mathfrak{h}_X = \bigcup_{m>0} N(ad(X)^m)$$

Then, we claim that $\mathfrak{h}$ is a Cartan algebra. First observe that $\mathfrak{h}$ is closed under linear combinations and under the Lie bracket operation. Indeed, $Y, Z \in \mathfrak{h}$ imply

$$ad(X)^m([Y, Z]) = \sum_{r=0}^{m} \binom{m}{r} [ad(X)^r(Y), ad(X)^{m-r}(Z)] = 0$$

for sufficiently large $m$. Here we have used that $ad(X)$ is a derivation on $\mathfrak{g}$ and $ad(X)^r(Y) = 0$ or $ad(X)^{m-r}(Y) = 0$ for all sufficiently large $m$ and all $r = 0, 1, ..., m$. Next we show that $\mathfrak{h}$ is a nilpotent algebra. Define $\mathfrak{h}'$ to be the set of all $Y \in \mathfrak{h}$ for which $\zeta_{\mathfrak{h}}(Y) \neq 0$. Clearly we have the direct sum decomposition

$$\mathfrak{g} = \bigcup_{m\geq 1} N(ad(X)^m) \oplus \bigcap_{m\geq 1} R(ad(X)^m) =$$

$$\mathfrak{h} \oplus q$$

Now $Y \in \mathfrak{h}$ implies that for some positive integer $m$, $ad(X)^m(Y) = 0$ and this implies that

$$(ad(adX))^m(ad(Y)) = 0$$

which implies that $ad(Y)$ leaves $\mathfrak{h} = \bigcup_{m\geq 1} N(ad(X)^m)$ invariant. Thus, $\mathfrak{h}$ is a Lie subalgebra of $\mathfrak{g}$. This also implies that $ad(Y)$ leaves $q = \bigcup_{m\geq 1} R(ad(X)^m)$ invariant as discussed above in the context of consequences of the primary decomposition theorem. $ad(X)$ is nilpotent on $\mathfrak{h}$ and non-singular on $q$. Thus, since $X$ has been assumed to be a regular element, it follows that $rk(\mathfrak{g}) = dim\mathfrak{h}$. Now suppose that $Y \in \mathfrak{h}'$. Then, $ad(Y)$ is non-singular on $\mathfrak{g}/\mathfrak{h}$ and hence also non-singular on $q$. Thus, $l(Y) \leq l(X) = rk(\mathfrak{g}) \leq l(Y)$. Thus, $l(Y) = rk(\mathfrak{g}) = dim\mathfrak{h}$. Since $ad(Y)$ is non-singular on $q$ and since $ad(Y)$ leaves $\mathfrak{h}$ invariant, it must necessarily follow that if $ad(Y)^m(Z) = 0$ for some positive integer $m$ and $Z \in \mathfrak{g}$, then $Z \in \mathfrak{h}$. In other words, we have shown that $\mathfrak{h}_Y \subset \mathfrak{h}$ and since

$$dim\mathfrak{h}_Y = l(Y) = dim\mathfrak{h}$$

it follows that

$$\mathfrak{h}_Y = \mathfrak{h}$$

It follows in particular that $ad(Y)$ is nilpotent on $\mathfrak{h}$ (since by definition, it is nilpotent on $\mathfrak{h}_Y$). Since $\mathfrak{h}'$ is a non-empty open subset of $\mathfrak{h}$ and since $Y$ is an arbitrary element of $\mathfrak{h}'$, it follows that $ad(Y)$ is nilpotent for every $Y \in \mathfrak{h}$. This, completes the proof that $\mathfrak{h} = \mathfrak{h}_X$ is a Cartan algebra for every regular $X \in \mathfrak{g}$.

Now let $\mathfrak{h}$ be a Cartan subalgebra of $\mathfrak{g}$. Let $\langle'$ denote the set of regular elements in $\mathfrak{h}$. Let $X \in \mathfrak{h}'$. We claim that $\mathfrak{h} = \mathfrak{h}_X$ where $\mathfrak{h}_X$ has been defined above as $\bigcup_{n\geq 1} N(ad(X)^n)$. Indeed, since $\mathfrak{h}$ is nilpotent, it follows that $ad(X)$ is nilpotent on $\mathfrak{h}$ and therefore $\mathfrak{h} \subset \mathfrak{h}_X$. On the other hand, we have seen above that $\mathfrak{h}_X$ is a Cartan algebra and hence maximally nilpotent. Since $\langle$ is also a Cartan algebra, it is also maximally nilpotent. Hence $\mathfrak{h} = \mathfrak{h}_X$. In other words, we have proved that every Cartan algebra is of the form $\mathfrak{h}_X$ for some regular element $X$ and in fact or argument shows that $\langle = \langle_X$ for any regular element $X$ of $\mathfrak{h}$.

Now let $\mathfrak{g}$ be any semisimple Lie algebra. Let $\mathfrak{h}$ be any Cartan algebra. Then, since $\mathfrak{h}$ is a nilpotent Lie algebra and hence also a solvable Lie algebra and $H \rightarrow ad(H)$ is a representation of $\mathfrak{h}$, in $\mathfrak{g}$, it follows that there is a basis for $\mathfrak{g}$ relative to which all the operators $ad(H), H \in \mathfrak{h}$ are upper-triangular (not necessarily upper triangular). Now, if $N\mathfrak{h}$ is such that $ad(N)$ is nilpotent on $\mathfrak{g}$, then its matrix relative to this basis will be strictly upper-triangular and hence

$$< N, H >= Tr(ad(N).ad(H)) = 0, H \in \mathfrak{h}$$

If we are able therefore to prove that $< .,. >$ is non-singular on $\mathfrak{h} \times \mathfrak{h}$, then it would follow from the above relation that $N = 0$, ie, $\mathfrak{h}$ does not have any nilpotent elements. To prove this claim, we choose $X \in \mathfrak{h}'$. Then, $\mathfrak{h} = \mathfrak{h}_X$. Let $X = S + N_1$ be the Jordan decomposition of $X$ into its semisimple and nilpotent components. Then, from the basic property of the Jordan decomposition, $[X, S] = 0$ and hence $S \in \mathfrak{h}_X = \mathfrak{h}$. further since $ad(X) = ad(S) + ad(N)$ and $ad(S)$ is semisimple while $ad(N)$ is nilpotent, it follows that $ad(X)$ and $ad(S)$ have the same characteristic polynomial and hence $S \in \mathfrak{h}'$. Hence, $\mathfrak{h} = \mathfrak{h}_S$. Since $ad(S)$ is semisimple, it then follows that for any $Y \in \mathfrak{h}$, we have $ad(S)(Y) = 0$, ie, $[S, Y] = 0$ and in fact, $\mathfrak{h} = N(ad(S))$. let $q = [S, \mathfrak{g}]$=R(ad(S)). Since $ad(S)$ is semisimple, we get

$$\mathfrak{g} = \mathfrak{h} \oplus q$$

Then,

$$< [Y, S], H >=< Y, [S, H] >= 0, H \in \mathfrak{h}, Y \in \mathfrak{g}$$

In other words,

$$< q, \mathfrak{h} >= 0$$

and hence, from the non-singularity of $< .,. >$ on $\mathfrak{g}$ (Cartan's criterion for semisimplicity of $\mathfrak{g}$), it must follow that $< .,. >$ is non-singular on $\mathfrak{h}$. This completes the proof of the claim.

Next, we show that if $\mathfrak{g}$ is semi-simple, then any Cartan algebra $\mathfrak{h}$ is maximal Abelian. Let $X \in \mathfrak{h}'$ and let $X = S + N$ be its Jordan decomposition. Then, $ad(X)(N) = [X, N] = 0$ and hence $N \in \mathfrak{h}_X = \mathfrak{h}$. Since $ad(N)$ is also nilpotent, by what we proved above, $N = 0$. Hence $X = S$ is semisimple. Thus $ad(X)^m(Y) = 0$ for some $Y, m > 0$ implies $[X, Y] = 0$. In other words, we have that $\mathfrak{h}'$ is Abelian and since $ad(X)$ is semisimple for all $X \in \mathfrak{h}'$, it follows that $ad(\mathfrak{h}')$ can be simultaneously diagonalized in $\mathfrak{g}$. By taking limits noting that $\mathfrak{h}'$ is a non-empty open subset of $\mathfrak{h}$ and hence dense in it it follows immediately that $ad(\mathfrak{h})$ is simultaneously diagonable and hence Abelian. From the faithfulness of the adjoint representation of a semisimple Lie algebra, it then follows that $\mathfrak{h}$ is also Abelian and since it is maximally nilpotent, it is also necessarily maximal Abelian. (Suppose $[\mathfrak{h}, Y] = 0$. Then, $ad(X)(Y) = 0$ for $X \in \mathfrak{h}'$ and hence $Y \in \mathfrak{h}_X = \mathfrak{h}$). Thus, we have proved the following fundamental result in the theory of semisimple Lie algebras:

Theorem: if $\mathfrak{g}$ is a semisimple Lie algebra then there exists a Lie subalgebra $\mathfrak{h}$ of $\mathfrak{g}$ such that $\mathfrak{h}$ is its own normalizer in $\mathfrak{g}$ and secondly, $\mathfrak{h}$ is Maximal Abelian, ie, $\mathfrak{h}$ is Abelian and $Y \in \mathfrak{g}, [\mathfrak{h}, Y] = 0$ implies $Y \in \mathfrak{h}$. $\mathfrak{h}$ is called a Cartan subalgebra of $\mathfrak{g}$.

Later on we shall prove that if the field is complex, $\mathfrak{g}$ has exactly one Cartan subalgebra upto conjugacy, ie, if $\mathfrak{h}_1, \mathfrak{h}_2$ are any two Cartan subalgebras of $\mathfrak{g}$, then there exists a $g \in G$ such that $\mathfrak{h}_2 = Ad(g).\mathfrak{h}_1$.

We have already shown assuming that $\mathfrak{g}$ is a semisimple Lie algebra, that if $\mathfrak{h}$ is a Cartan subalgebra, then there exists a regular element $X \in \mathfrak{g}$ such that $\mathfrak{h} = \mathfrak{h}_X$. Now we prove a slightly stronger version of this result namely: There exists a finite number $X_1, ..., X_r$ of regular elements in $\mathfrak{g}$ such that if $\mathfrak{h}$ is any Cartan algebra, then there exists an $i \in \{1, 2, ..., r\}$ and an $x \in G$ such that $\mathfrak{h} = \mathfrak{h}_i^x$ where $\mathfrak{h}_i = \mathfrak{h}_{X_i}$ and further that $\mathfrak{h}_i, i = 1, 2, ..., r$ are all mutually non-conjugate.

Note: Since we have made use of the Jordan decomposition on a semsimple Lie algebra in our construction of a Cartan subalgebra and proofs of some of these properties, we shall give a proof of this theorem in what follows.

Let $\mathfrak{g}$ be a semisimple Lie algebra and let $X \in \mathfrak{g}$. Write $ad(X) = T + U$ where $U$ is nilpotent on $\mathfrak{g}$ and $T$ is semisimple on $\mathfrak{g}$ and $[T, U] = 0$. This is the Jordan decomposition of a linear operator in a vector space. We claim that $T$ and $U$ are also derivations on $\mathfrak{g}$. In fact, we known from Chevalley's theory of replicas that $T$ and $U$ are replicas of $ad(X)$ and therefore since $ad(X)$ is a derivation, so are $T$ and $U$.

Remark: Let $V$ be a vector space and $\beta : V \times V \to V$ bilinear and let $D$ be a derivation on $V$ w.r.t $\beta$, ie,

$$D\beta(X, Y) = \beta(DX, Y) + \beta(X, DY), X, Y \in V$$

Then let $L$ be a replica of $D$. It is easy to see that $L$ is then also a derivation on $V$ w.r.t $\beta$. Indeed, consider the mapping $\eta : V \otimes V^* \otimes V^* \to B(V \times V, V)$ where $B(V \times V, V)$ is the space of $V$ valued bilinear forms on $V$. The map $\eta$ is defined by

$$\eta(X \otimes f_1 \otimes f_2)(U, V) = f_1(U) f_2(V) X$$

and then extending $\eta$ bilinearly w.r.t its first two arguments. It is then easy to see that $\eta$ is a vector space isomorphism. Note that if we choose a basis $\{e_1, ..., e_n\}$ for $V$, and if $\{e_1^*, ..., e_n^*\}$ denotes the corresponding dual basis, then

$$\beta(U, V) = \sum_{i,j} \beta(e_i, e_j) e_i^*(U) e_j^*(V)$$

$$= \eta(\sum_{i,j} \beta(e_i, e_j) \otimes e_i^* \otimes e_j^*)(U, V)$$

or equivalently,

$$\beta = \eta(\sum_{i,j} \beta(e_i, e_j) \otimes e_i^* \otimes e_j^*)$$

which proves that $\eta$ is surjective and therefore also injective since the dimension of $V \otimes V^* \otimes V^*$ equals $(dimV)^3$ which is also the dimension of the space of all $V$ valued bilinear forms on $V$. Writing

$$\beta_1(U,V) = \beta(DU,V), \beta_2(U,V) = \beta(U,DV)$$

We have

$$\beta_1(U,V) = \sum_{ij} \beta_1(e_i,e_j)e_i^*(U)e_j^*(V)$$

$$= \sum_{i,j} \beta(De_i,e_j)e_i^*(U)e_j^*(V)$$

$$= \sum_{i,j,k} [D]_{ki}\beta(e_k,e_j)e_i^*(U)e_j^*(V)$$

$$= \sum_{j,k} \beta(e_k,e_j)(D^T e_k^*)(U)e_j^*(V)$$

and therefore,

$$\beta_1 = \eta(\sum_{i,j} \beta(e_i,e_j) \otimes (D^T e_i^*) \otimes e_j)$$

Likewise,

$$\beta_2 = \eta(\sum_{i,j} \beta(e_i,e_j) \otimes e_i^* \otimes D^T e_j^*)$$

Further,

$$D(\beta(U,V)) = \sum_{i,j} (D\beta(e_i,e_j))e_i^*(U)e_j^*(V)$$

$$= \eta(\sum_{i,j} (D\beta(e_i,e_j)) \otimes e_i^* \otimes e_j^*)(U,V)$$

and hence the derivation property of $D$, namely

$$D(\beta(U,V)) = \beta_1(U,V) + \beta_2(U,V)$$

can equivalently be expressed as

$$\sum_{i,j} [(D\beta(e_i,e_j)) \otimes e_i^* \otimes e_j^* - \beta(e_i,e_j) \otimes D^T e_i^* \otimes e_j^* - \beta(e_i,e_j) \otimes e_i^* \otimes D^T e_j^*] = 0$$

which in the notation of replicas means that

$$D_{1,2}(\sum_{i,j,k} \beta(e_i,e_j) \otimes e_i^* \otimes e_j^*) = 0 - - - (1)$$

Now let $D$ be a derivation on $V$ w.r.t $\beta$ and let

$$D = T + U$$

be the Jordan decomposition of $D$. Then, it is known that $T$ and $U$ are also replicas of $D$ and in particular (1) implies that

$$T_{1,2}(\sum_{i,j,k} \beta(e_i, e_j) \otimes e_i^* \otimes e_j^*) = 0,$$

$$U_{1,2}(\sum_{i,j,k} \beta(e_i, e_j) \otimes e_i^* \otimes e_j^*) = 0$$

ie $T$ and $S$ are also derivations on $V$ w.r.t $\beta$. Now assuming that $\mathfrak{g}$ is a semisimple Lie algebra, we observe that $ad$ is faithful, ie, injective on $\mathfrak{g}$. Indeed, this follows from the fact that $ad(X) = 0$ for some non-zero $X \in \mathfrak{g}$ would imply that $[X, \mathfrak{g}] = 0$ with some non-zero $X$ and this would imply that $\mathfrak{g}$ has a non-zero centre which contradicts the semisimplicity of $\mathfrak{g}$. Thus, writing the Jordan decomposition of the derivation $ad(X)$ as

$$ad(X) = T + U$$

we get that the semisimple and nilpotent components of $ad(X)$, namely $T, U$ are also derivations and hence are inner, ie, there exist $S, N \in \mathfrak{g}$ such that $T = ad(S), U = ad(N)$. Then, from the faithfulness of $ad$, it follows that

$$X = S + N$$

with $ad(S)$ semisimple $ad(N)$ nilpotent and $[ad(S), ad(N)] = 0$ or equivalently, $ad([S, N]) = 0$ or equivalently, $[S, N] = 0$. This is the celebrated Jordan decomposition of a semisimple Lie algebra.

Remark: Let $D$ be a derivation on $\mathfrak{g}$ where $\mathfrak{g}$ is a semisimple Lie algebra. Then, $D$ is inner. Indeed, for $X, Y \in \mathfrak{g}$, we have

$$ad(DX)(Y) = [DX, Y] = D[X, Y] - [X, DY] = D[X, Y] - ad(X)(DY)$$

$$= Doad(X)(Y) - ad(X)(DY)$$

or equivalently,

$$ad(DX) = [D, ad(X)], \forall X \in \mathfrak{g}$$

Now, by non-degeneracy of the Cartan-Killing form $<.,.>$ on a semisimple Lie algebra, we have a unique $X \in \mathfrak{g}$ such that

$$Tr(D.ad(Y)) =< X, Y >= Tr(ad(X).ad(Y)) \forall Y \in \mathfrak{g}$$

since $Y \to Tr(D.ad(Y))$ is a linear functional on $\mathfrak{g}$. This shows that

$$Tr(D'.ad(Y)) = 0 \forall Y \in \mathfrak{g}$$

where

$$D' = D - ad(X)$$

Now, $D'$ is also a derivation since $D$ and $ad(X)$ and hence an application of the above formula to $D'$ in place of $D$, we get

$$ad(D'Y) = [D', ad(Y)], Y \in \mathfrak{g}$$

and hence

$$< D'Y, Z >= Tr(ad(D'Y)ad(Z)) = Tr([D', ad(Y)].ad(Z))$$

Now,

$$Tr([D', ad(Y)]ad(Z)) = Tr(D'ad(Y)ad(Z)) - Tr(ad(Y)D'ad(Z)) =$$

$$Tr(D'(ad(Y)ad(Z) - ad(Z)ad(Y)) = Tr(D'[ad(Y), a(Z)]) =$$

$$Tr(D'ad([Y, Z])) = 0$$

by what we just proved. Therefore,

$$< D'Y, Z >= 0 \forall Y, Z \in \mathfrak{g}$$

and hence by non-degeneracy of $< .,. >$, it follows that

$$D'Y = 0 \forall Y \in \mathfrak{g}$$

and hence

$$D' = 0$$

ie,

$$D = ad(X)$$

This completes the proof that $D$ is inner. This result has been used in the proof of the Jordan decomposition on a semisimple Lie algebra.

Remark: Let $T$ be a linear operator in a complex vector space $V$. Let $T'$ denote the commutant of $T$, ie, $T'$ is the set of all operators in $V$ that commute with $T$. Let $T''$ denote the double commutant of $T$, ie, the commutant of $T'$, ie, the set of all operators that commute with every operator in $T'$. Then, it is easy to show that $T''$ is precisely the set of all polynomials in $T$ with complex coefficients. In fact, this can be proved by restricting $T'$ and $T''$ to the space $W_k = N((T - c_k)^{m_k}) = R(E_k)$ where

$$I = E_1 + ... + E_r$$

or equivalently,

$$V = W_1 \oplus ... \oplus W_r$$

is the primary decomposition of $T$ with

$$p(t) = \Pi_{k=1}^{r}(t - c_k)^{m_k}$$

being the minimal polynomial of $T$. Note that $W_k$ is also $T'$ and $T''$-invariant since $T'' \subset T'$. Then by restricting to $W_k$ the claim reduces to proving that if $X \in (cI + N)''$ where $c$ is a complex scalar and $N$ nilpotent in $V$, then $X$ is a polynomial $cI + N$. Further reduction of this problem can be achieved by using the Jordan decomposition of $N$. In other words, proving the claim reduces to proving that if $J_c$ is a Jordan matrix in $V$, ie,

$$J_c = cI + Z$$

where $c \in \mathbb{C}$ and $Z$ has ones on the first superdiagonal and all the other entries as zero, then $J_c''$ is precisely the set of all polynomials in $J_c$.

Remark: Let $T$ be an operator in $V$. we claim that if $S$ a replica of $T$, then $S$ is a polynomial in $T$ with the constant term in the polynomial being zero, ie, $S = Tf(T)$ where $f$ is a polynomial. To see this, we first define an isomorphism $\mu : V \otimes V^* \to L(V)$, where $L(V)$ is the space of all linear operators on $V$ by

$$\mu(v \otimes w^*)(x) = w^*(x)v, x, v \in V, w^* \in V^*$$

and then extending $\mu$ by bilinearity. Then,

$$ad(\mu(v \otimes w^*))(T)(x) = [\mu(v \otimes w^*), T](x)$$

$$= w^*(Tx)v - w^*(x)Tv = (T^T w^*)(x)v - w^*(x)Tv = \mu(v \otimes T^T w^*)(x) - \mu(Tv \otimes w^*)(x)$$

$$= -\mu(T_{1,1}(v \otimes w^*))(x)$$

Equivalently,

$$ad(T)(\mu(v \otimes w^*))(x) = \mu(T_{1,1}(v \otimes w^*))(x)$$

or equivalently,

$$ad(T)o\mu = \mu oT_{1,1}$$

or equivalently,

$$ad(T) = \mu oT_{1,1}o\mu^{-1}, T_{1,1} = \mu^{-1}oad(T)o\mu$$

If $S$ is a replica of $T$, then $T_{r,s}\xi = 0$ implies $S_{r,s}\xi = 0$ for any $r, s \geq 0, r + s \geq 1$. Thus, in particular, $T_{1,1}\xi = 0$ implies $S_{1,1}\xi = 0$, or equivalently, in view of the above discussion $ad(T)(U) = 0$ implies $\mu oT_{1,1}o\mu^{-1}(U) = 0$ for an operator $U$ in $V$ implies $T_{1,1}o\mu^{-1}(U) = 0$ implies $S_{1,1}o\mu^{-1}(U) = 0$ implies $\mu oS_{1,1}o\mu^{-1}(U) = 0$ implies $ad(S)(U) = 0$. In other words, $S$ commutes with any operator $U$ that commutes with $T$, ie, $S \in T''$ and hence by the previous remark, $S$ is a polynomial in $T$, say $S = f(T)$ where $f$ is a polynomial. Further since $S$ is a replica of $T$ $T\xi = 0$ implies $S\xi = 0$ implies $f(T)\xi = 0$ implies that if $c$ is the constant term in $f(t)$, then $c\xi = 0$. In other words, if $T$ has a zero eigenvalue, then $c = 0$, ie, $S = Tg(T)$ where $g$ is a polynomial. Suppose that zero is not an eigenvalue of $T$. Then, $T$ is invertible and hence the minimal polynomial of $T$ has the form

$$p(t) = \Pi_{k=1}^{r}(t - c_k)^{m_k}, c_k \neq 0 \forall k$$

Thus,

$$p(t) = c + tq(t), c \neq 0$$

with $q$ a polynomial Then $p(T) = 0$ implies

$$cI + Tq(T) = 0$$

and therefore,

$$T^{-1} = -c^{-1}q(T)$$

ie, $T^{-1}$ is also a polynomial in $T$. It then follows that if $S$ is a replica of $T$, then

$$S = f(T) = c_0 I + Tg(T) = c_0 TT^{-1} + Tg(T) = T((-c_0/c)q(T) + g(T))$$

ie, $S$ can be expressed as a polynomial in $T$ where the polynomial has zero constant term. We shall now prove the following important result: $S$ is a replica of $T$ iff for all $r, s \geq 0, r + s \geq 1$, we have that $S_{r,s} = p_{r,s}(T_{r,s})$ where $p_{r,s}$ is a polynomial with zero constant term.

## 1.4 Exercises in Matrix Theory

:

### 1.4.1 The Polar decomposition

[1] Let $X$ be a square matrix of size $n \times n$. Prove that $X$ has a polar decomposition:

$$X = UP$$

where $U$ is unitary and $P$ is positive semidefinite. In case $X$ is non-singular, show that

$$P = \sqrt{X^*X}, U = X(X^*X)^{-1/2}$$

where $\sqrt{Q}$ denotes the unique positive semidefinite square root of a positive semidefinite matrix $Q$.

hint: Show that if $Q$ is positive definite of size $n \times n$, then $Q$ can have atmost $2^n$ distinct square roots out of which exactly one is positive semidefinite. Show that $X$, $X^*X$ and $|X| = \sqrt{X^*X}$ all have the same nullspace and hence the same nullity and hence the same rank and hence $R(|X|)^{\perp}|$ and $R(X)^{\perp}$ also have the same dimension. Show that the operator $U_1 : R(|X|) \to R(X)$ defined by $U_1|X|x = Xx \forall x \in \mathbb{C}^n$ is a well defined unitary operator. Do this by showing that the lengths of $|X|x$ and $Xx$ are the same. Hence show that there exists a unitary operator $U_2$ from $R(|X|)^{\perp} \to R(X)^{\perp}$. Let $V = \mathbb{C}^n$ and hence

$$V = R(|X|) \oplus R(|X|)^{\perp} = R(X) \oplus R(X)^{\perp}$$

Define a linear operator $U : V \to V$ by the relation that $U$ restricted to $R(|X|)$ equals $U_1$ and $U$ restricted to $R(|X|)^{\perp}$ equals $U_2$. Show that $U$ is unitary and

$$X = U_1|X| = U|X|$$

### 1.4.2 The singular value decomposition

[2] Deduce the singular value decomposition from the polar decomposition: If $X$ is a matrix of size $m \times n$ of rank $r$ (Note that $r \leq min(m,n)$), then there exist unitary matrices $U \in \mathbb{C}^{m\times m}, V \in \mathbb{C}^{n\times n}$ and a matrix $D \in \mathbb{C}^{m\times n}$ having the block structure

$$D = \begin{pmatrix} D_1 & 0 \\ 0 & 0 \end{pmatrix}$$

where

$$D_1 = diag[\sigma_1, ..., \sigma_r], \sigma_1, ..\sigma_r > 0$$

such that

$$X = UDV^*$$

### 1.4.3 The Riesz representation theorem

[3] Prove the Riesz representation theorem in an infinite dimensional Hilbert space $\mathcal{H}$: If $f : \mathcal{H} \to \mathbb{C}$ is a bounded linear functional, ie $\| f \| = sup_{x\neq 0}|f(x)|/ \| x \|$, then there exists a unique vector $z_f \in \mathcal{H}$ such that

$$f(x) = < z_f, x >, x \in \mathcal{H}$$

hint: Assume $\mathcal{H}$ to be separable which guarantees the existence of an orthonormal basis $\{e_n : n \geq 1\}$. Then show that for any $x \in \mathcal{H}$, we have

$$x = \sum_n e_n < e_n, x >$$

Deduce using the linearity and boundedness of $f$ that

$$f(x) = \sum_n f(e_n) < e_n, x >$$

Finally, show that

$$\infty > \| f \|^2 = \sum_n |f(e_n)|^2$$

and hence that

$$z_f = \sum_n \bar{f}(e_n) e_n \in \mathcal{H}$$

is well defined.

### 1.4.4 Transpose of a linear operator

[4] If $T : V_1 \rightarrow V_2$ is a linear transformation from one vector space $V_1$ to another vector space $V_2$, both assumed to be finite dimensional, then letting $V_1^*, V_2^*$ denote the vector space of linear functionals on $V_1$ and $V_2$ respectively, define the transpose $T'$ of $T$ as a transformation $T' : V_2^* \rightarrow V_1^*$ by

$$T'f(x) = f(Tx), x \in V_1, f \in V_2^*$$

Show that $T'$ is a well defined linear transformation and if we have a third vector space $V_3$ and a linear transformation $S : V_2 \rightarrow V_3$, then

$$(ST)' = T'S'$$

For these statements to be true, do we actually require $V_1, V_2, V_3$ to be finite dimensional or can we drop this condition ?

### 1.4.5 Differentiation of infinite dimensional vector valued functions

[5] This problem gives us some properties of the derivative of a function with values in a Hilbert space.

Let $\mathcal{H}$ be an infinite dimensional Hilbert space and $x : \mathbb{R} \rightarrow \mathcal{H}$ be a function such that

$$lim_{\delta \rightarrow 0}(x(t+\delta) - x(t))/\delta = y(t) \in \mathcal{H}$$

exists for $t \in (-a, +a)$ where the convergence is in the sense of the norm induced by the inner product in $\mathcal{H}$. Deduce that if $z \in \mathcal{H}$ is arbitrary, then

$$\frac{d}{dt} < x(t), z >=< y(t), z >, t \in (-a, a), z \in \mathcal{H}$$

where now the convergence is in the usual sense on the real line. We shall write $dx(t)/dt = y(t), t \in (-a, a)$ and say that $x(t)$ is differentiable in $(-a, a)$ with derivative $dx(t)/dt$ equal to $y(t)$ for $t \in (-a, a)$. Now prove that if $x_1(t)$ and $x_2(t)$ assume values in $\mathcal{H}$ and are differentiable in $(-a, a)$, then deduce that for any $c_1, c_2 \in \mathbb{C}$, $c_1x_1(t) + c_2x_2(t)$ is also differentiable in $(-a, a)$ with derivative given by

$$\frac{d}{dt}(c_1x_1(t) + c_2x_2(t)) = c_1\frac{dx_1(t)}{dt} + c_2\frac{dx_2(t)}{dt}, t \in (-a, a)$$

hint: Prove using the definitions

$$lim_{\delta \rightarrow 0} \| \frac{x_k(t+\delta) - x_k(t)}{\delta} - dx_k(t)/dt \|= 0, k = 1, 2$$

and the triangle inequality that

$$lim_{\delta \rightarrow 0} \| \frac{c_1x_1(t+\delta) + c_2x_2(t+\delta)}{\delta} - (c_1dx_1(t)/dt + c_2dx_2(t)/t) \|= 0$$

### 1.4.6 Invariant subspaces and primary decomposition

[6] Let $T$ be a linear operator on a finite dimensional complex vector space $V$ with minimal polynomial $p(t) = \Pi_{k=1}^{r}(t - c_k)^{m_k}$, $c_k's$ distinct and $m_k > 0$. The primary decomposition of $T$ is

$$V = \bigoplus_{k=1}^{r} W_k, I = \sum_{k=1}^{r} E_k, R(E_k) = W_k, E_k E_j = 0, k \neq j$$

Then if $S$ is another linear operator in $V$ such that $[T, S]$ leaves each of the $W_k$ invariant, then show that $S$ also shares the same property.

hint: Note that the $E_k's$ are all polynomials in $T$ and hence commute with $T$. Also note that an operator $L$ leaves each of the $W_k$ invariant iff

$$L = \sum_{k,j} E_k S E_j = \sum_k E_k S E_k$$

or equivalently, iff

$$E_k L E_j = 0 \forall k \neq j$$

Hence, for all $k \neq j$,

$$0 = E_k[T, S]E_j = [T, E_k S E_j]$$

and hence $E_k S E_j$ leaves every $W_i$ invariant and in particular, $W_j$ invariant. This means that $E_k S E_j = 0$ for all $k \neq j$ and hence

$$S = \sum_{k,j} E_k S E_j = \sum_k E_k S E_k$$

proving that $S$ leaves every $W_k$ invariant. Note that in the proof, we have used the easily verified fact that if $[T, U] = 0$, then $U$ leaves every $W_k$ invariant because

$$U E_k = E_k U, W_k = R(E_k)$$

Note that $U$ commutes with $E_k$ because $E_k$ is a polynomial in $T$.

### 1.4.7 Normalizer of a Lie subalgebra

[7] Let $G$ be a connected Lie group and let $H$ be a connected Lie subgroup of $G$. Thus if $\mathfrak{g}$ and $\mathfrak{h}$ are respectively the Lie algebras of $G$ and $H$, then the respective exponential maps are surjective, ie, $exp(\mathfrak{g}) = G$ and $exp(\mathfrak{h}) = H$. Show then that if $N(H)$ denotes the normalizer of $H$ in $G$, ie, the set of all $g \in G$ for which $gHg^{-1} \subset H$, and if $\mathfrak{n}(H)$ denotes the Lie algebra of $N(H)$, then

$$\mathfrak{n}(H) = \{X \in \mathfrak{g} : [X, \mathfrak{h}] \subset \mathfrak{h}\}$$

### 1.4.8 Finite dimensional Irreducible representations of $SL(2,\mathbb{C})$

[8] This problem discusses a method for obtaining all the finite dimensional irreducible representations of the Lie group $SL(2,\mathbb{C})$ or equivalently of its Lie algebra $sl(2,\mathbb{C})$. Let $G = SL(2,\mathbb{C})$, ie, the set of all $2 \times 2$ complex matrices having determinant one. Let $sl(2,\mathbb{C}) = \mathfrak{g}$, the Lie algebra of $G$. Show that $sl(2,\mathbb{C})$ is the set of all $2 \times 2$ complex matrices having trace zero. Show that a basis for $sl(2,\mathbb{C})$ is given by $\{H, X, Y\}$, where

$$H = \begin{pmatrix} 1 & 0 \\ 0 & -1 \end{pmatrix},$$

$$X = \begin{pmatrix} 0 & 1 \\ 0 & 0 \end{pmatrix},$$

$$Y = \begin{pmatrix} 0 & 0 \\ 1 & 0 \end{pmatrix}$$

Note prove the following identities:

[a]

$$[H, X] = 2X, [X, Y] = -2Y, [X, Y] = H$$

and hence if $\pi$ is any representation of $sl(2,\mathbb{C})$ in a vector space $V$, then

$$[\pi(H), \pi(X)] = 2\pi(X), [\pi(H), \pi(Y)] = -2\pi(Y), [\pi(X), \pi(Y)] = \pi(H)$$

Let now $\pi$ be in particular a representation of $sl(2,\mathbb{C})$ in a finite dimensional complex vector space $V$. Show that if $v$ is a vector in $V$ such that

$$\pi(H)v = \lambda v$$

for some $\lambda \in \mathbb{C}$, then

$$\pi(H)\pi(X)v = (\lambda + 2)\pi(X)v, \pi(H)\pi(Y)v = (\lambda - 2)\pi(Y)v$$

Hence, deduce from the finite dimensionality of $V$ that there exist a nonzero vector $v_0 \in V$ and a $\lambda_0 \in \mathbb{C}$ such that

$$\pi(X)v_0 = 0, \pi(H)v_0 = \lambda_0 v_0$$

and that there exists a smallest positive integer $l$ such that $\pi(Y)^l v_0 = 0$. By smallest, we mean that $\pi(Y)^{l-1} v_0 \neq 0$. Deduce that $V_0 = \{\pi(Y)^m v_0 : 0 \leq m \leq l-1\}$ is an invariant subspace for $\pi$ in $V$, ie,

$$\pi(sl(2,\mathbb{C}))(V_0) \subset V_0$$

and hence deduce that if $\pi$ is assumed to be irreducible, then $\{\pi(Y)^m v_0 : 0 \leq m \leq l-1\}$ is a basis for $V$. Now prove that if $n$ is any positive integer, then

$$[\pi(H), \pi(X)^n] = 2n.\pi(X)^n, [\pi(H), \pi(Y)] = -2n\pi(Y)^n$$

and

$$[\pi(X), \pi(Y)^n] =$$

$$\pi(H).\pi(Y)^{n-1} + \pi(Y)\pi(H).\pi(Y)^{n-2} + ... + \pi(Y)^{n-1}\pi(H)$$

$$= [-2(n-1)\pi(Y)^{n-1} + \pi(Y)^{n-1}\pi(H)] + [-2(n-2)\pi(Y)^{n-1} + \pi(Y)^{n-1}\pi(H)]$$

$$+ ... + [0.\pi(Y)^{n-1} + \pi(Y)^{n-1}\pi(H)]$$

$$= -2(0 + 1 + 2 + ... + (n-1)]\pi(Y)^{n-1} + n\pi(Y)^{n-1}\pi(H)$$

$$= -2n(n-1)\pi(Y)^{n-1} + n\pi(Y)^{n-1}\pi(H)$$

## 1.5 Conjugacy classes of Cartan sub-algebras

Consider a semisimple Lie algebra $\mathfrak{g}$ and let $\mathfrak{g}'$ denote the set of all of its regular elements. Let $\mathfrak{g}_i, i = 1, 2, ..., r$ denote all the connected components of $\mathfrak{g}$. For each $i = 1, 2, ..., r$, choose an $X_i \in \mathfrak{g}_i$. Then since $X_i$ is regular, it follows that $\mathfrak{h}_i = \mathfrak{h}_{X_i}$ is a Cartan algebra for each $i$. Let $\mathfrak{h}$ be any Cartan algebra. Then $\mathfrak{h} = \mathfrak{h}_X$ for some regular $X$ as we have already seen above. Since $X$ is regular, it follows that $X \in \mathfrak{g}_i$ for some $i$. Our aim is to show that $\mathfrak{h} = \mathfrak{h}_X$ is conjugate to $\mathfrak{h}_i$. We would then have established that any semisimple Lie algebra (finite dimensional) has a finite set of non-conjugate Cartan subalgebras such that any Cartan subalgebra is conjugate to one in this set. Let $\mathfrak{h}'_X$ denote the set of regular elements in $\mathfrak{h}_X$, ie, $\mathfrak{h}'_X = \mathfrak{h}_X \cap \mathfrak{g}'$. Let $\mathfrak{h}_{X+}$ denote the connected component of $\mathfrak{h}'_X$ that contains $X$ and define $\mathfrak{b}_X = (\mathfrak{h}_{X+})^G$. Then, it is clear that $\mathfrak{b}_X$ is connected. Choose any $Z \in \mathfrak{b}_X$. Then, $Z$ is regular and hence the Cartan algebra $\mathfrak{h}_Z$ defined. We claim that $\mathfrak{b}_Z = \mathfrak{b}_X$. Indeed, $Z^y$ is a regular element in $\mathfrak{h}_X$ for some $y \in G$ and hence, $\mathfrak{h}_Z^y = \mathfrak{h}_{Z^y} = \mathfrak{h}_X$ which implies that $\mathfrak{h}_Z'^y = \mathfrak{h}'_X$ and therefore

$$\mathfrak{b}_Z = (\mathfrak{h}'_Z)^G = (\mathfrak{h}'_X)^G = \mathfrak{b}_X$$

proving the claim. Now let $U, V \in \mathfrak{g}_i$. We claim that either $\mathfrak{b}_U = \mathfrak{b}_V$ or else $\mathfrak{b}_U \cap \mathfrak{b}_V = \phi$. Indeed, suppose $Z \in \mathfrak{b}_U \cap \mathfrak{b}_V$. Then, $Z \in \mathfrak{b}_U$ which implies as shown above that $\mathfrak{b}_Z = \mathfrak{b}_U$ and likewise, $\mathfrak{b}_Z = \mathfrak{b}_V$. Thus, $\mathfrak{b}_U = \mathfrak{b}_V$, thereby proving the claim. Further, $\mathfrak{b}_U$ is an open connected set of regular elements containing $U$ and hence $\mathfrak{b}_U \subset \mathfrak{g}_i$. In other words, we have proved that $\{\mathfrak{b}_U : U \in \mathfrak{g}_i\}$ is a family of connected open sets, each of which is contained in $\mathfrak{g}_i$ and two elements in this family are either disjoint or the same. Further, the union of this whole family is precisely $\mathfrak{g}_i$ since if $U \in \mathfrak{g}_i$, we have that $U \in \mathfrak{b}_U$. It follows from the connectedness of $\mathfrak{g}_i$ that $\mathfrak{b}_U = \mathfrak{g}_i \forall U \in \mathfrak{g}_i$. We have thus shown that $\mathfrak{b}_{X_i} = \mathfrak{g}_i, i = 1, 2, ..., r$. Now, let $\mathfrak{h} = \mathfrak{h}_X$ be as above with $X \in \mathfrak{g}_i$ (Recall that any Cartan subalgebra is of this form for some regular $X$, and some $i$). Then we have established that $\mathfrak{b}_X = \mathfrak{b}_{X_i}$. In other words, we have established that $\mathfrak{h}_{X+}^G = \mathfrak{h}_{X_i+}^G$ and this implies that $X = Z^y$ for some $y \in G$ and some $Z \in \mathfrak{h}_{X_i+}$

(Recall that $X \in \mathfrak{h}_{X+}$). Then, $\mathfrak{h} = \mathfrak{h}_X = \mathfrak{h}_{Z^y} = \mathfrak{h}_Z^y = \mathfrak{h}_{X_i}^y$ (since $Z, X_i$ are both regular elements in $\mathfrak{h}_{X_i}$). Therefore, $\mathfrak{h} = \mathfrak{h}_i^y$, ie, $\mathfrak{h}$ is conjugate to $\mathfrak{h}_i$. This completely proves our aim.

Remark: If the underlying field of $\mathfrak{g}$ is complex, then $\mathfrak{g}$ has just one Cartan subalgebra upto conjugacy, ie, any two of its Cartan subalgebras are conjugate.

## 1.6 Exercises

[1] Let $\mathfrak{h}$ be a Cartan subalgebra of any finite dimensional Lie algebra $\mathfrak{g}$, ie $\mathfrak{h}$ is a Lie subalgebra, nilpotent and its own normalizer. Show that there exists an $X \in \mathfrak{g}'$ ($\mathfrak{g}'$ is the set of regular elements in $\mathfrak{g}$) such that $\mathfrak{h}_X = \mathfrak{h}$ where

$$\mathfrak{h}_X = \bigcup_{m>0} N(ad(X)^m)$$

[2] With $\mathfrak{h}$ any Cartan subalgebra of a Lie algebra $\mathfrak{g}$, show that $X \in \mathfrak{h}$ is regular iff

$$\zeta(X) = det(ad(X)|_{\mathfrak{g}/\mathfrak{h}}) \neq 0$$

hint: $\zeta(X)$ is non-zero for $X \in \mathfrak{h}$ iff $ad(X)|_{\mathfrak{g}/\mathfrak{h}}$ has a zero eigenvalue iff when we write

$$det(tI - ad(X)) = c_k t^k + c_{k+1} t^{k+1} + ... + t^n, n = dim\mathfrak{g}$$

with $c_k \neq 0$, then $k > dim(\mathfrak{h}) = rk(\mathfrak{g})$. Do this by noting that for $X \in \mathfrak{h}$, all the eigenvalues of $ad(X)|_{\mathfrak{h}}$ are zero since by definition, $ad(X)$ is nilpotent on $\mathfrak{h}$ and therefore,

$$det(tI - ad(X)|_{\mathfrak{h}}) = t^l, l = dim(\mathfrak{h}) = rk(\mathfrak{g})$$

[2] With $\mathfrak{h}_X$ as in the problems [1,2] for $X$ regular (ie, $X \in \mathfrak{g}', \mathfrak{h}_X = \bigcup_{m>0} N(ad(X)^m)$), show without assuming that $\mathfrak{h}_X$ is a Cartan algebra that $ad(X)$ is non-singular on $\mathfrak{g}/\mathfrak{h}_X$ and hence $dim(\mathfrak{h}_X) = rk(\mathfrak{g})$(Note that to prove that $ad(X)$ is non-singular on $\mathfrak{g}/\mathfrak{h}_X$, you need not even assume that $X$ is regular since by the definition of $\mathfrak{h}_X$, $ad(X)|_{\mathfrak{g}/\mathfrak{h}_X}$ is non-singular). Now show that with $X$ regular, if $Y \in \mathfrak{h}_X$, then $ad(X)^m(Y) = 0$ for some $m > 0$ and hence $ad(Y)$ leaves $\mathfrak{h}_X$ invariant. Show further that if $\zeta(Y) = det(ad(Y)|_{\mathfrak{g}/\mathfrak{h}_X}) \neq 0$, then $Y$ is regular and hence $\mathfrak{h}_Y = \mathfrak{h}_X$. Deduce from this that for such a $Y$, $ad(Y)$ is nilpotent on $\mathfrak{h}_X$. Then, observe that if $Y \in \mathfrak{h}_X$ is arbitrary, we can write $Y = limY_n$ with $\zeta(Y_n) \neq 0$ and hence conclude that $ad(Y)$ is nilpotent on $\mathfrak{h}_X$,

ie $\mathfrak{h}_X$ is a nilpotent Lie algebra. Deduce from this that if $Y \in \mathfrak{h}_X$, then $Y$ is regular iff $\zeta(Y) = det(ad(Y)|_{\mathfrak{g}/\mathfrak{h}_X}) \neq 0$.

[3] This problem is a prerequisite for attempting the previous problems: let $V$ be a finite dimensional vector space and $T$ a linear operator on $V$. Let $W$ be a $T$ invariant $r$-dimensional subspace of $V$. Choose any basis $\{e_{r+1}+W, ..., e_n+W\}$ for $V/W$. Show that if $\{e_1, ..., e_r\}$ is any basis for $W$, then $B = \{e_1, ..., e_n\}$ is a basis for $V$. Show that $[T]_B$ has the following block structure:

$$[T]_B = \begin{pmatrix} A_{11} & A_{12} \\ 0 & A_{22} \end{pmatrix}$$

and hence deduce that

$$det(T) == det(A_{11}).det(A_{22}) = det(T|_W).det(T|_{V/W})$$

Note that $W$ is the zero element of the vector space $V/W$. Specialize this result to show that the characteristic polynomial of $T$ can be expressed as

$$f(t) = det(tI - T) = det(tI - T|_W).det(tI - T|_{V/W})$$

# 1.7 Appendix:Some applications of matrix theory to control theory problems

## 1.7.1 Controllability of the Yang-Mills non-Abelian field equations

The Lagrangian for the non-Abelian gauge fields $A^a_\mu(x), a = 1, 2, ..., N, \mu = 0, 1, 2, 3$ is

$$L = (-1/2)F^a_{\mu\nu}F^{\mu\nu a}$$

where

$$F^a_{\mu\nu} = [D_\mu, D_\nu]^a$$

with $D_\mu$ the gauge covariant derivative defined by

$$D_\mu = \partial_\mu + iA^a_\mu\tau_a = \partial_\mu + iA_\mu$$

where $\tau_a, a = 1, 2, ..., N$ are Hermitian generators of the gauge group Lie algebra $\mathfrak{g} = Lie(G)$. The structure constants associated with these generators are denoted by $C(abc)$:

$$[\tau_a, \tau_b] = -i\sum_{c=1}^{N} C(abc)\tau_c$$

Note that

$$F_{\mu\nu} = F^a_{\mu\nu} i\tau_a = [D_\mu, D_\nu] =$$

$$[\partial_\mu + iA_\mu, \partial_\nu + iA_\nu] =$$

$$i(A_{\nu,\mu} - A_{\mu,\nu}) - [A_\mu, A_\nu]$$

$$= (A^a_{\nu,\mu} - A^a_{\mu,\nu}) i\tau_a - A^b_\mu A^c_\nu [\tau_b, \tau_c]$$

$$= (A^a_{\nu,\mu} - A^a_{\mu,\nu} + C(abc) A^b_\mu A^c_\nu) i\tau_a$$

Thus,

$$F^a_{\mu\nu} = A^a_{\nu,\mu} - A^a_{\mu,\nu} + C(abc) A^b_\mu A^c_\nu$$

The field equations in the absence of current sources are obtained from the variational principle

$$\delta \int F^a_{\mu\nu} F^{\mu\nu a} d^4x = 0$$

and these give

$$[D_\nu, F^{\mu\nu}] = 0$$

or equivalently,

$$\partial_\nu F^{\mu\nu a} + C(abc) A^b_\nu F^{\mu\nu c} = 0$$

in the presence of interaction with a current source $J^{\mu a}$, with the interaction Lagrangian being

$$L_{int} = J^{\mu a} A^a_\mu$$

the field equations are

$$\partial_\nu F^{\mu\nu a} + C(abc) A^b_\nu F^{\mu\nu c} = J^{\mu a}$$

and these are non-Abelian $G$ generalizations of the Abelian $U(1)$ electromagnetic field equations where $G$ is a Lie subgroup of $U(M)$ with $dimG = N$, where by the dimension of a Lie group, we mean the dimension of its Lie algebra as a vector space. The above field equations be expanded as

$$\partial_\nu (A^{\nu a,\mu} - A^{\mu a,\nu} + C(abc) A^{\mu b} A^{\nu c})$$

$$+C(abc) A^b_\nu (A^{\nu c,\mu} - A^{\mu c,\nu} + C(cde) A^{\mu d} A^{\nu e}) = J^{\mu a}$$

or equivalently after attaching a perturbation parameter $\delta$ to keep track of the quadratic and cubic non-linear terms,

$$A^{\nu a,\mu}_{,\nu} - A^{\mu a,\nu}_{,\nu} + \delta C(abc)(A^{\mu b} A^{\nu c}))_{,\nu}$$

$$+\delta C(abc) A^b_\nu (A^{\nu c,\mu} - A^{\mu c,\nu}) + \delta^2 C(abc) C(cde) A^b_\nu A^{\mu d} A^{\nu e} = J^{\mu a} \quad --- (1)$$

The controllability problem for these field equations is then posed as follows: Let $U$ and $V$ be two disjoint subsets of $\mathbb{R}^3$. Then, given that at time $t = 0$, the potentials $A^a_\mu$ and their time derivatives $A^a_{\mu,0}$ have prescribed values on $U$, does there exist a control current field $J^{\mu a}(t, r), 0 \leq t \leq T, r \in \mathbb{R}^3$ such that at time $T$, these potentials have prescribed values on $V$ ?. More generally, we can ask the question that given two disjoint subsets $U, V$ of $\mathbb{R}^4$, does there exist a control current field $J^{\mu a}(x)$ on $\mathbb{R}^4$ and a solution $A^a_\mu$ to the above Yang-Mills field equations corresponding to this current source such that these potentials have prescribed values on both $U$ and $V$ ? We shall attempt to solve this controllability problem approximately by means of perturbation theory. First, we expand the solution in powers of $\delta$:

$$A^a_\mu = A^{a(0)}_\mu + \sum_{k\geq 1} \delta^k . A^{a(k)}_\mu \quad --- (2)$$

Since the gauge group $G$ has dimension $N$ equal to the number of possible values of the gauge index $a$, we can always gauge transform the gauge field so that the gauge conditions

$$\partial_\mu A^{\mu(a)} = 0$$

hold good. In that case, (1) reduces to

$$-A^{\mu a,\nu}_{,\nu} + \delta C(abc) A^{\mu b}_{,\nu} A^{\nu c}$$

$$+\delta C(abc) A^b_\nu (A^{\nu c,\mu} - A^{\mu c,\nu}) + \delta^2 C(abc) C(cde) A^b_\nu A^{\mu d} A^{\nu e} = J^{\mu a} \quad --- (3a)$$

or equivalently with

$$\Box = \partial^\nu \partial_\nu = \partial_0^2 - \nabla^2$$

denoting the D'Alembert wave operator,

$$-\Box A^{\mu a} + \delta C(abc) A^{\mu b}_{,\nu} A^{\nu c}$$

$$+\delta C(abc) A^b_\nu (A^{\nu c,\mu} - A^{\mu c,\nu}) + \delta^2 C(abc) C(cde) A^b_\nu A^{\mu d} A^{\nu e} = J^{\mu a} \quad --- (3b)$$

or on using the antisymmetry of the structure constants,

$$-\Box A^{\mu a} + \delta C(abc) A^{\nu c} (2A^{\mu b}_{,\nu} - A^b_{\nu,\mu})$$

$$+\delta^2 C(abc) C(cde) A^b_\nu A^{\mu d} A^{\nu e} = J^{\mu a} \quad --- (3b)$$

Substituting the perturbation expansion (2) into (3b) and equating equal powers of $\delta$ gives us (a), for $\delta^0 = 1$,

$$-\Box A^{\mu a(0)} = J^{\mu a},$$

for $\delta^1 = \delta$,

$$-\Box A^{\mu a(1)} + C(abc) A^{\nu c(0)} (2A^{\mu b(0)}_{,\nu} - A^{b(0)}_{\nu,\mu}) = 0,$$

for $\delta^2$,

$$-\Box A^{\mu a(2)} + C(abc)[A^{\nu c(0)} (2A^{\mu b(1)}_{,\nu} - A^{b(1)}_{\nu,\mu})$$

$$+A^{\nu c(1)} (2A^{\mu b(0)}_{,\nu} - A^{b(0)}_{\nu,\mu})]$$

$$+C(abc) C(cde) A^{b(0)}_\nu A^{\mu d(0)} A^{\nu e(0)} = 0$$

Let $G(x-x')$ denote the Green's function for the wave operator $\square$. Then, we can successively solve the above equations as

$$A^{\mu a(0)}(x) = -\int G(x-x')J^{\mu a}(x')d^4x'$$

$$A^{\mu a(1)}(x) = C(abc)\int G(x-x')A^{\nu c(0)}(x')(2A^{\mu b(0)}_{,\nu} - A^{b(0)}_{\nu,\mu})(x')d^4x',$$

$$A^{\mu a(2)}(x) =$$

$$C(abc)\int G(x-x')[A^{\nu c(0)}(x')(2A^{\mu b(1)}_{,\nu} - A^{b(1)}_{\nu,\mu})(x')$$

$$+A^{\nu c(1)}(x')(2A^{\mu b(0)}_{,\nu} - A^{b(0)}_{\nu,\mu})(x')]d^4x'$$

$$+C(abc)C(cde)\int G(x-x')A^{b(0)}_{\nu}(x')A^{\mu d(0)}(x')A^{\nu e(0)}(x')d^4x'$$

Remark: $G(x)$ satisfies the pde

$$\square G(x) = \delta^4(x)$$

which on Four dimensional Fourier transforming gives

$$\hat{G}(k) = \frac{1}{k^2}, k^2 = k_\mu k^\mu$$

It is easily deduced then that

$$G(x) = C\delta(x^2), x^2 = x_\mu x^\mu$$

is one such solution. That follows by four dimensional Fourier inversion. It is easily seen from the above formulas that upto $O(\delta^2)$, the solution can be expressed as

$$A(x) = \int G_0(x-x')J(x')d^4x' + \delta\int G_1(x-x', x-y')J(x')\otimes J(y')d^4x'd^4y' +$$

$$+\delta^2\int G_2(x-x', x-y', x-z')(J(x')\otimes J(y')\otimes J(z'))d^4x'd^4y'd^4z' \quad ---(4)$$

where $A(x), J(x)$ are appropriate vector space valued smooth functions on $\mathbb{R}^4$ and $G_0, G_1, G_2$ are appropriate matrix valued known functions on $\mathbb{R}^4, \mathbb{R}^4 \times \mathbb{R}^4$ and $\mathbb{R}^4 \times \mathbb{R}^4 \times \mathbb{R}^4$. Now the controllability problem is easily stated: Given an $\epsilon > 0$ and a function $A_g(x)$ on $\mathbb{R}^4$, does there exist a source current $J(x)$ such that the output of the above system is $A_d(x)$. Actually, the problem is more intricate if we take initial conditions into account. In that case, we absorb the intial conditions $A_i(0,r)$ into the solution for the zeroth order perturbation. Then,

$$A_0(x) = \int G(x-x')J(x')d^4x' + \int F(t,r|r')A_i(0,r')d^3r'$$

This initial condition then propagates into the higher order perturbations yielding finally upto second order in $\delta$ a solution of the form

$$A(x) = \int G_0(x-x')J(x')d^4x' + \delta \int G_1(x-x', x-y')(J(x') \otimes J(y'))d^4x'd^4y' +$$

$$+\delta^2 \int G_2(x-x', x-y', x-z')(J(x') \otimes J(y') \otimes J(z'))d^4x'd^4y'd^4z'$$

$$+ \int F_0(t, r|r')A_i(0, r')d^3r' + \delta \int F_1(t, r|r', r'')A_i(0, r') \otimes A_i(0, r'')d^3r'd^3r''$$

$$+\delta^2 \int F_2(t, r|r', r'')(A_i(0, r') \otimes A_i(0, r'') \otimes A_i(0, r''))d^3r'd^3r''d^3r''$$

and the question of approximate controllability is then the question of whether for a given input field $A_i(r)$ at time 0 and a given output field $A_f(r)$ at time $T$ and an $\epsilon > 0$, does there exist a control input current field $J(x)$ for which the output $A(T, r)$ in the above equation at time $T$ has a mean weighted square distance from $A_f(r)$ defined by

$$\int W(r) \parallel A(T, r) - A_f(r) \parallel^2 d^3r$$

smaller than $\epsilon$ ?

Remark: After discretization in the spatial variables, the Yang Mills field equations appears in state variable form as

$$x'(t) = A_0 x(t) + \delta.A_1(x(t) \otimes x(t)) + \delta^2.A_2(x(t) \otimes x(t) \otimes x(t)) + u(t)$$

A second order perturbative solution gives

$$x(t) = \int_0^t G_0(t-s)u(s)ds + \delta.\int_0^t \int_0^t G_1(t-s_1, t-s_2)(u(s_1) \otimes u(s_2))ds_1ds_2 +$$

$$\int_{[0,t]^3} G_2(t-s_1, t-s_2, t-s_3)u(s_1) \otimes u(s_2) \otimes u(s_3)ds_1ds_2ds_3$$

where

$$G_0(t) = exp(tA_0)$$

$$G_1(t_1, t_2) =$$

$$G_2(t_1, t_2, t_3) =$$

Using second order perturbation theory, with

$$x(t) = x_0(t) + \delta.x_1(t) + \delta^2.x_2(t) + O(\delta^3)$$

$$x_1'(t) = A_0 x_1(t) + A_1(x_0(t) \otimes x_0(t))$$

so

$$x_1(t) = \int_{0<s_1,s_2<s<t} G_0(t-s)A_1(G_0(s-s_1)\otimes G_0(s-s_2))(u(s_1)\otimes u(s_2))ds_1 ds_2 ds$$

and hence

$$G_1(t-s_1, t-s_2) = \int_{max(s_1,s_2)<s<t} G_0(t-s)A_1(G_0(s-s_1)\otimes G_0(s-s_2))ds$$

$$x_2'(t) = A_0 x_2(t) + A_1(x_0(t)\otimes x_1(t) + x_1(t)\otimes x_0(t)) + A_2(x_0(t)\otimes x_0(t)\otimes x_0(t))$$

so that

$$x_2(t) = \int_0^t G_0(t-s)(A_1(x_0(s)\otimes x_1(s)+x_1(s)\otimes x_0(s))+A_2(x_0(s)\otimes x_0(s)\otimes x_0(s))ds$$

Remark: If the input $u(t)$ is replaced by $Bu(t)$ where $B$ is a rectangular matrix, then

$$x_0(t) = \int_0^t G_0(t-s)Bu(s)ds,$$

$$x_1(t) = \int_0^t \int_0^t G_1(t-s_1, t-s_2)(B\otimes B)(u(s_1)\otimes u(s_2))ds_1 ds_2$$

It is clear from these expressions that for controllability upto $O(\delta)$, we must have that the matrix

$$C = [A^k B, , A_0^r A_1(A_0^s \otimes A_0^m)(B\otimes B), k, r, s, m \geq 0]$$

must have full row rank. Likewise a necessary condition for controllability upto $O(\delta^2)$ can be formulated.

Remark: It should be noted that in view of the Cayley-Hamilton equation satisfied by $A_0$, it is enough to restrict all the indices $k, r, s, m$ in the above controllability matrix $C$ to assume values in the range $0, 1, ..., n-1$ where $A_0 \in \mathbb{R}^{n\times n}$.

### 1.7.2 Appendix

Let $A_\mu \to A'_\mu$ be the transformation of the potential under a gauge group element $g(x) \in G$ dependent upon space-time in a differentiable way. Then, if $\psi(x)$ denotes the matter field which appears in the matter component of the Lagrangian either as a function of the $\psi, \psi^*$ and $D_\mu\psi, (D_\mu\psi)^*$, then it is clear that since under this local gauge transformation $\psi(x) \to g(x)\psi(x) = \psi'(x)$, in order to maintain invariance of the matter component of the Lagrangian under local gauge transformations, it must be of the form $L_M(\psi(x), D_\mu\psi(x))$ where $L_M$ is $G$-invariant, ie,

$$L_M(g.\psi, g.\xi_\mu) = L_M(\psi, \xi_\mu), g \in G$$

and hence for $L_M(\psi(x), D_\mu\psi(x))$ to be locally $G$-invariant, the potential $A_\mu$ must transform to $A'_\mu$ so that if $D'_\mu = \partial_\mu + A'_\mu$, then

$$D'_\mu g(x)\psi(x) = D'_\mu\psi'(x) = g(x).D_\mu\psi(x)$$

This happens only provided that

$$D'_\mu = g(x).D_\mu.g(x)^{-1}$$

or equivalently,

$$A'_\mu(x) = g(x)A_\mu(x).g(x)^{-1} + g(x)(\partial_\mu g(x)^{-1})$$

$$= g(x)A_\mu(x).g(x)^{-1} - (\partial_\mu g(x))g(x)^{-1}$$

In particular, for an infinitesimal local gauge transformation,

$$g(x) = 1 + \epsilon^a(x)\tau_a = 1 + \epsilon(x)$$

we get upto $O(\epsilon)$, the infinitesimal gauge transformation,

$$A'_\mu(x) - A_\mu(x) = \delta A_\mu(x) =$$

$$[\epsilon(x), A_\mu(x)] - \epsilon_{,\mu}(x)$$

or in component form,

$$\delta A^a_\mu(x) = -\epsilon^a_{,\mu}(x) + C(abc)\epsilon^b(x)A^c_\mu(x)$$

One candidate for the total action functional for the non-Abelian gauge and matter fields is given by

$$S[\psi, A_\mu] = (-1/4)\int F^a_{\mu\nu}F^{\mu\nu a}d^4x + c_1\int[\psi^*\gamma^0(i\gamma^\mu D_\mu - m)\psi]d^4x + c_2\int(D^\mu\chi)^*(D_\mu\chi)d^4x$$

where $\chi$ is a complex Klein-Gordon scalar field while $\psi$ is a non-Abelian Dirac matter field. This action is invariant under local gauge transformations. Note that the gauge group transformations commute with the Dirac Gamma matrices. The local gauge group symmetry can be broken by introducing control terms like $\int A^a_\mu J^{\mu a}d^4x$ and $\int B^{\mu a}\psi^*\gamma^0\gamma^\mu\psi d^4x$ with the control current $J^{\mu a}$ and the control gauge field potentials $B^{\mu a}$ being classical, ie, c-number fields. It should be noted here that $\psi^*\gamma^0\gamma^\mu\psi$ is the non-Abelian Dirac current density. Likewise, the Klein-Gordon current is

$$(-i/2)[\chi^* D_\mu\chi - \chi(D_\mu\chi)^*] = Im(\chi^* D_\mu\chi)$$

and we can also introduce an interaction term between this current and a control c-number four potential field $C^\mu$ as

$$\int C^\mu Im(\chi^* D_\mu\chi)d^4x$$

which also breaks the local gauge symmetry of the action.

Remark: This controllability problem is a generalization to the four dimensional continuum of the following problem in ordinary differential equations: Consider the vector valued ode

$$\frac{dx(t)}{dt} = A_0x(t) + A_1(x(t) \otimes x(t)) + .. + A_m(x(t)^{\otimes m}) + Bu(t), t \geq 0$$

where $x(t) \in \mathbb{R}^n$. Then under what conditions on the matrices $A_0, .., A_m, B$ is this system controllable ?, ie, when does there exist a control input $u(t), 0 \leq t \leq T$ that takes any given initial state $x(0)$ to a final state $x(T)$ ? If $A_k = 0, k = 1, 2, ..., m$, then Kalman proved that the system is controllable iff the matrix

$$[B, AB, A^2B, ..., A^{n-1}B]$$

has full row rank.

Symmetry breaking in non-Abelian gauge field theory. Let the gauge group $G$ be spontaneously broken into the subgroup $H$. Thus, if $\psi(x)$ transforms according to $G$, and if we write $g(x) = \gamma(x)h(x)$ where $h(x) \in H$ and $\gamma(x)$ is a representative element for the coset space $G/H$, then we can express the local gauge transformation of $\psi(x)$ as follows: First write

$$\psi(x) = \gamma(x)\tilde{\psi}(x)$$

where $\tilde{\psi}(x)$ transforms according to $H$. Then if $g \in G$, the transformation law of $\psi$ under $g$ can be expressed as

$$\psi(x) \to \psi'(x) = g\psi(x) = g\gamma(x)\tilde{\psi}(x) =$$

$$\gamma(g, x)h(g, x)\tilde{\psi}(x)$$

where

$$h(g, x) \in H,$$

and $\gamma(g, x)$ is the representative element in the coset space $G/H$ corresponding to the element $g\gamma(x) \in G$. Thus, if we write

$$\psi'(x) = \gamma'(g, x)\tilde{\psi}'(x)$$

where $\gamma'(g, x)$ is a representative element in the coset space $G/H$ and $\tilde{\psi}'(x)$ transforms according to $H$, then we have

$$\gamma'(g, x)\tilde{\psi}'(x) = \gamma(g, x)h(g, x)\tilde{\psi}(x)$$

Note that

$$g\gamma(x) = \gamma(g, x)h(g, x)$$

Remark: We can look upon $\gamma(x)$ as that component of the wave function $\psi(x)$ that corresponds to the broken symmetry degrees of freedom and $\tilde{\psi}(x)$ as that component of $\psi(x)$ that transforms according to the unbroken subgroup $H$. Note that the broken degrees of freedom are represented by the coset space $G/H$. Consider now the gauge transformation of the gauge field $A_\mu(x)$. $D_\mu = \partial_\mu + iA_\mu$ transforms under local gauge transformations according to the adjoint representation of the gauge group. Now define $\tilde{A}_\mu(x)$ by the equation

$$\tilde{A}_\mu(x) = \gamma(x)^{-1}A_\mu(x)\gamma(x)$$

or equivalently,

$$A_\mu(x) = \gamma(x)\tilde{A}_\mu(x)\gamma(x)^{-1}$$

This transformation may be viewed as the removal of the broken symmetry degrees of freedom from the gauge potentials $A_\mu$. We therefore expect $\tilde{A}_\mu(x)$ to transform under local gauge transformations $g(x) \in G$ in accordance with an unbroken subgroup $H$ element derived from $g(x)$. more precisely, for a given gauge group transformation $g(x) \in G$, we expect that $\tilde{D}_\mu = \partial_\mu + i\tilde{A}_\mu(x)$, after perhaps the addition of some space-time element in the Lie algebra of $H$, to transform according to $Ad(h(g,x))$. To make these matters precise, we compute:

## 1.8 Controllability of supersymmetric field theoretic problems

The Lagrangian of the super-Yang-Mills field is

$$L = F^a_{\mu\nu}F^{\mu\nu a} + c.\bar{\chi}^a\gamma^\mu D_\mu\chi^a - - - (A.1)$$

where $F^a_{\mu\nu}$ is the antisymmetric field tensor corresponding to the Yang-Mills gauge potentials $A^a_\mu$ and $\chi^a$ is the gaugino field. $D_\mu$ acts on the gaugino field in the adjoint representation. In other words, by $D_\mu\chi^a$ we actually mean $[D_\mu, \chi]^a$ where $\chi = \chi^a\tau_a$ and $D_\mu = \partial_\mu + iA^a_\mu\tau_a$. Thus,

$$[D_\mu, \chi]^a = \chi^a_{,\mu} + C(abc)A^b_\mu\chi^c$$

It is well known that the action corresponding to the Lagrangian (1) is invariant under the infinitesimal local supersymmetric transformations

$$\delta A^a_\mu(x) = \bar{\epsilon}(x)\gamma_\mu\chi^a(x),$$

$$\delta\chi^a(x) = \gamma_{\mu\nu}\epsilon(x)F^{\mu\nu a}(x)$$

where $\epsilon(x)$ is an infinitesimal Majorana parameter field and

$$\bar{\epsilon}(x) = \epsilon(x)^T\gamma^0$$

provided that the dimension of space-time is 10. This critical dimension is needed for the term involving a product of three gaugino Majorana spinor fields $\chi^a$ in

$$\delta_A(\bar{\chi}^a\gamma^\mu D_\mu\chi^a) = \bar{\chi}^a\gamma^\mu(\delta D_\mu)\chi^a$$

$$= \bar{\chi}^a \gamma^\mu (C(abc)(\delta A^b_\mu)\chi^c$$

$$= C(abc)\bar{\chi}^a \gamma^\mu \chi^c.\bar{\epsilon}(x)\gamma_\mu \chi^b(x),$$

$$= C(abc)\bar{\chi}^a \gamma^\mu \chi^c \bar{\chi}^b \gamma_\mu \epsilon$$

to vanish for any $\epsilon(x)$. This is equivalent to require the vanishing of

$$C(abc)\bar{\chi}^a \gamma^\mu \chi^c \bar{\chi}^b \gamma_\mu$$

Then, to the locally supersymmetric Lagrangian (1), we can add supersymmetry breaking terms by coupling the gauge field $A^a_\mu$ to classical current source $J^{\mu a}$ and its superpartner, the gaugino field $\chi^a$ to a classical vector potential $C^\mu$ so that the perturbing Lagrangian is

$$\Delta L = J^{\mu a} A^a_\mu + C^\mu \bar{\chi}^a \gamma_\mu \chi^a$$

and then address the controllability question.

Controllability of Yang-Mills gauge fields in the quantum context using Feynman's path integral approach to quantum field theory. Consider the action functional

$$I_T[A|J] = \int_{0 \le x_0 \le T, (x^1, x^2, x^3) \in \mathbb{R}^3} [(-1/4)F^a_{\mu\nu} F^{\mu\nu a} + A^a_\mu J^{\mu a}] d^4x$$

where $J^{\mu a}(x)$ is a classical current field. Let $\psi_\alpha(A), \alpha \in I$ be a family of wave functionals of the gauge field $A$ and let $S_g$ be a given scattering matrix in gauge field space. We compute the scattering matrix elements w.r.t these states:

$$< \psi_\alpha | S_g | \psi_\beta > = \int \psi_\alpha(A)^* S_g \psi_\beta(A) > dA, \alpha, \beta \in I$$

and ask the question, when does there exist a control current field $J$ so that the actual scattering matrix elements over a given time duration $[0, T]$ defined by

$$S_{\alpha\beta}(J) = \int \psi_\alpha(A)^* S(J) \psi_\beta(A) dA$$

where

$$S(J)\psi(A_f) = \int K(A_f, A_i | J) \psi(A_i) dA_i$$

with

$$K(A_f, A_i | J) = \int_{A(0,.) = A_i, A(T,.) = A_f} exp(i I_T[A|J]) \Pi_{0 \le x^0 \le T, \mathbf{x} \in \mathbb{R}^3} dA(x^0, \mathbf{x})$$

being the Feynman path integral evolution kernel are close to the given ones ?

## 1.9 Large deviations and control theory

Let $p(\partial) = L$ be a linear partial differential operator in $n$ variables $x = (x_1, ..., x_n)$ such that $L$ transforms a vector valued signal $[f_1(x), ..., f_n(x]^T$ into another vector valued signal $[g_1(x), ..., g_m(x)]^T$. Thus, $L$ can be looked upon as an $m \times n$ matrix partial differential operator:

$$L = ((p_{ij}(\partial)))_{1 \leq i \leq m, 1 \leq j \leq n}$$

We consider the special case when $m = n$ and then consider the stochastic pde

$$Lf(x) = s(x) + \sqrt{\epsilon} w(x), x \in \mathbb{R}^n$$

where $s(x)$ is an input signal field and $w(x)$ is a zero mean Gaussian noise field with covariance

$$\mathbb{E}(w(x)w(y)^T) = K_w(x, y)$$

In the absence of noise, assuming $L$ to be invertible, we write the general solution as $f_0(x) = L^{-1}s(x) + h(x)$ where $h(x)$ is any solution to the homogeneous pde

$$Lh(x) = 0$$

Now we ask the following question: When weak noise is present, then what the the approximate probability for the solution

$$f(x) = L^{-1}(s(x) + \sqrt{\epsilon} w(x)) + h(x) = f_0(x) + \sqrt{\epsilon} L^{-1}(w)(x)$$

to deviate by an amount greater than $\delta$ from the noiseless solution $f_0(x)$ over a domain $D \subset \mathbb{R}^n$ in the $L^2$ sense ? Using the theory of large deviations for Gaussian random variables, this probability is approximately given by

$$P(\int_D \parallel f(x) - f_0(x) \parallel^2 > dx\delta^2) \approx$$

$$exp(-(1/2(\epsilon))inf_{g: \int_D \|g(x)\|^2 > \delta^2} \int_{D \times D} g(x)^T Q(x, y) g(y) dx dy)$$

where

$$\begin{aligned} Q = R^{-1}, R(x, y) &= \mathbb{E}((L^{-1}w(x)).(L^{-1}w)(y)^T) \\ &= \int_{D \times D} L^{-1}(x, u) K_w(u, v) L^{-1}(y, v) du dv \end{aligned}$$

This approximate probability evaluates to

$$exp(-\delta^2/2\epsilon\lambda_{max}(R))$$

where $\lambda_{max}(R)$ is the maximum eigenvalue of the kernel $R$ on $D \times D$ and $Q$ is the inverse kernel of $R$ on $D \times D$:

$$\int_D R(x, u) Q(u, y) du = \delta^n(x - y), x, y \in D$$

Now we wish to reduce this deviation probability by adding non-random control terms which would reduce the effects of noise. Specifically, we modify the system to

$$Lf(x) = s(x) + \sqrt{\epsilon}w(x) + u(x,\theta)$$

where $u(x,\theta)$ is an input depending upon a control parameter vector $\theta$. We write

$$L^{-1}u(x,\theta) = v(x,\theta)$$

and then the controlled system output can be expressed as

$$f(x) = f_0(x) + v(x,\theta) + \sqrt{\epsilon}L^{-1}w(x)$$

and the corresponding deviation probability then gets modified to on defining the output noise field

$$d(x) = L^{-1}w(x)$$

to

$$P(\int_D \| f(x) - f_0(x) \|^2 > dx\delta^2)$$

$$= P(\int_D \| v(x,\theta) + \sqrt{\epsilon}d(x) \|^2 \, dx >\geq \delta^2) =$$

$$exp(-(2\epsilon)^{-1} inf_{g:\int_D \|v(x,\theta)+g(x)\|^2 > dx\delta^2} \int_{D\times D} g(x)^T Q(x,y)g(y)dxdy)$$

Finally, the parameter vector $\theta$ must be chosen so that this deviation probability is as small as possible, or equivalently, such that

$$inf_{g:\int_D \|v(x,\theta)+g(x)\|^2 > dx\delta^2} \int_{D\times D} g(x)^T Q(x,y)g(y)dxdy)$$

is as large as possible.

Some additional remarks on controllability: Suppose we have a nonlinear vector sde of the form

$$dx(t) = f(t, x(t)|\theta)dt + \sqrt{\epsilon}g(t, x(t), \theta)dB(t)$$

Then, we wish to design the control parameters $\theta$ such that the probability of the trajectory falling in a set $D \subset \mathbb{C}[0,T]^n$ over the time duration $[0,T]$ is minimized. Then, the large deviation solution to this problem will be to maximize

$$E(\theta) = \inf_{\{x\in D, \xi\in C^1[0,T]: dx(t)/dt = f(t,x(t),\theta)+g(t,x(t),\theta)\xi(t)\}} \int_0^T \| \xi(t) \|^2 \, dt$$

## 1.10 Approximate controllability of the Maxwell equations

The wave equations for the vector and scalar potentials are

$$\square A^{\mu} = J^{\mu} + \sqrt{\epsilon} w^{\mu}$$

where

$$J^{\mu}_{,\mu} = 0, w^{\mu}_{,\mu} = 0$$

and $w^r(x)$ are jointly Gaussian noise fields with zero mean. Thus,

$$w^0_{,0} = -w^r_{,r}, w^0 = -\int_0^{x^0} w^r_{,r} dx^0$$

We now incorporate control terms in the current, ie, replace $J^{\mu}$ by $J^{\mu}(x) + K^{\mu}(x|\theta)$ where $\theta$ are control parameters and $K^{\mu}$ satisfies

$$K^{\mu}_{,\mu}(x|\theta) = 0 \forall \theta \in \Theta$$

The solution is

$$A^{\mu}(x) = \int G(x - x')(J^{\mu}(x') + K^{\mu}(x'|\theta) + \sqrt{\epsilon} w^{\mu}(x'|\theta))d^4x'$$

and the problem is to use large deviation theory for Gaussian processes to mininimize the probability

$$P(\int_D \parallel A^{\mu}(x) - A^{\mu}_d(x) \parallel^2 W(x)d^4x > \epsilon)$$

by instead minimizing the inifimum of the rate function of $A^{\mu}$ over the given set indicated in the probability w.r.t the control parameters $\theta$.

## 1.11 Controllability problems in quantum scattering theory

Let $H_0$ denote the free projectile Hamiltonian and $H_1(\theta)$ the Hamiltonian when the projectile interacts with the scattering centre. $\theta$ is a control parameter vector for the scattering potential. The wave operators are

$$\Omega_+(\theta) = lim_{t \to \infty} exp(-iH_1(\theta)).exp(itH_0),$$

$$\Omega_-(\theta) = lim_{t \to -\infty} exp(-itH_1(\theta)).exp(itH_0)$$

These wave operators can be computed using the Lippmann-Schwinger equations. The scattering matrix is then

$$S(\theta) = \Omega_+(\theta)^* \Omega_-(\theta)$$

The controllability problem is then to determine whether for each given scattering operator $S_g$ (ie, a unitary operator) in a given family whether there exists a $\theta$ for which $S(\theta)$ has a distance smaller than $\epsilon$ w.r.t the spectral norm from $S_d$.

Kalman's notion of controllability and its extension to pde's: Consider the state equations

$$X'(t) = AX(t) + BU(t)$$

This can also be expressed as

$$[Id/dt - A, B] \begin{pmatrix} X(t) \\ U(t) \end{pmatrix} = 0$$

The controllability problem then involves determining whether a solution $(X(t), U(t))$ to this equation exists over the time interval $[t_1, t_2]$ such that the value of $X(t)$ at $t_1$ and $t_2$ are given. Generalizing this to pde's, we ask the question, given a matrix partial differential operator $p(\partial)$, there exists a solution $f$ to it, ie, $p(\partial)f(x) = 0$ such that $\Psi(f(x))$ has specified values on two disjoint Borel sets $U$ and $V$. Note that in the above special case considered by Kalman, the two disjoint open sets are $\{t_1\}$ and $\{t_2\}$.

## 1.12 Controllability in the context of representations of Lie groups

Let $\pi$ be a representation of a Lie group $G$ that acts on a manifold $M$. Let an image field $f_1(x)$ be given on $M$. After transforming it by a $G$-action and adding noise to it, the image field becomes

$$f_2(x) = f_1(g^{-1}.x) + w(x), x \in M$$

We can regard $f_1$ as the input image field and $f_2$ as the output image field with the system being defined by $g \in G$.

We can write $g(t) = exp(tX)$ for a one parameter sub-group $t \to g(t)$ of $G$ where $X$ is an element of the Lie algebra $\mathfrak{g}$ of $G$. Then the initial image field $f(x)$ after time $t$ transforms to

$$f_2(t,x) = f_1(exp(-tX).x), t \geq 0, x \in M$$

Its rate of change at time $t$ is given by

$$\partial f_2(t,x)/\partial t = -\xi_X(x).f_2(t,x)$$

where $\xi_X$ is the vector field induced on $M$ by the infinitesimal action of the one parameter group $g(t)$ on $M$, ie,

$$\xi_X(x) = \frac{d}{dt} exp(tX).x|_{t=0}$$

Formally, we can express the solution to the above pde as

$$f_2(t,x) = exp(-t\xi_X(x)).f_1(x)$$

and now we can pose the controllability question: Given a family of smooth functions $\mathcal{F}$ on $M$, and two elements $f_a, f_b$ in $M$ does there exist an element $X \in \mathfrak{g}$ such that

$$f_b(x) = exp(-T.\xi_X(x))f_a(x)$$

for some fixed $T > 0$. Note that this is equivalent to the existence of a one parameter group $g(t)$ such that

$$f_b(x) = f_a(g(T)^{-1}x)$$

Now the representation $\pi$ of $G$ induces a representation $d\pi$ of $\mathfrak{g}$ such that

$$\pi(exp(tX)) = exp(t\pi(X)), t \geq 0$$

If $G$ acts transitively on $M$, we can define a Fourier transform of $f(x)$ at $\pi$ formally by choosing a point $x_0 \in M$ and defining

$$\hat{f}(\pi) = \int_G f(gx_0)\pi(g)dg$$

After time $t$, $f(x)$ evolves to $f(t,x) = f(exp(-tX)x)$ and its Fourier transform at $\pi$ is then given by

$$\hat{f}(t,\pi) = \int_G f(t,gx_0)\pi(g)dg = \int_G f(exp(-tX)gx_0)\pi(g)dg$$

$$= \int_G f(gx_0)\pi(exp(tX)g)dg = \pi(exp(tX))\hat{f}(\pi)$$

which is equivalent to saying that

$$\partial\hat{f}(t,\pi)/\partial t = d\pi(X).\hat{f}(t,\pi)$$

the right side being interpreted in terms of ordinary matrix multiplication. Thus, we have

$$\hat{f}(t,\pi) = exp(t.d\pi(X))\hat{f}(\pi)$$

and hence we can pose the controllability problem of when this operation will carry the Fourier transform of an initial signal field evaluated at $\pi$ to the Fourier transform of another signal field at $\pi$ when both the signal fields are taken from a given family ?

## 1.13 Irreducible representations and maximal ideals

Let $A$ be an algebra with a unit 1 and $\pi$ a representation of this algebra in a vector space $V$. Assume that $\pi$ has a cyclic vector $v$, ie,

$$\pi(A)v = V$$

Define

$$I_0 = \{x \in A : \pi(x)v = 0\}$$

Clearly, $I_0$ is a left ideal of $A$ and $A/I_0$ is isomorphic to $V$ as a vector space via the mapping $x + I_0 \to \pi(x)v, x \in A$. Now, let $W$ be a $\pi$-invariant subspace of $V$ and define

$$I_W = \{x \in A : \pi(x)v \in W\}$$

Clearly since $W$ is $\pi$ invariant, it follows that $I_W$ is a left ideal in $A$ containing $I_0$. We can define a representation $\bar{\pi}$ of $A$ in $A/I_0$ by $\bar{\pi}(x)(y+I_0) = xy+I_0, x, y \in A$. Then, it is clear that $\bar{\pi}$ is isomorphic to $\pi$ and that $I_W/I_0$ is a $\bar{\pi}$-invariant subspace of $A/I_0$. This argument shows that there is a one-one correspondence, ie, bijection between all $\pi$-invariant subspaces of $V$ and all $\bar{\pi}$ invariant subspaces of $A/I_0$. Equivalently, there is a one-one correspondence between all $\pi$-invariant subspaces of $V$ and all ideals of $A$ containing $I_0$. In particular, if $\pi$ is irreducible, then $I_0$ is a maximal ideal of $A$ and conversely if $I_0$ is any maximal ideal of $A$, then the representation $\bar{\pi}$ of $A$ in $A/I_0$ is irreducible with $1+I_0$ as a cyclic vector. In other words, there is a one-one correspondence between equivalence classes of irreducible representations of $A$ and maximal ideals in $A$. This fact was used by Harish-Chandra with great power in constructing all the finite dimensional irreducible representations of a semisimple Lie algebra using dominant integral weights.

## 1.14 Controllability of the Maxwell-Dirac equations using external classical current and field sources

Suppose that we are given a second quantized Dirac electron-positron field and and a second quantized Maxwell photon field inside a cavity having perfectly conducting boundary. We wish to control these fields so that the far field radiation pattern has a given set of quantum moments in a given state of the cavity field, say the tensor product of a Bosonic and Fermionic coherent state. Let $A^c_\mu$ and $J^c_\mu$ denote the external classical field and current sources into the cavity. The perturbed Maxwell-Dirac equations are then given by

$$\square A_\mu = -e\psi^* \alpha^\mu \psi + J^c_\mu,$$

$$\gamma^\mu(i\partial_\mu - m)\psi = -e\gamma^\mu(A_\mu + A^c_\mu)\psi$$

Let $D(x-x')$ and $S(x-x')$ denote respectively the photon and electron propagator kernels:

$$D = \Box^{-1}, S = (i\gamma^\mu\partial_\mu - m + i0)^{-1}$$

The approximate solutions to these equations is

$$A_\mu(x) = A^{(0)}_\mu(x) + \int G(x-x')(-e\psi_0(x')^*\alpha_\mu\psi_0(x') + J^c_\mu(x'))d^4x'$$

$$= A^{(0)}_\mu(x) + \delta A_\mu(x)$$

$$\psi(x) = \psi_0(x) + \int S(x-x')(-e\gamma^\nu(A^{(0)}_\nu(x') + A^c_\nu(x'))d^4x'$$

$$= \psi_0(x) + \delta\psi(x),$$

where $\psi_0$ free second quantized Dirac field that satisfies

$$[i\gamma^\mu\partial_\mu - m]\psi_0 = 0$$

and $A^{(0)}_\mu$ is the free second quantized photon field that satisfies

$$\Box A^{(0)}_\mu = 0$$

$\psi_0$ is expressed as a linear superposition of plane waves with dispersion relation $p^0 = \sqrt{m^2 + p_1^2 + p_2^2 + p_3^2}$ with coefficients being the electron annihilation and positron creation operators in momentum space while $A^{(0)}_\mu$ is expressed as a linear superposition of plane waves with dispersion relation $k^0 = \sqrt{k_1^2 + k_2^2 + k_3^2}$ with coefficients being photon annihilation and creation operators in momentum space. Using these expressions, if $|\Phi>$ is any state of the electrons, positrons and photons within the cavity, then we can calculate in principle all the moments of the radiation and Dirac field in this state. electromagnetic radiation from the cavity comes from the current of electron and positrons within the cavity as well as from the surface current density induced by the the quantum magnetic field on the cavity boundary. Applying the retarded potential formula to these two currents, it follows that the total quantum electromagnetic field radiated from the cavity will have the form

$$A_{rad,\mu}(x) = \int_{cavity} G_1(x,x')\psi(x')^*\alpha_\mu\psi(x')d^4x'$$

$$+ \int_{cavity} G^\nu_{2\mu}(x,x')A_\nu(x')d^4x'$$

In this expansion, we retain terms only upto linear orders in $\delta\psi$ and $\delta A_\mu$. It follows then that we can express the radiated electromagnetic field as

$$A_{rad,\mu}(x) = F_{1\mu}(x) + \int F_{2,\mu}(x,x')\delta\psi(x')d^4x'$$

$$+\int F_{3\mu}^{\nu}(x,x')\delta A_{\nu}(x')d^4x'$$

where $F_{2\,mu}(x,x')$ is a linear functional of $\psi_0$ and hence of the electron-positron creation and annihilation operators while $F_{3\mu}^{\nu}$ are c-number functions. $F_{1\mu}$ is the retarded Maxwell potential produced by the free Dirac quantum current density $-e\psi_0^*\alpha_\mu\psi_0$ and is therefore a quadratic functional of the electron and positron creation and annihilation operator fields. It should be noted that the classical current and potential sources are contained in the terms $\delta\psi$ and $\delta A_\mu$.

The controllability issue can then be posed as follows: For a given $\epsilon,\delta>0$ and a given electromagnetic field $A_{g,\mu}(x')$ in a region $V$ of space-time, do there exist classical control fields $J_\mu^c$ and $A_\mu^c$ so that the quantum average of $A_{rad,\mu}(x)$ in the given coherent state of the electrons, positrons and photons has a distance smaller than $\epsilon$ from the given electromagnetic field in the sense of a weighted integral of the error square over $V$ and simultaneously, this field has a fluctuation mean square value smaller than $\delta^2$ over this region in the coherent state ? By fluctuation mean square value, we mean the quantity

$$\int_{V\times V} <\Phi|A_{rad,\mu}(x)A_{rad,\nu}(x')|\Phi> W^{\mu\nu}(x,x')d^4xd^4x'$$

$$-\int_{V\times V} <\Phi|A_{rad,\mu}(x)|\Phi><\Phi|A_{rad,\nu}(x')|\Phi> W^{\mu\nu}(x,x')d^4xd^4x'$$

We require that this must be smaller than $\delta^2$ and likewise, we require that

$$\int_{V\times V}(<\Phi|A_{rad,\mu}(x)|\phi>-A_{g,\mu}(x))(<\Phi|A_{rad,\nu}(x')|\phi>-A_{g,\nu}(x'))W^{\mu\nu}(x,x')d^4xd^4x'$$

must be smaller than $\epsilon^2$.

**Controllability of the EEG signals on the brain surface modeled as a spherical surface by influencing the infinitesimal dipoles in the cells of the brain cortex to vary in accord to sensory perturbations**

If $U(r)$ is the potential generated on the brain surface by infinitesimal dipoles $p_1,...,p_N$ present at the locations $r_1,...,r_N$ in the cortex, then $U$ satisfies Poisson's equation

$$\nabla^2U(r)=-\rho(r)/\epsilon,\rho(r)=\sum_k p_k.\nabla\delta^3(r-r_k)$$

More generally, if we assume the presence of stochastic perturbation terms in the charge distribution, we obtain the following stochastic pde

$$\nabla^2U(r)=-\sum_k p_k.\nabla\delta^3(r-r_k)+w(r)$$

Assume that there are no charges present on the brain surface, ie, if $\hat{n}$ is the unit normal at any point on the brain surface, then

$$\partial U(r)/\partial\hat{n}=0$$

Then if $G(r,r')$ is the Green's function for the Neumann boundary value problem, we have

$$\nabla^2 G(r,r') = \delta^3(r-r'), \partial G(r,r')/\partial \hat{n} = 0$$

and we get as solution

$$U(r) = \int G(r,r')(-\sum_k p_k.\nabla\delta^3(r'-r_k) + w(r'))d^3r'$$

Assume that $w(r)$ is weak zero mean Gaussian noise with known autocorrelation

$$R_w(r,r') = \mathbb{E}(w(r)w(r'))$$

Then, the controllability problem is to add additional charge sources to the brain cortex, defined by a control charge density $\rho_c(r)$, so that the controlled potential on the brain surface

$$U_c(r) = \int G(r,r')(-\sum_k p_k.\nabla\delta^3(r'-r_k) + w(r') + \rho_c(r'))d^3r'$$

has a minimal probability of determining an electric field $E_c(r) = \nabla U_c(r)$ on the brain surface $S$ that deviates from a given surface electric field $E_g(r)$ on $S$ by an amount $> \epsilon$. This is the classical control problem based on large deviation theory.

## 1.15 Application of the representation theory of $SL(2,\mathbb{C})$ as an alternative way of characterizing Lorentz transformations to control problems

We use the irreducible representations of $SL(2,\mathbb{C})$ to estimate a Lorentz group transformation element on a time varying three dimensional image field and using this estimate, to design an error feedback controller in the group domain so that the transformed image field is as close as possible in the sense of some distance measure to a given 3-D time varying image field. This idea can be compared to the extended Kalman filter based state observer to design a controller based on output error feedback so that the state tracks a given trajectory.

## 1.16 Construction of irreducible representations

Given a Lie group or more generally an algebraic group over a field, the question is how to construct an appropriate basis for an irreducible representation of the group or an irreducible representation of a module. The standard Verma-Module or Borel-Weil method involves starting with a formal vector, called the highest weight vector and operating on it freely by the negative root vectors of the semisimple Lie algebra and then extracting out a Maximal ideal from this free module and using the fact that the quotient of the universal enveloping algebra by a maximal ideal is an irreducible module for the universal enveloping algebra of the Lie algebra of the group. Standard monomials based on Schubert varieties of the Grassmannian provide nice bases for irreducible modules of algebraic groups like $SL(n, K), SO(n, K)$ where $K$ is any algebraic field. Such fields naturally appear in the construction of classical codes and we can look upon the elements of these algebraic groups as linear transformations on the space of code vectors and formulate code pattern recognition problems for the same via the irreducible representations of these groups.

Acknowledgements: I am grateful to Professor Shiva-Shankar for suggesting this problem to me and for providing me with his lecture notes on controllability, partial differential equations and the vector potential delivered at the Steklov Institute, Moscow.

References:

[1] Shiva Shankar, "Six lectures at the Steklov Institute, Moscow on Controllability and the vector potential.

[2] Amir Dembo and Ofer Zeitouni, "Large deviations, Techniques and Applications", Springer.

[3] V.S.Varadarajan, "Lie groups, Lie algebras and their Representations, Springer, 1984.

[4] C.S.Seshadri, "An Introduction to Standard Monomials", Hindustan Book Agency.

## 1.17 More on root space decomposition of a semisimple Lie algebra

Let $\mathfrak{g}$ be a semisimple Lie algebra and let $\mathfrak{h}$ be a Cartan subalgebra of it. Then, $\mathfrak{h}$ is Abelian, $ad(\mathfrak{h})$ is an Abelian family of semisimple linear operators on $\mathfrak{g}$ and hence can be simultaneously diagonalized. Thus we get the root space decomposition of $\mathfrak{g}$ as

$$\mathfrak{g} = \mathfrak{h} \oplus \bigoplus_{\alpha \in \Delta} \mathfrak{g}_\alpha$$

where $\Delta$ is a finite subset of $\mathfrak{h}^*$, none of which is zero (The zero eigenspace of $ad(\mathfrak{h})$ is precisely $\mathfrak{h}$ itself since $\mathfrak{h}$ is maximal Abelian) and

$$ad(H)(X) = [H, X] = \alpha(H)X, iffX \in \mathfrak{g}_\alpha \forall H \in \mathfrak{h}$$

In other words, for any $\alpha \in \Delta$, $\mathfrak{g}_\alpha$ is the eigenspace of $ad(H)$ corresponding to the eigenvalue $\alpha(H)$ for every $H \in \mathfrak{h}$. An element of $\Delta$ is called a root. We now claim that

$$\Delta = -\Delta$$

In fact, we have that for any $X \in \mathfrak{g}_\alpha, Y \in \mathfrak{g}_\beta$ the identity

$$B([H,X],Y) = -B(X,[H,Y]), H \in \mathfrak{h}$$

and hence

$$(\alpha(H) + \beta(H))B(X,Y) = 0$$

Further,

$$\alpha(H')B(H,X) = B(H,[H',X]) = -B([H,H'],X) = 0, H, H' \in \mathfrak{h}$$

and since $\alpha \in \mathfrak{h}^*$ is nonzero, it follows that

$$B(H,X) = 0 \forall H \in \mathfrak{h}$$

In other words, we have proved two things: One, that $\mathfrak{h} \perp \mathfrak{g}_\alpha \forall \alpha \in \Delta$ and two that $\beta \neq -\alpha$ implies that $\mathfrak{g}_\alpha \perp \mathfrak{g}_\beta$. From these two and the above root space decomposition of $\mathfrak{g}$, it easily follows that if there is an $\alpha \in \Delta$ such that $-\alpha \notin \Delta$, then $g_\alpha \perp \mathfrak{g}$ which contradicts the non-degeneracy of $B(.,.)$. This proves that $\Delta = -\Delta$. Next, we prove that $dim\mathfrak{g}_\alpha = 1 \forall \alpha \in \Delta$ and $\mathfrak{g}_{k\alpha} = 0, k > 1 \alpha \in \Delta$. Indeed consider the subspace $V$ of $\mathfrak{g}$ defined by

$$V = span\{Y\} \oplus \mathfrak{h}\mathfrak{g}_\alpha \oplus \mathfrak{g}_{2\alpha} \oplus ... \oplus \mathfrak{g}_{k\alpha} \oplus ...$$

the series terminating after a finite number of steps since $\mathfrak{g}$ and hence $V$ are finite dimensional. Here, $Y \in \mathfrak{g}_{-\alpha}$ is any non-zero element. Note that for any $\alpha, \beta \in \Delta$, we have

$$[\mathfrak{g}_\alpha, \mathfrak{g}_\beta] \subset \mathfrak{g}_{\alpha+\beta}$$

and

$$[\mathfrak{g}_\alpha, \mathfrak{g}_{-\alpha}] \subset \mathfrak{h} \forall \alpha, \beta \in \Delta$$

the first follows by Jacobi's identity and the second by Jacobi's identity and maximal Abelian property of $\mathfrak{h}$. Now choose $0 \neq X \in \mathfrak{g}_\alpha$ and Then $ad(X)$ leaves $V$ invariant while $ad(Y)$ also leaves $V$ invariant. Then, $ad(H)$ with $H = [X,Y] \in \mathfrak{h}$ also leaves $V$ invariant. But

$$ad(H) = [ad(X), ad(Y)]$$

and hence $Tr(ad(H)|_V) = 0$. Thus we get

$$0 = -\alpha(H) + \alpha(H)dim(\mathfrak{g}_\alpha) + 2\alpha(H)dim(\mathfrak{g}_{2\alpha}) + ...$$

Choosing $H \in \mathfrak{h}$ so that $\alpha(H) \neq 0$, we get

$$0 = -1 + dim(\mathfrak{g}_\alpha) + 2dim(\mathfrak{g}_{2\alpha}) + ...$$

and this easily results in

$$dim(\mathfrak{g}_\alpha) = 1, dim(\mathfrak{g}_{k\alpha}) = 0, k \geq 2$$

Now let $\alpha, \beta \in \Delta$. We look at the root spaces $\mathfrak{g}_{\alpha+k\beta}, k = -p, -p+1, ..., q-1, q$ where $\{\alpha + k\beta : k = -p, -p+1, ..., q-1, q\}$ is a maximal chain, ie, $\alpha + k\beta, k = -p, p+1, ..., q-1, q$ are all roots, ie, elements of $\Delta$ but $\alpha + (q+1)\beta$ and $\alpha - (p+1)\beta$ and $\alpha + (q+1)\beta$ are not roots. By maximality, it is clear that no other chain of this sort can overlap with this chain. Note that we have already proved that $dim(\mathfrak{g}_{\alpha+k\beta}) = 1, k = -p, ..., q$. Our immediate aim is to show that $V = \bigoplus_{k=-p}^{q} \mathfrak{g}_{\alpha+k\beta}$ is a vector space such that the restriction of the $sl(2, \mathbb{C})$ Lie algebra generated by $\{H_\alpha, X_\alpha, X_{-\alpha}\}$ has its adjoint representation restricted to $V$ an irreducible representation of $sl(2, \mathbb{C})$. Here, $X_\alpha \in \mathfrak{g}_\alpha$ and $X_{-\alpha} \in \mathfrak{g}_{-\alpha}$ are nonzero elements and their normalizations are chosen so that $B(X_\alpha, X_{-\alpha}) = 1$. Then, we have

$$[X_\alpha, X_{-\alpha}] = c\bar{H}_\alpha$$

for some $\bar{H}_\alpha \in \mathfrak{h}$ and hence,

$$cB(\bar{H}_\alpha, H) = B([X_\alpha, X_{-\alpha}], H) = B(X_\alpha, [X_{-\alpha}, H]) =$$

$$\alpha(H)B(X_\alpha, X_{-\alpha}), H \in \mathfrak{h}$$

We choose $\bar{H}_\alpha$ so that

$$B(\bar{H}_\alpha, H) = \alpha(H), H \in \mathfrak{h}$$

This is possible in view of the above equation by taking

$$c = B(X_\alpha, X_{-\alpha})$$

Note that since $B$ is non-singular on $\mathfrak{g}$, and $X_\alpha$ is orthogonal to both $\mathfrak{h}$ and to $\mathfrak{g}_\beta \forall \beta \neq -\alpha$, and since $X_{-\alpha}$ is a non-zero element of the one dimensional vector space $\mathfrak{g}_{-\alpha}$, it follows that $B(X_\alpha, X_{-\alpha}) \neq 0$. Thus with the above choice of $c$, we get that

$$[X_\alpha, X_{-\alpha}] = B(X_\alpha, X_{-\alpha})\bar{H}_\alpha$$

Now we define

$$H_\alpha = B(X_\alpha, X_{-\alpha})\bar{H}_\alpha$$

and obtain

$$[X_\alpha, X_{-\alpha}] = H_\alpha$$

and then,

$$B(H_\alpha, H_\alpha) = B(H_\alpha, [X_\alpha, X_{-\alpha}]) =$$
$$B([H_\alpha, X_\alpha], X_{-\alpha}) =$$
$$\alpha(H_\alpha)B(X_\alpha, X_{-\alpha})$$

If we choose the normalizations of $X_\alpha, X_{-\alpha}$ so that

$$B(X_\alpha, X_{-\alpha}) = 2/B(\bar{H}_\alpha, \bar{H}_\alpha) = 2/\alpha(\bar{H}_\alpha)$$

then we get

$$H_\alpha = 2\bar{H}_\alpha/\alpha(\bar{H}_\alpha) = 2\bar{H}_\alpha/B(\bar{H}_\alpha, \bar{H}_\alpha)$$

and then we get

$$[H_\alpha, X_\alpha] = \alpha(H_\alpha)X_\alpha = 2X_\alpha,$$
$$[H_\alpha, X_{-\alpha}] = -\alpha(H_\alpha)X_{-\alpha} = -2X_{-\alpha}$$

and as before,

$$[X_\alpha, X_{-\alpha}] = H_\alpha$$

In other words, the triplet $\{H_\alpha, X_\alpha, X_{-\alpha}\}$ form a standard set of generators of an $sl(2, \mathbb{C})$ Lie algebra.

## 1.18 A problem in Lie group theory

Let $H, X, Y$ be the standard generators of $sl(2, \mathbb{C})$, ie, $[H, X] = 2X, [H, Y] = -2Y, [X, Y] = H$. Then define for $t, x, y \in \mathbb{R}$,

$$g(t, x, y) = exp(tH + xX + yY)$$

and express

$$\partial_t g(t, x, y) = g(t, x, y)(a(1)H + a(2)X + a(3)Y),$$
$$\partial_x g(t, x, y) = g(t, x, y)(b(1)H + b(2)X + b(3)Y),$$
$$\partial_y g(t, x, y) = g(t, x, y)(c(1)H + c(2)X + c(3)Y)$$

Evaluate $a(k), b(k), c(k), k = 1, 2, 3$.

Hint: Use the the following formula for the differential of the exponential map:

$$\partial_t exp(A + tB) = exp(A + tB)(f(ad(A + tB))(B)$$

where

$$f(z) = (1 - exp(-z))/z$$

Hence, express the Haar measure on $SL(2,\mathbb{R})$ in terms of $t, x, y$, ie if $\mu$ is the Haar measure, then

$$d\mu(g(t,x,y)) = F(t,x,y)dtdxdy$$

and evaluate the Haar density $F$. For doing this, you must use the formulas obtained to express

$$g(t,x,y)H, g(t,x,y)X, g(t,x,y)Y$$

as linear combinations of $\partial_t g(t,x,y), \partial_x g(t,x,y), \partial_y g(t,x,y)$

## 1.19 More problems in linear algebra and functional analysis

### 1.19.1 Riesz' representation theorem

[1] Prove Riesz' representation theorem in the following form: Let $X = C[0,1]$, the space of continuous functions on $[0,1]$ with sup-norm, ie,

$$\| x \| = sup(|x(t)| : t \in [0,1]\}, x \in X$$

Show that $(X, \| . \|)$ is a Banach space.

[2] Let $(X, d)$ be a compact metric space and let $C(X)$ denote the space of all continuous functions on $X$. Prove that $C(X)$ under the sup norm is a Banach space. Let $\psi : C(X) \to \mathbb{R}$ be a continuous linear functional on $C(X)$. Then, show that there exists a unique signed measure $\mu$ on the Borel subsets of $X$ such that

$$\psi(f) = \int_X f(x)d\mu(x), f \in C(X)$$

Show further that

$$\| \psi \| = sup(|\psi(f)| : \| f \| \leq 1) = |\mu|(X)$$

where

$$\| f \| = sup(|f(x)| : x \in X)$$

and $|\mu|$ is the variation norm of $\mu$, ie, corresponding to the Hahn decomposition

$$X = X_+ \cup X_-, X_+ \cap X_- = \phi$$

so that $\mu$ is non-negative on all the Borel subsets of $X_+$ and non-positive on all the Borel subsets of $X_-$, we define

$$|\mu|(E) = \mu(X_+ \cap E) - \mu(X_- \cap E)$$

### 1.19.2 Lie's theorem on solvable Lie algebras

Let $\mathfrak{g}$ be a solvable Lie algebra, ie for some finite positive integer $p$, we have

$$\mathcal{D}^{p+1}\mathfrak{g} = 0, \mathfrak{D}^p\mathfrak{g} \neq 0$$

where if $S$ is any subset of $\mathfrak{g}$, then $DS$ denotes the linear span of the elements $[X,Y], X,Y \in S$. Let $\rho$ be any finite dimensional representation of $\mathfrak{g}$ in a complex vector space $V$. Then, we claim that there exists a non-zero vector $v \in V$ and a linear functional $\lambda$ on $\mathfrak{g}$, ie, $\lambda \in \mathfrak{g}^*$ such that

$$\rho(X)v = \lambda(X)v, \forall X \in \mathfrak{g}$$

We prove this theorem by induction. $\mathfrak{g}$ is solvable, it is clear that $D\mathfrak{g}$ is a proper subspace of $\mathfrak{g}$ (In fact $D\mathfrak{g}$ is an ideal in $\mathfrak{g}$). Thus, we can choose a subspace $\mathfrak{h}$ of $\mathfrak{g}$ such that $dim\mathfrak{h} = dim\mathfrak{g} - 1$, or equivalently, $dim(\mathfrak{g}/\mathfrak{h}) = 1$ and such that

$$D\mathfrak{g} \subset \mathfrak{h} \subset \mathfrak{g}$$

It is clear that $\mathfrak{h}$ is an ideal in $\mathfrak{g}$ since

$$[X, \mathfrak{h}] \subset D\mathfrak{g} \subset \mathfrak{h}, X \in \mathfrak{g}$$

In particular,

$$D\mathfrak{h} \subset D\mathfrak{g} \subset \mathfrak{h}$$

or in other words, that $\mathfrak{h}$ is a proper Lie subalgebra of $\mathfrak{g}$ having codimension one in $\mathfrak{g}$. Hence, $\rho$ restricted to $\mathfrak{h}$ is a representation. By the induction hypothesis (induction is on $dim\mathfrak{g}$), it follows that there is a nonzero $w_0 \in V$ such that

$$\rho(Y)w_0 = \lambda(Y)w_0, \forall Y \in \mathfrak{h}$$

for some $\lambda \in \mathfrak{h}^*$. Now since $\mathfrak{h}$ has codimension one in $\mathfrak{g}$, it is clear that we can choose an $X_0 \in \mathfrak{g}, X_0 \notin \mathfrak{h}$ and then it follows that

$$\mathfrak{g} = \{cX_0 + Y : c \in \mathbb{C}, Y \in \mathfrak{h}\} = < X_0 > + \mathfrak{h}$$

Now let $q$ be the smallest non-negative integer for which $\{\rho(X_0)^k w_0 : 0 \leq k \leq q\}$ forms a linearly independent set. Then, $\rho(X_0)^{q+1}w_0$ is linearly dependent on $\{\rho(X_0)^k w_0 : 0 \leq k \leq q\}$ and by successive application of $\rho(X_0)$, it follows that for any $m > q$, $\rho(X_0)^m w_0$ is also linearly dependent upon $\{\rho(X_0)^k w_0 : 0 \leq k \leq q\}$. Define the subspaces

$$W_m = span\{w_0, w_1, ..., w_m\}, m = 0, 1, ..., q, w_k = \rho(X_0)^k w_0, k = 0, 1, ..., q$$

Since then $\rho(X_0)w_k = w_{k+1}, k = 0, 1, ..., q-1$ and $\rho(X_0)w_q \in W_q$, it follows that $W_q$ is $\rho(X_0)$ invariant. We have further,

$$\rho(Y)\rho(X_0)w_k = [\rho([Y, X_0]) + \rho(X_0)\rho(Y)]w_k, Y \in \mathfrak{h} - - - (1)$$

Now, $\mathfrak{h}$ is an ideal in $\mathfrak{g}$ and hence, if we assume that the proposition $P_k$ defined by

$$\rho(Y)w_k = \lambda(Y)w_k mod W_{k-1}, Y \in \mathfrak{h}$$

is valid for some $k$, then $P_0$ is true with $W_{-1} = 0$ and $P_k$ implies in view of (1) that

$$\rho(Y)w_{k+1} = \lambda([Y, X_0])w_k + \lambda(Y)w_{k+1}, Y \in \mathfrak{h} --- (2)$$

so that $P_{k+1}$ is also true. Note that since $[Y, X_0] \in \mathfrak{h}, Y \in \mathfrak{h}$ since $\mathfrak{h}$ is an ideal. Thus, by induction, it follows that $P_k$ is true for $k = 0, 1, ..., q$, ie,

$$\rho(Y)w_k = \lambda(Y)w_k mod W_{k-1}, k = 0, 1, ..., q, Y \in \mathfrak{h}$$

In particular, $W_q$ is $\rho(\mathfrak{h})$-invariant and further,

$$Tr(\rho(Y)|_{W_q}) = (q+1)\lambda(Y), Y \in \mathfrak{h} --- (3)$$

Now, we have seen that $W_q$ is $\rho(X_0)$ invariant and also $\rho(Y)$ invariant and hence, since $[Y, X_0] \in \mathfrak{h}$, we have

$$Tr(\rho([Y, X_0])|_{W_q}) = Tr([\rho(Y), \rho(X_0)]|_{W_q}) = 0, Y \in \mathfrak{h}$$

Thus from (3), with $[Y, X_0]$ in place of $Y$, we have

$$\lambda([Y, X_0]) = 0, Y \in \mathfrak{h}$$

and hence, we get from (2) that

$$\rho(Y)w_{k+1} = \lambda(Y)w_{k+1}, k = 0, 1, ..., q-1, Y \in \mathfrak{h}$$

and hence

$$\rho(Y)w_k = \lambda(Y)w_k, k = 0, 1, ..., q, Y \in \mathfrak{h}$$

In other words, for any $Y \in \mathfrak{h}$ $\rho(Y)$ acts on $W_q$ as multiplication by $\lambda(Y)$ times the identity operator. Further, since $W_q$ is $\rho(X_0)$-invariant, and the field is the $\mathbb{C}$ which is algebraically closed, we can choose a nonzero vector $v \in W_q$ such that

$$\rho(X_0)v = c_1 v$$

for some complex number $c_1$. Now, we extend the domain of definition of $\lambda$ from $\mathfrak{h}^*$ to the whole of $\mathfrak{g}$ by setting

$$\lambda(cX_0 + Y) = cc_1 + \lambda(Y), Y \in \mathfrak{h}$$

and then the proof of the result is complete. Now from this result we deduce the result that if $\mathfrak{g}$ is a solvable Lie algebra and $\rho$ is a finite dimensional representation of $\mathfrak{g}$ in a complex vector space $V$, then we can choose a basis $B$ for $V$ such that for each $X \in \mathfrak{g}$, $[\rho(X)]_B$ is an upper triangular matrix. In fact, using Lie's theorem, we first choose a non-zero vector $v_1 \in V$ and a $\lambda \in \mathfrak{g}^*$ so that

$$\rho(X)v_1 = \lambda(X)v_1, X \in \mathfrak{g}$$

Then define

$$V_1 =< v_1 >$$

and observe that since $V_1$ is $\rho$-invariant, $\rho$ induces in a natural way a representation $\rho_1$ of $\mathfrak{g}$ on $V/V_1$ and again applying Lie's theorem to this representation, we get an element $v_2 \in V, v_2 \notin V_1$ such that $\rho(X)v_2 - \lambda_2(X)v_2 \in V_1, X \in \mathfrak{g}$ for some $\lambda \in \mathfrak{g}^*$. Define

$$V_2 =< v_1, v_2 >$$

Then we have

$$\rho(X)v_2 = \lambda_2(X)v_2 mod V_1, V_1 \subset V_2$$

In general, suppose, we have constructed linearly independent vectors $v_1, ..., v_k$ and linear functionals $\lambda_j \in \mathfrak{g}^*, j = 1, 2, ..., k$ such that

$$\rho(X)v_j = \lambda_j(X)v_j mod < v_1, ..., v_{j-1} >, j = 1, 2, ..., k, X \in \mathfrak{g}$$

Then, the vector space

$$V_k =< v_1, ..., v_k >$$

is clearly $\rho$-invariant and if $V_k = V$, the proof is complete while if $V_k \neq V$, then $\rho$ induces in a natural way a representation of $\mathfrak{g}$ on the non-zero vector space $V/V_k$ and hence by application of Lie's theorem, we can select a $v_{k+1} \notin V_k$ and a $\lambda_{k+1} \in \mathfrak{g}^*$ such that

$$\rho(X)v_{k+1} - \lambda_{k+1}(X)v_{k+1} = 0 mod V_k$$

Then set,

$$V_{k+1} =< v_1, ..., v_k, v_{k+1} >$$

and the induction proceeds further. This process will terminate in a finite number of steps since $V$ is finite dimensional.

### 1.19.3 Engel's theorem on nil-representations of a Lie algebra

Let $\mathfrak{g}$ be a (finite dimensional) Lie algebra and let $\rho$ be a nil-representation of $\mathfrak{g}$ in a finite dimensional complex vector space $V$, ie, $\rho(X)$ is nilpotent for all $X \in \mathfrak{g}$. Then, there exists a non-zero vector $v \in V$ such that $\rho(X)v = 0 \forall X \in \mathfrak{g}$. Hence, there exists a basis $B$ for $V$ such that $[\rho(X)]_B$ is strictly upper triangular for all $X \in \mathfrak{g}$. In particular, if $\mathfrak{g}$ is a nilpotent Lie algebra, ie, $ad(X)$ is nilpotent for all $X \in \mathfrak{g}$, then, there is a basis $B$ for $\mathfrak{g}$ such that $[ad(X)]_B$ is strictly upper triangular for every $X \in \mathfrak{g}$ and consequently, if $n = dim\mathfrak{g}$ and $X_1, ..., X_n \in \mathfrak{g}$ are arbitrary then

$$ad(X_1).ad(X_2)...ad(X_n) = 0$$

Solution:

Let

$$\mathfrak{a} = Ker(\rho) = \{X \in \mathfrak{g} : \rho(X) = 0\}$$

Then, $\mathfrak{a}$ is an ideal in $\mathfrak{g}$ since if $X \in \mathfrak{g}, Y \in \mathfrak{a}$, then $\rho(Y) = 0$ and therefore,

$$\rho([X, Y]) = [\rho(X), \rho(Y)] = 0$$

implying that $[X, Y] \in \mathfrak{a}$. So it is meaningful to speak of the Lie algebra

$$\mathfrak{g}' = \mathfrak{g}/\mathfrak{a}$$

with the bracket defined by

$$[X + \mathfrak{a}, Y + \mathfrak{a}] = [X, Y] + \mathfrak{a}, X, Y \in \mathfrak{g}$$

Note that the bracket is well defined since

$$X' + \mathfrak{a} = X + \mathfrak{a}, Y' + \mathfrak{a} = Y + \mathfrak{a}$$

imply

$$U = X' - X, V = Y' - Y \in \mathfrak{a}$$

which imply

$$[X', Y'] = [X + U, Y + V] = [X, Y] + Z$$

where

$$Z = [X, V] + [U, Y] + [U, V] \in \mathfrak{a}$$

since $\mathfrak{a}$ is an ideal. Thus,

$$[X', Y'] + \mathfrak{a} = [X, Y] + \mathfrak{a}$$

proving thereby that the bracket on the vector space $\mathfrak{g}/\mathfrak{a}$ is well defined. The necessary properties of the bracket, namely bilinearity, skew-symmetry and the Jacobi identity immediately follows from the same properties of the bracket on $\mathfrak{g}$. Now, clearly the Lie algebra $\mathfrak{g}/\mathfrak{a}$ is isomorphic with the Lie algebra $\rho(\mathfrak{g})$ via the isomorphism $\rho_1$ that maps $X + \mathfrak{a}$ to $\rho(X)$ for $X \in \mathfrak{g}$. For example, the injectivity of $\rho_1$ follows from the fact that $X + \mathfrak{a} = \mathfrak{a}$ implies $X \in \mathfrak{a}$ implies $\rho(X) = 0$. Surjectivity of $\rho_1$ is obvious. That $\rho_1$ preserves the bracket follows from

$$\rho_1([X + \mathfrak{a}, Y + \mathfrak{a}]) = \rho_1([X, Y] + \mathfrak{a})$$
$$= \rho([X, Y]) = [\rho(X), \rho(Y)] = [\rho_1(X + \mathfrak{a}), \rho_1(Y + \mathfrak{a})]$$

Note that $\rho_1$ is well defined because $X + \mathfrak{a} = X' + \mathfrak{a}$ implies $X' - X \in \mathfrak{a}$ implies $\rho(X') - \rho(X) = \rho(X' - X) = 0$. In other words, $\rho_1$ is a faithful representation of $\mathfrak{g}'$ in $V$. Since the elements of $\rho(\mathfrak{g})$ are all nilpotent, it follows therefore from the above isomorphism that $\mathfrak{g}' = \mathfrak{g}/\mathfrak{a}$ is a nilpotent Lie algebra, ie, all the elements of $ad(\mathfrak{g}')$ are nilpotent. To see this clearly, we observe that if $U \in \mathfrak{g}'$, then

$$\rho_1 o adU(V) = \rho_1([U, V]) = [\rho_1(U), \rho_1(V)] = ad(\rho_1(U))(\rho_1(V))$$

or equivalently,

$$\rho_1 o ad(U) = ado\rho_1(U)$$

or equivalently,

$$\rho_1 o ad = ado\rho_1$$

on $\mathfrak{g}'$ or equivalently,

$$ad|_{\mathfrak{g}'} = \rho_1^{-1} o ad|_{\rho(\mathfrak{g})} o \rho_1$$

on $\mathfrak{g}'$ and since $ad|_{\rho(\mathfrak{g})}$ is nilpotent on $L(V)$ because $\rho(\mathfrak{g})$ is nilpotent on $V$, it follows that $ad|_{\mathfrak{g}'}$ is also nilpotent on $\mathfrak{g}'$.

Remark: if $A$ is a linear nilpotent operator on a vector space $W$, then $ad(A)$ is nilpotent on $L(W)$. This follows from the identity

$$ad(A)^n(B) = (L_A - R_A)^n(B) = \sum_{r=0}^{n} \binom{n}{r} L_A^r(-R_A)^{n-r}(B)$$

$$= \sum_{r=0}^{n} \binom{n}{r} (-1)^{n-r} A^r B A^{n-r} \forall B \in L(W)$$

In view of the above remarks, we can assume without any loss of generality that $\mathfrak{g}$ is a nilpotent Lie algebra and $\rho$ is a nil-representation of $\mathfrak{g}$ in $V$ in order to prove the existence of a non-zero vector $v$ such that $\rho(\mathfrak{g})v = 0$.

Let $\mathfrak{h}$ be a maximal proper Lie subalgebra of $\mathfrak{g}$. We claim that $dim\mathfrak{h} = dim\mathfrak{g} - 1$ or equivalently, that $dim(\mathfrak{g}/\mathfrak{h}) = 1$. In fact, $ad(\mathfrak{h})$ leaves $\mathfrak{h}$ invariant since $\mathfrak{h}$ is a Lie subalgebra and hence $ad(\mathfrak{h})$ acting on $\mathfrak{g}/\mathfrak{h}$ defines a non-trivial representation of $\mathfrak{h}$ in the vector space $\mathfrak{g}/\mathfrak{h}$. Further, this representation is nilpotent since $\mathfrak{g}$ is a nilpotent Lie algebra and $\mathfrak{h}$ is a subset of $\mathfrak{g}$. It follows therefore from the induction hypothesis that there exists an $X_0$ in $\mathfrak{g}$ that is not contained in $\mathfrak{h}$ such that $ad(\mathfrak{h})(X_0) \subset \mathfrak{h}$, ie, $[X_0, \mathfrak{h}] \subset \mathfrak{h}$. But then $span < \mathfrak{h}, X_0 >$ is a Lie subalgebra of $\mathfrak{g}$ that properly contains $\mathfrak{h}$. Hence, by maximality of $\mathfrak{h}$ as a proper subalgebra, it follows that $< \mathfrak{h}, X_0 >= \mathfrak{g}$. In fact, this argument shows that $\mathfrak{h}$ is a ideal in $\mathfrak{g}$. This proves the claim. Now by the induction hypothesis, the vector space

$$W = \{v \in V : \rho(Y)v = 0 \forall Y \in \mathfrak{h}\}$$

has at least one non-zero vector. Further, $W$ is $\rho(X_0)$-invariant since $u \in W$ implies

$$\rho(Y)\rho(X_0)u = \rho([Y, X_0)])u + \rho(X_0)\rho(Y)u = 0, \forall Y \in \mathfrak{h}$$

since $[Y, X_0] \in \mathfrak{h}$. Since $W$ is therefore a non-zero complex $\rho(X_0)$-invariant subpspace, and since $\rho(X_0)$ is nilpotent on $V$ and hence also on $W$, it follows that $\rho(X_0)$ contains an eigenvector $v_0$ in $W$ and this eigenvector must have zero as its eigenvalue. Obivously since $v_0 \in W$, we have that $\rho(Y)v_0 = 0 \ \forall Y \in \mathfrak{h}$ and since $\mathfrak{h}$ and $X_0$ together span the whole of $V$, it follows that $\rho(X)v_0 = 0 \forall X \in \mathfrak{g}$. This completes the proof of Engel's theorem.

## 1.20 Spectral theory in Banach Algebras

[8] Prove the spectral theorem for bounded normal operators on a Hilbert space using the Gelfand-Naimark theorem on spectrum of a commutative Banach algebra combined with the Riesz representation theorem for continuous functionals on the space of continuous functions on a compact metric space.

hint: Let $\mathcal{H}$ be a Hilbert space and $A$ a commutative Banach subalgebra of $B(\mathcal{H})$. Let $\Delta$ denote the space of continuous homomorphisms from $A$ into $\mathbb{C}$. Thus, $h \in \Delta$ means that $h : A \to \mathbb{C}$ is a continuous map such that

$$h(TS) = h(T)h(S), T, S \in A, h(I) = 1,$$

$$h(aT + bS) = ah(T) + bh(S), a, b \in \mathbb{C}, T, S \in A$$

Note that if $T \in A$ is invertible, then it follows that

$$h(T^{-1}) = h(T)^{-1}$$

If $T \in A$, then $\lambda \in \sigma(T)$ (the spectrum of $T$) iff $\lambda - T$ is non-invertible in $A = B(\mathcal{H})$. Note that if $T \in A$, then for $\lambda \in \mathbb{C}$, $\lambda - T \in A$, in particular, this is in $B(\mathcal{H})$ and hence defined on the whole of $\mathcal{H}$, so if it is invertible, then by the open mapping theorem, $(\lambda - T)^{-1}$ is bounded, ie, in $A$ which means that $\lambda \notin \sigma(T)$. Thus, for $T \in A$, we have that $\lambda \in \sigma(T)$ iff $(\lambda - T)$ is non-invertible. If $\lambda \notin \sigma(T)$ and $h \in \Delta$, then

$$h((\lambda - T)^{-1})h(\lambda - T) = 1$$

implies that $h(T) \neq h(\lambda) = \lambda$ ($h(\lambda)$ is an abbreviation for $h(\lambda.e) = \lambda.h(e) = \lambda$). Equivalently, if $h(T) = \lambda$ for some $h \in \Delta$, then $\lambda \in \sigma(T)$ and this implies that $(\lambda - T)$ is not invertible. Note that if we write $\hat{\lambda}(h) = h(\lambda), \lambda \in \mathbb{C}$, then $\hat{\lambda}(h) = \lambda \forall h \in \Delta \forall \lambda \in \mathbb{C}$ and $\hat{T}(h) = h(T), h \in \Delta, T \in A$. Then, it follows that $\lambda \in \sigma(T)$ iff

$$h(T) = \hat{T}(h) = \lambda$$

for some $h \in \Delta$.

Remark: If $\lambda \in \sigma(T)$, then $\lambda - T$ is not invertible and hence $(\lambda - T)^{-1}$ is not defined. Choose a sequence $\lambda_n \to \lambda$ such that $\lambda_n - T$ is invertible for every $n$ and then we get

$$h((\lambda_n - T)^{-1})h(\lambda_n - T) = 1$$

implies

$$h((\lambda_n - T)^{-1})(\lambda_n - h(T)) = 1 \forall n$$

Now since $\lambda - T$ is not invertible and $\lambda_n \to \lambda$, it follows that there must exist an $h \in \Delta$ such that $limsup_n |h((\lambda_n - T)^{-1}| = \infty$ and therefore

$$h(T) = lim \lambda_n = \lambda$$

for such an $h$.

Remark: Another way to see this is as follows: Suppose $X = \lambda - T$ is not invertible. Then, $I = \{YX : Y \in A\}$ is an ideal in $A$ that does not contain the identity element $e$ and hence $I$ is contained in a maximal proper ideal of $A$. Thus, there exists an $h \in \Delta$ that vanishes on this ideal and in particular, $h(X) = 0$. From this, it follows that $h(T) = h(\lambda) = \lambda$. In other words, we have proved that

$$\sigma(T) = \{h(T) : h \in \Delta\} = \hat{T}(\Delta)$$

Note that if $h \in \Delta$, then $\lambda\mu = h(\lambda\mu) = h(\lambda)h(\mu), \lambda, \mu \in \mathbb{C}$.

Now let $T$ be a normal operator in $B(\mathcal{H})$, ie, $T$ is bounded and it commutes with $T^*$. Consider the commutative Banach algebra $A$ generated by $\{T, T^*\}$. Choose $x, y \in \mathcal{H}$ and consider the mapping $\hat{T} \to < x, Ty >$ from $C(\Delta)$ into

$\mathbb{C}$. This mapping is well defined since if $T, S \in A$ and $\hat{T} = \hat{S}$, then $h(T) = h(S) \forall h \in \Delta$ and hence $h(T - S) = 0 \forall h \in \Delta$ which implies that $\| T - S \| = 0$, ie, $T = S$.

Remark: if $A$ is a commutative Banach algebra and $x \in A$, then with $\Delta$ denoting the space of continuous homomorphisms on $A$, we have that

$$\| x \| = sup\{h(x) : h \in \Delta\} = sup\{\hat{x}(h) : h \in \Delta\} = \| \hat{x} \|$$

Remark: We have seen that the spectrum $\sigma(T)$ of $T$ can be identified with the continuous function $\hat{T}$ on $\Delta$. In fact, we have seen that if $A$ is a commutative Banach algebra, then

$$\sigma(T) = \hat{T}(\Delta) = \{h(T) : h \in \Delta\}$$

For example, if $T$ is a bounded Hermitian operator in the Hilbert space $\mathfrak{H}$ and if $A$ is the Banach algebra of all bounded functions of $T$, then from the spectral representation of $T$, then for any bounded function $f$ on $\mathbb{R}$, we have

$$f(T) = \int f(\lambda) dE(\lambda) = \int \lambda dE(f^{-1}(-\infty, \lambda]) - -(1)$$

We note that if $S_1, S_2 \in A$,

$$S_k = f_k(T) = \int f_k(\lambda) dE(\lambda), k = 1, 2$$

for some bounded measurable functions $f_k, k = 1, 2$ on $\mathbb{R}$ and therefore, if $h \in \Delta$ where $\Delta$ is the space of continuous homomorphisms from $A$ into $\mathbb{C}$, then

$$h(S_1 S_2) = \int f_1(\lambda) f_2(\lambda) h(dE(\lambda)) =$$

$$\int f_1(\lambda) f_2(\lambda) dh(E(\lambda))$$

on the one hand and on the other,

$$h(S_1 S_2) = h(S_1) h(S_2) = \int f_1(\lambda) f_2(\mu) h(dE(\lambda)).h(dE(\mu))$$

$$= \int f_1(\lambda) f_2(\mu) dh(E(\lambda)) dh(E(\mu))$$

and comparing these two expressions, we can infer that

$$dh(E(\lambda)).dh(E(\mu)) = dh(E(\lambda)) \delta_{\lambda,\mu}$$

which is equivalent to saying that

$$h(E(B_1)).h(E(B_2)) = h(E(B_1 \cap B_2)), B_1, B_2 \in \mathcal{B}(\mathbb{R})$$

This is in agreement with the homomorphism property of $h : A \rightarrow \mathbb{C}$:

$$h(E(B_1 \cap B_2)) = h(E(B_1).E(B_2)) = h(E(B_1)).h(E(B_2)), B_1, B_2 \in \mathcal{B}(\mathbb{R})$$

Now, suppose, we pick a $\lambda \in \sigma(T)$ and define

$$h_\lambda(T) = \lambda,$$

and more generally, for

$$S = \int f(\lambda)dE(\lambda)$$

we define

$$h_\lambda(S) = f(\lambda)$$

then

$$h_\lambda(S_1 S_2) = f_1(\lambda)f_2(\lambda) = h_\lambda(S_1)h_\lambda(S_2),$$

$$h_\lambda(c_1 S_1 + c_2 S_2) = c_1 f_1(\lambda) + c_2 f_2(\lambda) =$$

$$c_1 h_\lambda(S_1) + c_2 h_\lambda(S_2)$$

and hence $h_\lambda : A \rightarrow \mathbb{R}$ is a homomorphism, ie,

$$h_\lambda \in \Delta$$

It is clear from the above formulas that

$$sup\{|h_\lambda(S)| : \lambda \in \sigma(T)\} = sup\{|f(\lambda)| : \lambda \in \sigma(T)\} = \| S \|$$

More generally, if $h \in \Delta$ is arbitrary, ie, $h : A \rightarrow \mathbb{C}$ is a homomorphism, then for

$$S = \int f(\lambda)dE(\lambda)$$

we find that

$$h(S) = \int f(\lambda)dh(E(\lambda))$$

where the integral is taken over $\lambda \in \sigma(T)$. We now observe that for any Borel function $g$ on $\mathbb{R}$, we have that

$$h(g(T)) = g(h(T)), h \in \Delta$$

where $g(T)$ is defined as an operator via the spectral theorem for this result follows from

$$h(T^n) = h(T)^n, n = 0, 1, ...$$

and hence by taking linear combinations,

$$h(g(T)) = g(h(T))$$

In particular, taking $g$ as the function which maps $T$ to $E(\lambda)$, ie, $g(x) = \chi_{(-\infty,\lambda]}(x)$, we get

$$h(E(\lambda)) = \chi_{(-\infty,\lambda]}(h(T))$$

and hence,

$$h(S) = \int f(\lambda)d_\lambda\chi_{(-\infty,\lambda]}(h(T))$$

$$= \int f(\lambda)d\theta(\lambda - h(T)) = f(h(T))$$

which is consistent with our above observation. Thus,

$$\| S \| = sup\{|f(\lambda)| : \lambda \in \sigma(T)\}$$

$$= sup\{|f(h(T))| : h \in \Delta\}$$

which is once again, consistent with our previous observations regarding the definition of the spectrum of $T$ as the set of all $\lambda$ for which $\lambda - T$ is not invertible and our result that this spectrum is also equal to the set of all $h(T)$ as $h$ varies over $\Delta$.

Now we return to the commutative Banach algebra $A$ generated by $T, T^*$ where $T$ is a bounded normal operator. Then, we get from the Riesz representation theorem, in view of the fact that $\hat{S} \to < x, Sy >$ from $C(\Delta)$ into $\mathbb{C}$ is a bounded linear functional (in fact,

$$\| S \| = sup\{|h(S)| : h \in \Delta\} =$$

$$= sup\{|\hat{S}(h)| : h \in \Delta\} = \| \hat{S} \|$$

where the last norm occurs because it is the standard *sup*-norm on the space $C(\Delta)$)

$$< x, Sy > = \int_\Delta \hat{S}(h)d\mu_{x,y}(h), S \in A$$

for some measure $\mu_{x,y}$ on $\Delta$. Replacing $S$ by $f(T)$ gives us

$$< x, f(T)y > = \int_\Delta \Phi(f)(h)d\mu_{x,y}(h)$$

where we have defined

$$\Phi(f) = \hat{S} \in C(\Delta), S = f(T)$$

where $f$ is a bounded Borel function on $\mathbb{R}$. Now writing $S_1 = f_1(T), S_2 = f_2(T)$, we get

$$\Phi(f_1f_2)(h) = (S_1S_2)\hat{}(h) = h(S_1S_2) = h(S_1)h(S_2)$$

$$= \hat{S}_1(h)\hat{S}_2(h) = \Phi(f_1)(h)\Phi(f_2)(h) = (\Phi(f_1)\Phi(f_2))(h), h \in \Delta$$

and hence

$$\Phi(f_1f_2) = \Phi(f_1)\Phi(f_2)$$

for all bounded Borel functions $f_1, f_2$ on $\mathbb{R}$. In particular, if $B_1, B_2$ are bounded Borel subsets of $\mathbb{R}$, then we get

$$\Phi(\chi_{B_1 \cap B_2}) = \Phi(\chi_{B_1}).\Phi(\chi_{B_2})$$

We also have the following useful identities:

$$\int \Phi(f_1)(h) d\mu_{f_2(T)x,y}(h) =< f_2(T)x, f_1(T)y >=< x, f_2(T)^* f_1(T)y >$$

$$=< f_1(T)^* x, f_2(T)^* y >= \mu_{f_1(T)^*x, f_2(T)^*y}(\Delta)$$

In fact, suppose $S \in A$, ie, $S$ is a Borel function of $T, T^*$. Then,

$$< x, STy >=< S^* x, Ty >=< x, TSy >=< T^* x, Sy >$$

and hence,

$$\mu_{x,STy}(\Delta) = \mu_{S^*x,Ty}(\Delta) = \mu_{x,TSy}(\Delta) = \mu_{T^*x,Sy}(\Delta)$$

For

$$S = f(T, T^*)$$

we define $\Psi(f) = \hat{S} \in C(\Delta)$ and then get

$$< x, Sy >= \int_\Delta h(S) d\mu_{x,y}(h)$$

Replacing $S$ by $\chi_B(S)$ where $B$ is a bounded Borel set in $\mathbb{R}$ gives us

$$< x, \chi_B(S)y >= \int_\Delta h(\chi_B(S)) d\mu_{x,y}(h)$$

and thus, if $B_1, B_2$ are two bounded Borel subsets of $\mathbb{R}$, we have

$$< x, \chi_{B_1 \cap B_2}(S)y >= \int_\Delta h(\chi_{B_1}(S)) h(\chi_{B_2}(S)) d\mu_{x,y}(h)$$

Remark: To relate this circle of ideas to the spectral theorem for normal operators, we must transform the integral over $\Delta$ to an integral over $\sigma(T) = \hat{T}(\Delta)$ Since

$$\hat{T}(\Delta) = \{h(T) : h \in \Delta\} = \sigma(T),$$

it follows that

$$< x, Ty >= \int_\Delta \hat{T}(h) d\mu_{x,y}(h) = \int_{\sigma(T)} \lambda d\mu_{x,y} o \hat{T}^{-1}(\lambda)$$

and more generally,

$$< x, f(T)y >= \int_\Delta \Phi(f)(h) d\mu_{x,y}(h)$$

with

$$\Phi(f_1 f_2) = \Phi(f_1)\Phi(f_2), \Phi(cf_1 + f_2) = c\Phi(f_1) + \Phi(f_2)$$

More importantly,

$$< x, f(T)y >= \int_\Delta h(f(T))d\mu_{x,y}(h)$$

Now if $f(\lambda) = \lambda^n$, then

$$h(f(T)) = h(T^n) = h(T)^n = f(h(T))$$

and since $h$ is linear, it follows that if $f$ is any polynomial, then

$$h(f(T)) = f(h(T))$$

and by taking limits and using the continuity of $h$ we get that this identity is true even if $f$ is any bounded Borel function. Thus,

$$< x, f(T)y >= \int_\Delta f(h(T))d\mu_{x,y}(h)$$

$$= \int_\Delta fo\hat{T}(h)d\mu_{x,y}(h) = \int_{\sigma(T)} f(\lambda)d\mu_{x,y}o\hat{T}^{-1}(\lambda)$$

which is precisely the content of the spectral theorem once we are able to show that for any bounded Borel set $B$ in $\mathbb{R}$, we can write

$$\mu_{x,y}(\hat{T}^{-1}(B)) =< x, E_T(B)y >$$

where $E_T$ is a spectral measure on the Borel subsets of $\mathbb{R}$ with values in the space of orthogonal projection operators on $\mathcal{H}$.

## 1.21 Atiyah-Singer index theorem:A supersymmetric proof

Prerequisites:

[1] The Dirac operator in curved space-time with a Yang-Mills connection term.

[2] Lichnerowicz' formula for the square of the Dirac operator in curved space time with a Yang-Mills connection term.

[3] The index of a linear operator.

[4] Expressing the index of a linear operator $D$ using the difference trace of the heat kernels $exp(-tD^*D)$ and $exp(-tDD^*)$ by observing that the non-zero eigenvalues of $D^*D$ and $DD^*$ are identical inclusive of multiplities and hence the contribution to the index comes only from the multiplicities of their zero eigenvalues.

[5] Write

$$T = \begin{pmatrix} 0 & D^* \\ D & 0 \end{pmatrix}$$

and observe that

$$T^2 = \begin{pmatrix} D^*D & 0 \\ 0 & DD^* \end{pmatrix}$$

and hence with $str$ denoting supertrace, we have

$$str(exp(-tT^2)) = Tr(exp(-tD^*D)) - Tr(exp(-tDD^*)) = Index(D), t \in \mathbb{R}$$

## 1.22 Jordan decomposition on a semisimple Lie algebra

[a] Choose any $X \in \mathfrak{g}$ and consider the derivation

$$D = ad(X)$$

of $\mathfrak{g}$.

[b] Write down the Jordan decomposition of $D$ viewed as a linear operator on $\mathfrak{g}$:

$$D = U + V$$

where $U, V$ are respectively the semisimple and nilpotent parts of $D$. Thus $[U, V] = 0$ and hence $[X, U] = [X, V] = 0$ and in fact, we know from basic linear algebra that $U$ and $V$ are expressible as polynomial functions of $D$ having constant coefficient zero and that $U, V$ with this constaint on semisimplicity and nilpotency and mutual commutativity are uniquely determined by $D$.

[c] Hence, it is obvious that

$$D_{r,s} = U_{r,s} + V_{r,s}, r, s \geq 0, r + s \geq 1$$

is the Jordan decomposition of $D_{r,s}$ on $W_{r,s} = W^{\otimes r} \otimes W^{*s}$ with $V = \mathfrak{g}$.

[d] Thus $U_{r,s}, V_{r,s}$ are expressible as polynomial functions of $D_{r,s}$ with these polynomials having zero constant coefficient. Hence

$$N(D_{r,s}) \subset N(U_{r,s}), N(V_{r,s}), \forall r, s$$

ie, $U, V$ are replicas of $D$. In particular, taking $r = 1, s = 2$ gives us $N(D_{1,2}) \subset N(U_{1,2}), N(V_{1,2})$ which means that $U, V$ are also derivations on $V = \mathfrak{g}$. Since every derivation of a semisimple Lie algebra is inner (this is proved using the non-singularity of the Cartan-Killing form), it follows that $U = ad(S), V = ad(N)$ where $S, N \in \mathfrak{g}$. Since $U$ and $V$ commute and $ad([S, N]) = [ad(S), ad(N)]$, it follows from the faithfulness of the $ad$ map on $\mathfrak{g}$ that $[S, N] = 0$ and hence, we obtain the Jordan decomposition on the semisimple Lie algebra $\mathfrak{g}$:

$$X = S + N, S, N \in \mathfrak{g},$$

$ad(S)$ is semisimple, $ad(N)$ is nilpotent and $[S, N] = 0$. Furthermore, this decomposition is unique as follows from the Jordan decomposition on a vector space and the faithfulness of the $ad$ map on semisimple Lie algebras.

## 1.23 Construction of a Cartan subalgebra

[7] Show that if $\mathfrak{g}$ is a semisimple Lie algebra, and $\mathfrak{h}$ is a Cartan subalgebra, then there exists a regular element $X \in \mathfrak{g}$ such that

$$\mathfrak{h} = \mathfrak{h}_X = \bigcup_{m \geq 1} N(ad(X)^m)$$

hint: Let $X \in \mathfrak{h}$ be a regular element. Since $ad(X)$ is nilpotent on $\mathfrak{h}$ and since $X$ is regular, it follows that $ad(X)$ is non-singular on $\mathfrak{g}/\mathfrak{h}$. Then, it is easy to see that $\mathfrak{h}_X \subset \mathfrak{h}$ and by regularity of $X$, we get that $\mathfrak{h}_X = \mathfrak{h}$.

[8] Show that if $\mathfrak{g}$ is a semisimple Lie algebra, and $\mathfrak{h}$ is a Cartan subalgebra, then $\mathfrak{h}$ is maximal Abelian. hint:The steps involved in the proof ar outlined below:

[a] Let $X$ be regular in $\mathfrak{g}$ such that $\mathfrak{h} = \mathfrak{h}_X$. Let

$$X = S + N$$

be its Jordan decomposition. Then since $ad(X)(S) = [X, S] = 0, ad(X)(N) = [X, N] = 0$, it follows that $S, N \in \mathfrak{h}_X = \mathfrak{h}$. It is also clear that $S$ is regular and hence $\langle_S = \langle$. Regularity of $S$ can be seen as follows. Clearly, since $ad(S)$ and $ad(N)$ commute, $ad(N)$ leaves each eigensubspace of $ad(S)$ invariant and on such a subspace, $ad(N)$ is nilpotent. Thus, there is a basis for $\mathfrak{g}$ relative to which $ad(S)$ is diagonal and $ad(N)$ is strictly upper triangular. It follows that relative to such a basis $ad(X)$ is upper triangular with the same diagonal entries as those of $ad(S)$. Hence, the characteristic polynomial of $ad(X)$ is the same as that of $ad(S)$ and since $X$ is regular, it follows that $S$ is also regular. Now we wish to show that $N = 0$ and that will prove that $ad(\mathfrak{h})$ consists of only semisimple elements.

Another proof of the fact that $\mathfrak{h}_S = \mathfrak{h}_X$. First observe that $ad(X)(S) = 0$ and hence $S \in \mathfrak{h}_X$. Suppose that $Y \in \mathfrak{h}_S$. Then, $ad(S)^m(Y) = 0$ for some positive $m$ and since $ad(S)$ is semisimple, it follows that $ad(S)(Y) = 0$. It follows that $ad(S)^m(Y) = 0$ for all $m > 0$ and therefore,

$$ad(X)^m(Y) = \sum_{r=0}^{m} \binom{m}{r} ad(N)^{m-r} ad(S)^r(Y) = ad(N)^m(Y) = 0$$

for sufficiently large $m$ since $ad(N)$ is nilpotent. This implies that

$$Y \in \mathfrak{h}_X$$

and therefore,

$$\mathfrak{h}_S \subset \mathfrak{h}_X$$

Conversely suppose $Y \in \mathfrak{h}_X$. Then,

$$ad(S)^m(Y) = (ad(X) - ad(N))^m(Y) = \sum_{r=0}^{m} (-1)^{m-r}(ad(X))^r(ad(N))^{m-r}(Y)$$

for sufficiently large $m$ since $ad(X)$ is nilpotent on $\mathfrak{h}_X$ while $ad(N)$ is nilpotent. This proves that

$$Y \in \mathfrak{h}_S$$

and hence,

$$\mathfrak{h}_X = \mathfrak{h}_S$$

Another way to see that $\mathfrak{h}_X = \mathfrak{h}_S$ is simply to prove that $S$ is a regular element of $\mathfrak{h}_X$. But this would follow immediately if we are able to prove that the characteristic polynomial of $ad(S)$ equals that of $ad(X)$. This can be seen in the following way:

$$ad(X) = ad(S) + ad(N), [ad(S), ad(N)] = 0$$

is the Jordan decomposition of $ad(X)$. Now $ad(S)$ is semisimple on $\mathfrak{g}$ and hence we can choose a basis $B$ for $\mathfrak{g}$ such that $[ad(S)]_B$ is diagonal. Let $c_1, .., c_r$ be the distinct eigenvalues of $ad(S)$, then we can write the spectral decomposition of $ad(S)$ as

$$ad(S) = \sum_{k=1}^{r} c_k E_k$$

where

$$I = \sum_{k=1}^{r} E_k$$

is a resolution on $I$ (not necessarily orthogonal). Since $ad(N)$ commutes with $ad(S)$, it follows that it also commutes with every $E_k$, ie, it leaves $R(E_k) = N(ad(N) - c_k I)$ invariant. Now we can choose a basis for $R(E_k)$ so that $ad(N)$ restricted to $R(E_k)$ is strictly upper triangular since $ad(N)$ restricted to $E_k$ is nilpotent. If we pool up all these bases for $R(E_k)$ then we get a basis for $\mathfrak{g}$ such that $ad(S)$ in this basis is diagonal and $ad(N)$ in this basis is strictly upper triangular. In other words, we have shown that $ad(X) = ad(S) + ad(N)$ in this basis is upper-triangular with same diagonal entries as that of $ad(S)$. This immediately proves that the characteristic polynomials of $ad(X)$ and $ad(S)$ are the same and therefore since $X$ is regular, it follows that $S$ is also regular.

## 1.24 A criterion for regularity of an element in a Lie algebra

Problem: Show that if $X$ is regular then $Y \in \mathfrak{h}_X = \mathfrak{h}$ is regular iff $det(ad(Y)|_{\mathfrak{g}/\mathfrak{h}}) \neq 0$.

hint: The above determinant is zero iff there exists a $Z_1 \in \mathfrak{g} - \mathfrak{h}$ such that

$$ad(Y)(Z_1) = 0$$

Now choose elements $Z_2, ..., Z_r$ such that $\{Z_k + \mathfrak{h} : k = 1, 2, ..., r\}$ form a basis for $\mathfrak{g}/\mathfrak{h}$ and show that

$$\mathfrak{g} = \mathfrak{h} \oplus \mathfrak{q}$$

where

$$\mathfrak{q} = span\{Z_1, ..., Z_m\}$$

Now choose a basis $\{H_1, ..., H_l\}$ for $\mathfrak{h}$ such that $ad(Y)|_{\mathfrak{h}}$ in this basis is strictly upper triangular (This is possible because $\mathfrak{h}$ is a nilpotent Lie algebra). Then compute the characteristic polynomial of $ad(Y)$ in the basis $\{H_1, ..., H_l, Z_1, ..., Z_r\}$ for $\mathfrak{g}$ and observe that if $ad(Y)(Z_1) = 0$, then this characteristic polynomial is of the form

$$t^{l+1} f(t)$$

where $f$ is a polynomial and hence $Y$ cannot be regular. Further, if there is no such vector $Z_1$, then show that the characteristic polynomial of $ad(Y)$ relative to this basis must be of the form $t^l f(t)$ where $f$ is a polynomial with $f(0) \neq 0$ or in other words, $Y$ is regular. To show this, we must make the fact that $\mathfrak{h}_X$ is a nilpotent Lie algebra for regular $X$. To prove this fact, let $Y \in \mathfrak{h}_X$ be such that the determinant of $ad(Y)$ on $\mathfrak{g}/\mathfrak{h}_X$ is non-zero. Then, it is clear that the characteristic polynomial of $ad(Y)$ must be of the form $t^k f(t)$ where $k \leq dim\mathfrak{h}_X$ and $f(0) \neq 0$. Hence, $\mathfrak{h}_Y \subset \mathfrak{h}_X$. This is possible only if $\mathfrak{h}_Y = \mathfrak{h}_X$ because $X$ is regular and is nilpotent on $\mathfrak{h}_X$ and non-singular on $\mathfrak{g}/\mathfrak{h}_X$ and further that $\mathfrak{h}_X$ is $ad(Y)$-invariant (since $ad(X)^m(Y) = 0$ for some positive $m$, it follows that $ad(ad(X))^m(ad(Y)) = 0$ and therefore from elementary linear algebra, $ad(Y)$ leaves $\mathfrak{h}_X = \bigcup_{m \geq 1} N(ad(X)^m)$ invariant). In particular, $ad(Y)$ is nilpotent on $\mathfrak{h}_X$. Since the elements $Y$ of $\mathfrak{h}_X$ for which $ad(Y)$ is non-singular on $\mathfrak{g}/\mathfrak{h}_X$ are dense in $\mathfrak{h}_X$, it follows that every $Y \in \mathfrak{h}_X$ is such that $ad(Y)$ is nilpotent on $\mathfrak{h}_X$. Note that this proof also shows that if $Y$ is a regular element of $\mathfrak{h}_X$, then $\mathfrak{h}_Y = \mathfrak{h}_X$.

## 1.25 Cartan subalgebras of a semisimple Lie algebra are maximal Abelian

Now coming back to the proof that every Cartan algebra $\mathfrak{h}$ is maximal abelian, when $\mathfrak{g}$ is semisimple we choose a regular $X$ in $\mathfrak{h}$ and observe that $\mathfrak{h} = \mathfrak{h}_X$ because, $ad(X)$ is nilpotent on $\mathfrak{h}$ (by that part in the definition of a Cartan algebra which states that a Cartan algebra is nilpotent) and hence $\mathfrak{h} \subset \mathfrak{h}_X$. Further, consider the Lie algebra $\mathfrak{h}_X/\mathfrak{h}$. By Engel's theorem applied to the adjoint representation of the nilpotent algebra $\mathfrak{h}$ on this space, we deduce that there exists a $Y \in \mathfrak{h}_X - \mathfrak{h}$ such that $[Y, \mathfrak{h}] \subset \mathfrak{h}$ and since by that part in the definition of a Cartan algebra which states that it is its own normalizer, it follows that $Y \in \mathfrak{h}$, a contradiction unless $\mathfrak{h} = \mathfrak{h}_X$. (This argument does not require $\mathfrak{g}$ to be semisimple Now, for semisimple $\mathfrak{g}$, we let

$$X = S + N$$

be the Jordan decomposition of $X$ where $X$ is a regular element in $\mathfrak{h}$. We wish to show that $N = 0$. (Note that the existence of the Jordan decomposition depends on the semisimplicity of $\mathfrak{g}$. Since $ad(S)$ is semisimple, we have the decomposition

$$\mathfrak{g} = R(ad(S)) \oplus N(ad(S))$$

Since $S$ is a regular element of $\mathfrak{h} = \mathfrak{h}_X$, it follows that $\mathfrak{h}_S = \mathfrak{h}$ and since $ad(S)$ is semisimple, it follows that

$$\mathfrak{h} = \mathfrak{h}_S = N(ad(S))$$

So we can write

$$\mathfrak{g} = R(ad(S)) \oplus \mathfrak{h}$$

Now since $[S, N] = 0$, it follows that for any $Y \in \mathfrak{g}$, we have

$$< N, ad(S)(Y) >=< N, [S, Y] >=< [N, S], Y >= 0$$

and hence by the nonsingularity of $< .,. >$ for semisimple Lie algebras, it follows that

$$N \perp R(ad(S))$$

To prove that $N \perp \mathfrak{h}$, we take any $Y \in \mathfrak{h} = N(ad(S))$. We must show that

$$Tr(ad(N).ad(Y)) =< N, Y >= 0$$

Recall that $ad(X)(S) = ad(S)(N) = 0$ and hence $S, N \in \mathfrak{h}_X = \mathfrak{h}$. Now every nilpotent Lie algebra is also solvable. (This can be seen by making use of Engel's theorem for nil-representations. Indeed let $\mathfrak{g}_0$ be a nilpotent Lie algebra. Then the adjoint representation of $\mathfrak{g}_0$ on itself is a nil representation and hence according to Engel's theorem, there exists a basis for $\mathfrak{g}_0$ relative to which all the operators $ad(Z), Z \in \mathfrak{g}_0$ are strictly upper triangular. Hence, by working in this basis, we deduce that for all sufficiently large $n$, we have

$$ad(Z_1).ad(Z_2)..ad(Z_n) = 0 \forall Z_1, ..., Z_n \in \mathfrak{g}_0$$

which is the same as saying that

$$[Z_1, [Z_2, .., [Z_{n-1}, [Z_n, Z]]...,] = 0, forall Z_1, ..., Z_n, Z \in \mathfrak{g}_0$$

and in particular,

$$D^n \mathfrak{g}_0 = 0$$

thereby establishing solvability of $\mathfrak{g}_0$). Thus, $\mathfrak{h}$ is a solvable Lie algebra and hence by applying Lie's theorem to the adjoint representation of $\mathfrak{h}$ in $\mathfrak{g}$, we deduce that there exists a basis for $\mathfrak{g}$ relative to which all the operators $ad(H), H \in$

$\mathfrak{h}$ are upper-triangular matrices. In particular, relative to this basis $ad(N)$ and $ad(Y)$ are also upper-triangular. But since $ad(N)$ is nilpotent on $\mathfrak{g}$, it must necessarily follow that relative to this basis, $ad(N)$ is strictly upper-triangular. Since the product of an upper-triangular matrix and a strictly upper-triangular matrix is strictly upper-triangular, it follows that $ad(N).ad(Y)$ is strictly upper-triangular and hence its trace is zero, ie,

$$< N, Y >= 0$$

Therefore, we have proved that $N \perp R(ad(S)) \oplus N(ad(S)) = \mathfrak{g}$ and therefore, $N = 0$ by non-degeneracy of the Cartan-Killing symmetric bilinear form on $\mathfrak{g}$. Hence $X = S$. In other words, we have proved that if $X$ is a regular element of $\mathfrak{h}$, then $X$ is semisimple (ie, $ad(X)$ is semisimple on $\mathfrak{g}$). Now for such an $X$, we have that since $\mathfrak{h} = \mathfrak{h}_X$ the result that $ad(X)^m(\mathfrak{h}) = 0$ for sufficiently large $m$ and therefore (by expressing the semisimple operator $ad(X)$ on $\mathfrak{h}$ relative to basis which makes it diagonal) that $ad(X)(\mathfrak{h}) = 0$, ie, $[X, \mathfrak{h}] = 0$. Now if $Y \in \mathfrak{h}$ is arbitrary, we can write $Y = limX_n$ where $X_n \in \mathfrak{h}$ are semisimple because the set of regular elements of $\mathfrak{h}$ is dense in $\mathfrak{h}$ and we have shown that every regular element of $\mathfrak{h}$ is semisimple in $\mathfrak{g}$. Hence,

$$[Y, \mathfrak{h}] = lim[X_n, \mathfrak{h}] = 0$$

proving that $\mathfrak{h}$ is Abelian. Another way to see this is to start with the equation

$$[X, \mathfrak{h}] = 0 \forall X \in \mathfrak{h}'$$

and hence,

$$[\mathfrak{h}', \mathfrak{h}'] = 0$$

with $ad(\mathfrak{h}')$ consisting only of semisimple elements. Thus, $ad(\mathfrak{h}')$ is simultaneously diagonable and hence there is a basis $B$ for $\mathfrak{g}$ so that $[ad(\mathfrak{H})]_B$ is diagonal for all $H \in \mathfrak{h}'$. Then taking the limit points of this set we get that $[ad(H)]_B$ is diagonal for all $H \in \mathfrak{h}$.

Remark: We are making use of the fact that the space of regular elements of any Lie algebra $\mathfrak{g}$ is dense in $\mathfrak{g}$. To prove this result, we choose an element $Y \in \mathfrak{g}$ that is non-regular, ie, if $l = rk(\mathfrak{g})$ (then,

$$det(tI - ad(Y)) = c(l+1)t^{l+1} + ... + t^n$$

Let $Z \in \mathfrak{g}$ be arbitrary and consider

$$f(t, s) = det(t - ad(X + sZ)) = det(tI - ad(X) - s.ad(Z))$$

This is a polynomial in $t, s$ and hence it can be expressed as

$$f(t, s) = c_l(s)t^l + c_{l+1}(s)t^{l+1} + ... + c_n(s)t^n$$

where $c_l(s)$ is a polynomial in $s$. Suppose that $c_l(s)$ vanishes for all $|s| < \delta$. Then it vanishes for all $s$ since $c_l(s)$ being a polynomial has only a finite number of zeroes. In this case, it would follow that $X + sZ$ is an irregular element for all $s$. By choosing a basis $\{Z_1, ..., Z_n\}$ for $\mathfrak{g}$ and applying this argument, we would deduce that any element of the form $X + s_1Z_1 + ... + s_nZ_n$ is irregular and hence that there exists no regular element in $\mathfrak{g}$ which is a contradiction to the definition of $l = rk(\mathfrak{g})$. This means that there exists an element $Z \in \mathfrak{g}$ such that for each $\delta > 0$ there is an $s$ with $|s| < \delta$ such that $X + sZ$ is regular. In other words, there is a sequence $s_n \to 0$ such that $X + s_nZ$ is regular for all $n = 1, 2, ....$ Writing $X_n = X + s_nZ$, we deduce that $X_n$ is regular for each $n$ and $lim X_n = X$. The proof is complete.

## 1.26 Maximal nilpotency of a Cartan subalgebra

Remark: If $\mathfrak{g}$ is any Lie algebra and $\mathfrak{h}$ is a CSA (ie, $\mathfrak{h}$ is a nilpotent Lie subalgebra of $\mathfrak{g}$ and is also its own normalizer in $\mathfrak{g}$, then $\mathfrak{h}$ is also maximally nilpotent, ie, every CSA is maximally nilpotent. Note that when we say that $\mathfrak{h}$ is a nilpotent sublgebra, we mean that $[\mathfrak{h}, \mathfrak{h}] \subset \mathfrak{h}$ and secondly that $ad(H)^m(\mathfrak{h}) = 0 \forall H \in \mathfrak{h}$ where $m$ some finite positive integer. Note that since we are dealing only with finite dimensional Lie algebras, nilpotency of $\mathfrak{h}$ is equivalent to saying that for each $H_1 \in \mathfrak{h}$ there is a finite positive integer $m$ such that $ad(H_1)^m(H_2) = 0$ $\forall H_2 \in \mathfrak{h}$.

To see the maximal nilpotency of a CSA $\mathfrak{h}$, suppose $\mathfrak{h} \subset \mathfrak{h}_1$ is a proper inclusion where $\mathfrak{h}_1$ is a nilpotent subalgebra. Then since $\mathfrak{h}$ is $ad(\mathfrak{h})$-invariant, it follows that $ad : \mathfrak{h} \to L(\mathfrak{h}_1/\mathfrak{h})$ is a well defined representation and since $ad(\mathfrak{h}_1)$ is nilpotent on $\mathfrak{h}_1$ and $\mathfrak{h} \subset \mathfrak{h}_1$, it follows that $ad(\mathfrak{h})$ is nilpotent on $\mathfrak{h}_1/\mathfrak{h}$. In other words, this representation is a nil-representation and hence by Engel's theorem, there exists a $Y \in \mathfrak{h}_1 - \mathfrak{h}$ such that $ad(\mathfrak{h})(Y + \mathfrak{h}) = \mathfrak{h}$ which is equivalent to saying that $[Y, \mathfrak{h}] \subset \mathfrak{h}$ and since $\mathfrak{h}$ is its own normalizer in $\mathfrak{g}$, it follows that $Y \in \mathfrak{h}$, a contradiction.

## 1.27 Lecture Plan, Matrix Theory, M.Tech, Instructor:Harish Parthasarathy

**Note:Two lectures per topic, totally fifty lectures**

[1] Vector spaces, basis, linear transformations

[2] Matrix representation of a linear transformation relative to a basis, similarity transformations.

[3] Rank and nullity of a matrix.

[4] Decomposition theorems for matrices.

[a] QR factorization based on the Gram-Schmidt orthonormalization process.

[b] LDU decomposition of a positive definite matrix.

[e] Spectral theorem for normal operators in finite and infinite dimensional Hilbert spaces.

[f] Polar decomposition in a Hilbert space.

[g] Singular value decomposition in finite dimensional Hilbert spaces.

[5] Primary decomposition theorem of a matrix in a finite dimensional complex vector space.

[6] Jordan decomposition of a matrix.

[7] Canonical representation of nilpotent matrices.

[8] The Jordan canonical form.

[9] Computational algorithms for calculating the Jordan canonical form.

[10]Perturbation theory for computing the eigenvalues and eigenvectors of Hermitian and diagonable matrices.

[11] Calculating functions of matrices using contour integrals in the complex plane.

[12] Computing functions of matrices using the Jordan canonical form.

[13] Matrix norms and Banach algebras.

[14] Homomorphisms and spectra of commutative Banach algebras.

[15] Numerical methods for computing eigenvalues and eigenvectors of a matrix.

[16] Tensor product of vector spaces and of linear transformations.

[17] Application of the tensor product to describing nonlinear systems.

[18] Quotient vector spaces and linear transformations on them.

[19] Lie groups, Lie algebras and their representations.

[19.1] Definition of a Lie group and its Lie algebra.

[19.2] Jacobi identity on a Lie algebra.

[19.3] Representation of a Lie group and the associated representation of its Lie algebra.

[19.4] The adjoint representation of a Lie algebra.

[19.5] Solvable and semisimple Lie algebras.

[19.6] Nilpotent Lie algebras and nilpotent representations of a Lie algebra.

[19.7] Lie's theorem on reprsentations of solvable Lie algebras.

[19.8] Engel's theorem on nilpotent representations of a Lie algebra.

[19.9] Regular elements of a Lie algebra.

[19.10] Cartan subalgebras of a Lie algebra and their construction.

[19.11] Properties of Cartan subalgebras.

[19.12] Jordan decomposition on a Lie algebra, proofs based on Chevalley's theory of replicas.

[19.13] Cartan subalgebras of a semisimple Lie algebra and their properties.

[19.14] The root space decomposition of a semisimple Lie algebra.

[19.15] Cartan's classification of all the complex simple Lie algebras.

[20] Some notions in the theory of unbounded operators in a Hilbert space.

[20.1] The uniform boundedness principle.

[20.2] The Hahn-Banach theorem.

[20.3] The closed graph theorem.

[20.4] Closed operators.
[20.5] Adjoint of an operator.
[21] Applications of unbounded operators to quantum mechanics.

## 1.28 Some other remarks on Lie algebras

### 1.28.1 Root space decomposition of a semisimple Lie algebra and some of its properties

Let $\mathfrak{g}$ be a complex SSLA(Semisimple Lie algebra) and let $\mathfrak{h}$ be a CSA (Cartan subalgebra). Consider the root space decomposition

$$\mathfrak{g} = \mathfrak{h} \oplus \bigoplus_{\alpha \in \Delta} \mathfrak{g}_\alpha$$

Note that $\Delta$ is a finite subset of $\mathfrak{h}^*$. We have that

$$[H, X] = \alpha(H)X, H \in \mathfrak{h}, X \in \mathfrak{g}_\alpha$$

$$[H, H'] = 0, H, H' \in \mathfrak{h}$$

and $ad(H), H \in \mathfrak{h}$ are semisimple operators on $\mathfrak{g}$ which is what makes the above root space decomposition possible. It is clear that

$$[\mathfrak{g}_\alpha, \mathfrak{g}_\beta] \subset \mathfrak{g}_{\alpha+\beta}, \alpha, \beta \in \Delta$$

where $\mathfrak{g}_{\alpha+\beta}$ is to be taken as zero if $\alpha + \beta \neq 0$ and $\alpha + \beta$ is not a root and as $\mathfrak{h}$ if $\alpha + \beta = 0$.

This is because if $X \in \mathfrak{g}_\alpha, Y \in \mathfrak{g}_\beta$, then by the Jacobi identity,

$$[H, [X, Y]] = -([X, [Y, H]] + [Y, [H, X]] = (\alpha + \beta)(H)[X, Y], H \in \mathfrak{h}$$

Further,

$$[\mathfrak{h}, \mathfrak{g}_\alpha] \subset \mathfrak{g}_\alpha$$

as again follows by use of the Jacobi identity and the Abelian character of $\mathfrak{h}$.

Note that if $X \in \mathfrak{g}_\alpha, Y \in \mathfrak{g}_{-\alpha}$, then for any $H \in \mathfrak{h}$, we have by the Jacobi identity that

$$[H, [X, Y]] = 0$$

and hence since $\mathfrak{h}$ is maximal Abelian, it follows that

$$[X, Y] \in \mathfrak{h}$$

In other words,

$$[\mathfrak{g}_\alpha, \mathfrak{g}_{-\alpha}] \subset \mathfrak{h}, \alpha \in \Delta$$

It is clear that $< .,. >$ is non-singular on $\mathfrak{h} \times \mathfrak{h}$ as well as on $\mathfrak{g}_\alpha \times \mathfrak{g}_{-\alpha}$ for the following reasons:

$$< X, [H, Y] >=< [X, H], Y >= \beta(H) < X, Y >= -\alpha(H) < X, Y >, X \in \mathfrak{g}_\alpha, Y \in \mathfrak{g}_\beta, H \in \mathfrak{h}$$

If $\beta + \alpha \neq 0$, it follows that there exists an $H \in \mathfrak{h}$ for which $(\beta + \alpha)(H) \neq 0$ and therefore $< X, Y >= 0$. In other words, we have proved that if $\alpha, \beta \in \Delta$ and $\beta \neq -\alpha$, then $\mathfrak{g}_\beta \perp \mathfrak{g}_\alpha$. Further $\mathfrak{h} \perp \mathfrak{g}_\alpha \forall \alpha \in \Delta$. This follows from the identity

$$\alpha(H') < H, X >=< H, [H', X] >=< [H, H'], X >= 0, H, H' \in \mathfrak{h}, X \in \mathfrak{g}_\alpha$$

which implies the $< H, X >= 0, H \in \mathfrak{h}$ since $\alpha \neq 0$ in $\mathfrak{h}^*$. Since therefore, for a given $\alpha \in \Delta$, $\mathfrak{g}_\alpha$ is orthogonal to $\mathfrak{h}$ as well as to $\mathfrak{g}_\beta$ for all roots $\beta \neq -\alpha$, the root space decomposition implies that for any $\alpha \in \Delta$, $X \in \mathfrak{g}_\alpha$ cannot be orthogonal to $\mathfrak{g}_{-\alpha}$ for other wise, it would be orthogonal to $\mathfrak{g}$ which is false since for a SSLA, the Cartan-Killing form is non-singular. Likewise since $\mathfrak{h}$ is orthogonal to $\mathfrak{g}_\alpha \forall \alpha \in \Delta$, it follows from the root space decomposition and the non-singularity of the Cartan-Killing form that $\mathfrak{h}$ cannot be orthogonal to itself, ie, the Cartan-Killing form is non-singular on $\mathfrak{h} \times \mathfrak{h}$.

Remark:$\Delta = -\Delta$, ie, if $\alpha \in \Delta$, then $-\alpha \in \Delta$. For suppose that $\alpha \in \Delta$ but $-\alpha \notin \Delta$. Then $\mathfrak{g}_\alpha$ is orthogonal to $\mathfrak{h}$ as well to all the $\mathfrak{g}_\beta, \beta \in \Delta$ since $-\alpha \notin \Delta$. Thus by the root space decomposition, $\mathfrak{g}_\alpha$ is orthogonal to $\mathfrak{g}$ which contradicts the non-degeneracy of the Cartan-Killing form on $\mathfrak{g}$.

Now we show that $dim \mathfrak{g}_\alpha = 1 \forall \alpha \in \Delta$. In fact, since the Cartan-Killing form is non-singular on $\mathfrak{g}_\alpha \times \mathfrak{g}_{-\alpha}$, it follows that we can select $X_\alpha \in \mathfrak{g}_\alpha$ and $X_{-\alpha} \in \mathfrak{g}_{-\alpha}$ so that

$$< X_\alpha, X_{-\alpha} >\neq 0$$

and hence, we can define $H_\alpha \in \mathfrak{h}$ so that

$$[X_\alpha, X_{-\alpha}] =< X_\alpha, X_{-\alpha} > H_\alpha, \alpha \in \Delta$$

Then, it is clear that

$$\begin{aligned} < H_\alpha, H >&= \frac{< [X_\alpha, X_{-\alpha}], H >}{< X_\alpha, X_{-\alpha} >} \\ &= \frac{< X_\alpha, [X_{-\alpha}, H] >}{< X_\alpha, X_{-\alpha} >} \\ &= \alpha(H), H \in \mathfrak{h} \end{aligned}$$

Now choose complex numbers $c_\alpha, c_{-\alpha}$ such that

$$c_\alpha c_{-\alpha} < X_\alpha, X_{-\alpha} >= 2/\alpha(H_\alpha)$$

and define

$$\bar{X}_\alpha = c_\alpha X_\alpha, \bar{X}_{-\alpha} = c_{-\alpha} X_{-\alpha}$$

Also define

$$\bar{H}_\alpha = 2H_\alpha/\alpha(H_\alpha)$$

(We shall soon be showing that $\alpha(H_\alpha) \neq 0$ so that division by it is justified).

Then, it is easy to see that

$$[\bar{H}_\alpha, \bar{X}_\alpha] = 2\bar{X}_\alpha, [\bar{H}_\alpha, X_{-\alpha}] = -2\bar{X}_{-\alpha},$$

$$[\bar{X}_\alpha, \bar{X}_{-\alpha}] = \bar{H}_\alpha$$

or equivalently, $\{\bar{H}_\alpha, \bar{X}_\alpha, \bar{X}_{-\alpha}\}$ forms a standard basis for an $\mathfrak{sl}(2,\mathbb{C})$ Lie algebra. Hence, we shall denote $\bar{X}_\alpha$ by $X_\alpha$ and $\bar{X}_{-\alpha}$ by $X_{-\alpha}$ for notational convenience. Thus, in our new notation, $\{\bar{H}_\alpha, X_\alpha, X_{-\alpha}\}$ forms a standard $\mathfrak{sl}(2,\mathbb{C})$ triplet.

Result: $span\{H_\alpha : \alpha \in \Delta\} = \mathfrak{h}^*$. To see this, it suffices to show that (in view of the non-singularity of the Cartan-Killing form on $\mathfrak{h} \times \mathfrak{h}$ that if $H \in \mathfrak{h}$ is such that $\alpha(H) =< H_\alpha, H >= 0 \forall \alpha \in \Delta$, then $H = 0$. But from the definition of the Cartan-Killing form,

$$< H', H >= \sum_{\beta\in\Delta} dim(\mathfrak{g}_\beta)\beta(H')\beta(H) = 0 \forall H' \in \mathfrak{h}$$

since by hypothesis, $\beta(H) = 0 \forall \beta \in \Delta$. But then since $< .,. >$ is non-singular on $\mathfrak{h} \times \mathfrak{h}$, it follows that $H = 0$ and we are done.

Result: For any $\alpha, \beta \in \Delta$, there exists a real rational number $q_{\beta,\alpha}$ such that

$$\beta(H_\alpha) = q_{\beta,\alpha}\alpha(H_\alpha)$$

To see this, we first construct a maximal chain of roots $\beta + k\alpha, k = -p, -p + 1, ..., q-1, q$ where $p, q$ are non-negative integers. By maximal chain, we mean that $\beta - (p+1)\alpha$ and $\beta + (q+1)\alpha$ are not roots. Then, it follows that

$$ad(X_{-\alpha})(\mathfrak{g}_{\beta-p\alpha}) = 0,$$

$$ad(X_\alpha)(\mathfrak{g}_{\beta+q\alpha}) = 0,$$

$$ad(X_{-\alpha})(\mathfrak{g}_{\beta+k\alpha}) \subset \mathfrak{g}_{\beta+(k-1)\alpha}$$

for $-p < k \leq q$,

$$ad(X_\alpha)(\mathfrak{g}_{\beta+k\alpha}) \subset \mathfrak{g}_{\beta+(k+1)\alpha}$$

for $-p \leq k < q$ and

$$ad(\bar{H}_\alpha)(\mathfrak{g}_{\beta+k\alpha}) \subset \mathfrak{g}_{\beta+k\alpha}$$

for $-p \leq k \leq q$. In particular, we see that the vector space

$$V_{\beta,\alpha} = \bigoplus_{k=-p}^{q} \mathfrak{g}_{\beta+k\alpha}$$

is invariant under the $sl(2,\mathbb{C})$ adjoint algebra $\{ad(\bar{H}_\alpha), ad(X_\alpha), ad(X_{-\alpha})\}$. Thus,

$$Tr(ad(\bar{H}_\alpha)|_{V_{\beta,\alpha}} = Tr([ad(X_\alpha), ad(X_{-\alpha})]|_{V_{\beta,\alpha}}) = 0$$

Evaluating this trace gives us

$$\sum_{k=-p}^{q} dim(\mathfrak{g}_{\beta+k\alpha})(\beta+k\alpha)(\bar{H}_\alpha)=0$$

or equivalently since $H_\alpha$ is a scalar times $\bar{H}_\alpha$, we get

$$\beta(H_\alpha).(\sum_{k=-p}^{q} dim(\mathfrak{g}_{\beta+k\alpha})+(\sum_{k=-p}^{q} kdim(\mathfrak{g}_{\beta+k\alpha}))\alpha(H_\alpha)=0$$

which completes the proof once we note that

$$q_{\beta,\alpha}=-\frac{\sum_{k=-p}^{q} k.dim(\mathfrak{g}_{\beta+k\alpha})}{\sum_{k=-p}^{q} dim(\mathfrak{g}_{\beta+k\alpha}}$$

is a real rational number.

Result:$\alpha(H_\alpha)\neq 0$ for any $\alpha\in\Delta$. To see this, suppose that for some $\alpha\in\Delta$, we have that $\alpha(H_\alpha)=0$. Then by the previous result, $\beta(H_\alpha)=q_{\beta,\alpha}\alpha(H_\alpha)=0\forall\beta\in\Delta$. This implies that $H_\alpha=0$ since as already noted above span $\{\Delta\}=\mathfrak{h}^*$. Thus, $\alpha=0$ since $\alpha(H)=<H_\alpha,H>, H\in\mathfrak{h}$. This contradiction proves the result.

Result:For any $\alpha\in\Delta$, $\alpha(H_\alpha)$ is a real, positive rational number. To prove this, we use the two results proved above, namely that $\alpha(H_\alpha)$ is a nonzero complex number and that $\beta(H_\alpha)=q_{\beta,\alpha}\alpha(H_\alpha)$ for any $\beta\in\Delta$ where $q_{\beta,\alpha}$ is a real number (in fact a real rational number). Then, we get

$$\alpha(H_\alpha)=<H_\alpha,H_\alpha>=\sum_{\beta\in\Delta}\beta(H_\alpha)^2 dim(\mathfrak{g}_\beta)$$

$$=\sum_{\beta\in\Delta} dim(\mathfrak{g}_\beta)q_{\beta,\alpha}^2\alpha(H_\alpha)^2$$

(Note that $dim\mathfrak{g}_\beta\geq 1\forall\beta\in\Delta$). Since $\alpha(H_\alpha)\neq 0$, we can cancel it from both the sides to get

$$\alpha(H_\alpha)=[\sum_{\beta\in\Delta} dim(\mathfrak{g}_\beta)q_{\beta,\alpha})^2]^{-1}$$

which proves the claim.

Result: $dim\mathfrak{g}_\alpha=1\ \forall\alpha\in\Delta$.
Let $\alpha\in\Delta$ and define the subspace

$$V=span\{X_{-\alpha}\}\oplus\mathfrak{h}\oplus\mathfrak{g}_\alpha\oplus\mathfrak{g}_{2\alpha}\oplus...\oplus\mathfrak{g}_{m\alpha}\oplus..$$

It is clear that this series terminates after a finite number of steps because $\mathfrak{g}$ is finite dimensional. Further, it is clear that $V$ is invariant under $\{ad(X_{-\alpha}),ad(X_\alpha),ad(\bar{H}_\alpha)\}$ and therefore since

$$ad(\bar{H}_\alpha)=[ad(X_\alpha),ad(X_{-\alpha})]$$

it follows that

$$Tr(ad(\bar{H}_\alpha)|_V) = 0$$

## 1.29 Question Paper on Matrix Theory

Attempt any four questions. Each question carries five marks.

[1] Let $\mathbf{A}$ be an $m \times n$ matrix with $m > n$ and having rank $n$. Write down the general structure of the singular value decomposition of $\mathbf{A}$. Explain how using this SVD, you will obtain the least squares solution to the problem of calculating $\theta \in \mathbb{R}^n$ so that

$$(\mathbf{x} - \mathbf{A}\theta)^T(\mathbf{x} - \mathbf{A}\theta)$$

is minimum for a given $\mathbf{x} \in \mathbb{R}^n$. How will you modify your method if we have to solve the weighted least squares problem of minimizing

$$(\mathbf{x} - \mathbf{A}\theta)^T\mathbf{W}(\mathbf{x} - \mathbf{A}\theta)$$

where $\mathbf{W}$ is a positive definite $n \times n$ matrix.

[2] Write down the root space decomposition of the Lie algebra $\mathfrak{sl}(n, \mathbb{C})$, ie, the Lie algebra of all $n \times n$ complex matrices having zero trace. Identify clearly the Cartan subalgebra and the root vectors. Prove that this Lie algebra is indeed semisimple by calculating it Cartan-Killing form and showing that this form is non-singular.

[3] Let $\mathbf{A}$ be an $n \times n$ matrix having exactly two Jordan blocks of sizes $n_1 \times n_1$ with eigenvalue $c_1$ and $n_2 \times n_2$ with eigenvalue $c_2$ where $c_1 \neq c_2$. Write down the explicit forms of $exp(t\mathbf{A})$ and $(\lambda I - \mathbf{A})^{-1}$ in terms of these Jordan blocks and hence deduce that

$$\int_0^\infty exp(t\mathbf{A})exp(-\lambda t)dt = (\lambda I - \mathbf{A})^{-1}$$

provided that $Re(\lambda) > Re(c_k), k = 1, 2$.

[4] Using the primary decomposition theorem, prove that if $T, S$ are complex $n \times n$ matrices such that $ad(T)^m(S) = 0$ for some positive integer $m$, then $S$ leaves both the subspaces $\bigcup_{k \geq 1} \mathcal{N}(ad(T)^k)$ and $\bigcap_{k \geq 1} \mathcal{R}(ad(T)^k)$ invariant.

[5] Write short notes on the following:

[a] The primary decomposition theorem.

[b] Construction of a Cartan subalgebra of a semisimple Lie algebra.

[c] LDU and UDL decomposition of a positive definite matrix with applications to signal prediction theory.

[d] Gram-Schmidt orthonormalization with application to the QR decomposition of a full column rank matrix.

## 1.30 Root space decompositions of the complex classical Lie algebras

[a] $\mathfrak{g} = \mathfrak{sl}(n, \mathbb{C})$. This Lie algebra consists of all $n \times n$ complex matrices having trace zero. It has the root space decomposition

$$\mathfrak{g} = \mathfrak{h} \oplus \bigoplus_{k>l} \mathfrak{g}_{kl} \oplus \bigoplus_{k<l} \mathfrak{g}_{kl}$$

where $\mathfrak{h}$ consists of all complex diagonal matrices of the form

$$H = diag[h_1, h_2, ..., h_n], h_1 + h_2 + ... + h_n = -0$$

and for $k \neq l$, $\mathfrak{g}_{kl} = \mathbb{C}.E_{kl}$. Note that we can write

$$H = \sum_{k=1}^{n-1} h_k(E_{kk} - E_{nn})$$

and hence

$$\dim \mathfrak{h} = n - 1$$

and $\mathfrak{h}$ is spanned by the linearly independent elements $E_{k,k} - E_{k+1,k+1}, k = 1, 2, ..., n-1$.

## 1.31 More on the root space decomposition of a semisimple Lie algebra

We've sequentially established that if $\mathfrak{g}$ is a complex SSLA and $\mathfrak{h}$ a CSA, then $\mathfrak{h}$ is a maximal Abelian subalgebra of $\mathfrak{g}$ whose elements are all semisimple and we have the root space decomposition of the form

$$\mathfrak{g} = \mathfrak{h} \oplus \bigoplus_{\alpha \in \Delta} \mathfrak{g}_\alpha$$

where $\Delta$ is a finite subset of $\mathfrak{h}^*$ not containing the zero element and

$$\mathfrak{g}_\alpha = \{X \in \mathfrak{g} : [H, X] = \alpha(H)X \forall H \in \mathfrak{h}\}$$

and further that using the obvious identity

$$< H, H' >= Tr(ad(H).ad(H')) = \sum_{\alpha \in \Delta} dim(\mathfrak{g}_\alpha)\alpha(H)\alpha(H'), H, H' \in \mathfrak{h}$$

and the fact that (a) $< .,. >$ is non-singular on $\mathfrak{h} \times \mathfrak{h}$ and (b) that

$$< \beta, \alpha >= \beta(H_\alpha) = q_{\beta,\alpha}\alpha(H_\alpha) = q_{\beta,\alpha} < \alpha, \alpha >$$

where $q_{\beta,\alpha}$ are real rational numbers that $span_{\mathbb{C}}\Delta = \mathfrak{h}^*$ and hence that $\alpha(H_\alpha) =< \alpha, \alpha >$ is a nonzero complex number and hence also a positive rational number for any $\alpha \in \Delta$. In order to establish (b), we had defined

$$V = V_{\beta,\alpha} = \bigoplus_{k=-p}^{q} \mathfrak{g}_{\beta+k\alpha}$$

where the direct sum is over a maximal string of root spaces so that $V$ is invariant under

$$\mathfrak{sl}(2,\mathbb{C}) = \{\bar{H}_\alpha, X_\alpha, X_{-\alpha}\}$$

and hence that

$$0 = Tr([ad(X_\alpha), ad(X_{-\alpha})]|_V) = Tr(ad(\bar{H}_\alpha)|_V)$$

and hence that

$$0 = Tr[ad(H_\alpha)|_V] = \sum_{k=-p}^{q} dim(\mathfrak{g}_{\beta+k\alpha}) < \beta + k\alpha, \alpha >$$

Fundamental to all this was our observation that $-\Delta = \Delta$. After that, we had taken a connected component $C$ of $\mathfrak{h}'$(all the regular elements in $\mathfrak{h}$) and defined $P(C) = P$ to be the set of all roots $\alpha$ which assume positive values on $C$. (We had noted that on each connected component of $\mathfrak{h}'$, every root assumes non-zero values). Then we had defined the simple system $S(C) = S$ associated to $P$ to be the set of all roots in $P$ which could not be expressed as a sum of two roots in $P$. Based on this definition, we had noted that $\Delta = P \cup -P$ and that every root in $P$ could be expressed as a non-negative integer linear combination of the elements of $S$. We had then observed that since, in particular, any element of $P$ is a linear combination of elements of $S$ it followed that every root (ie element of $\Delta$) is also in particular a linear combination of elements of $S$ and hence since $span_{\mathbb{C}}\Delta = \mathfrak{h}^*$, that $dim\mathfrak{h} \leq \mu(S)$ and hence in order to show that $\mu(S) = dim\mathfrak{h}$, it would suffice to prove the linear independence of $S$. This we had established by first proving that $S$ is linearly independent over $\mathbb{R}$ which was settled using the theorem that if $v_1, ..., v_n$ is a set of vectors in a vector space $V$ such that for any $i \neq j$, $< v_i, v_j >< 0$ then $\sum_{i=1}^n c(i)v_i = 0$ for some real numbers $\{c(i)\}$ implied $c(i) = 0$ for all $i$, ie, $v_1, ..., v_n$ was linearly independent over $\mathbb{R}$. Then we were in fact able to show that the $v_i's$ were in fact linearly independent over $\mathbb{C}$ by noting that if $d(i)$ were complex numbers such that $\sum_i d(i)v_i = 0$, then $\sum_i d(i) < v_i, v_j >= 0$ for all $j$ and hence $\sum_i Re(d(i)) < v_i, v_j >= 0$ and $\sum_i Im(d(i)) < v_i, v_j >= 0 \forall j$ which implied by considering appropriate linear combinations over $j$ using $Re(d(j))$ and $Im(d(j))$ that $\sum_i Re(d(i))v_i = \sum_i Im(d(i))v_i = 0$ and therefore using the linear independence of the $v_i's$ over $\mathbb{R}$, we got that $d(i) = 0 \forall i$. For this result to be valid, it was noted that we did not even require $< .,. >$ to be a positive definite inner product on $V$. It was sufficient to require that $< .,. >$ be a positive definite bilinear form over the real vector space spanned by the $v_i's$. In order to apply this result to prove the linear independence of $S$, we had first shown that if $\alpha, \beta \in S$ were distinct, then $< \alpha, \beta >< 0$. This in turn was seen to be a consequence of the fact that if $\alpha, \beta \in S$, then $\beta - \alpha$ could not be a root and hence the maximal chain of roots $\beta + k\alpha, -p \leq k \leq q$ had to have $p = 0$, implying that the least eigenvalue of $ad(\bar{H}_\alpha)$ on this irreducible subspace for the Lie algebra $sl(2, \mathbb{C})$ whose standard generators are $\{\bar{H}_\alpha, X_\alpha, X_{-\alpha}\}$ is $\beta(\bar{H}_\alpha) = 2 < \beta, \alpha > / < \alpha, \alpha >$ and hence from the general theory of irreducible representations of $\mathfrak{sl}(2, \mathbb{C})$ it follows that this is a negative integer and in particular, $< \beta, \alpha >$ is a negative rational number. Further, in order to apply this result, we had first considered the real vector space $\mathfrak{h}_R = \sum_{\alpha \in \Delta} \mathbb{R}.H_\alpha$ and noted that the Cartan-Killing form defined a positive definite inner product on this real vector space. In order to show this, we had taken an $H \in \mathfrak{h}_R$ and computed

$$< H, H >= \sum_{\alpha \in \Delta} dim(\mathfrak{g}_\alpha)\alpha(H)^2$$

and noted that since $\alpha(H_\beta)$ was a real number (in fact a rational number) for any pair of roots $\alpha, \beta$, it follows that $\alpha(H)$ was real and hence $< H, H >$ was non-negative and was zero iff $\alpha(H) = 0 \forall \alpha \in \Delta$ which implied that $H = 0$ since $span_{\mathbb{C}} \Delta = \mathfrak{h}^*$.

Some other remarks: We had stated that if for two roots $\alpha, \beta, \beta + k\alpha, -p \leq k \leq q$ was a maximal chain of roots, then the vector space $V = \bigoplus_{k=-p}^{q} \mathfrak{g}_{\beta+k\alpha}$ was invariant and infact even irreducible under the adjoint representation of the $\mathfrak{sl}(2, \mathbb{C})$ Lie algebra $\{\bar{H}_\alpha, X_\alpha, X_{-\alpha}\}$. Invariance is obvious and irreducibility followed from the following argument: Suppose $W_1$ and $W_2$ were two invariant irreducible subspaces of $V$ under this representation with zero intersection. Then, from the basic properties of irreducible representations of $\mathfrak{sl}(2, \mathbb{C})$, $ad(\bar{H}_\alpha)$ has a maximal eigenvalue of $j_k$ and a minimal eigenvalue of $-j_k$ on $W_k$ for $k = 1, 2$ where $j_k, k = 1, 2$ are positive integers. We may assume without loss of generality that $j_1 \geq j_2$. Then $W_1$ is spanned by eigenvectors of $ad(\bar{H}_\alpha)$ with eigenvalues $j_1, j_1 - 2, ..., -j_1$ and $W_2$ by eigenvectors of $ad(\bar{H}_\alpha)$ with eigenvalues $j_2, j_2 - 2, .., -j_2$ and for each eigenvalue of $ad(\bar{H}_\alpha)$ in $W_1$, there is only one eigenvector upto a proportionality constant because of irreducibility and likewise for $W_2$. We had then observed that $dim\mathfrak{g}_{\beta+k\alpha} = 1, -p \leq k \leq q$. We next observe that the only eigenvectors of $ad(\bar{H}_\alpha)$ are precisely those in $\mathfrak{g}_{\beta+k\alpha}, k = -p, ..., q$ and the corresponding eigenvalues differ by even integers since these eigenvalues are $(\beta + k\alpha)(\bar{H}_\alpha) = \beta(\bar{H}_\alpha) + 2k, k = -p, ..., q$. For $W_1$ and $W_2$ to be disjoint apart from the zero vector, it was necessary that $j_1$ and $j_2$ differ by an odd integer for otherwise they would have a common non-zero element. But this would mean that $ad(\bar{H}_\alpha)$ has two eigenvalues that differ by an odd integer which is not possible and thereby establishing the irreducibility of $V$. Note that for this argument to work, we had to prove that $dim\mathfrak{g}_\alpha = 1$ forall $\alpha \in \Delta$. But this follows by considering the finite dimensional vector space

$$V_1 = span\{X_{-\alpha}) \oplus \mathfrak{h} \oplus \mathfrak{g}_\alpha \oplus \mathfrak{g}_{2\alpha} \oplus ...$$

noting that $V_1$ is invariant under $ad\{\bar{H}_\alpha, X_\alpha, X_{-\alpha}\}$ and hence that

$$0 = Tr([adX_\alpha, adX_{-\alpha}]|_{V_1}) = Tr(ad\bar{H}_\alpha|_{V_1})$$

$$= -alpha(\bar{H}_\alpha) + 0 + \sum_{k\geq 1} k\alpha(\bar{H}_\alpha)dim(\mathfrak{g}_{k\alpha})$$

$$= -2 + \sum_{k\geq 1} 2.dim(\mathfrak{g}_{k\alpha})$$

and therefore,

$$dim\mathfrak{g}_\alpha = 1, dim\mathfrak{g}_{k\alpha} = 0, k > 1$$

Note that this argument works only because we have been able to show that $< \alpha, \alpha >= \alpha(H_\alpha) \neq 0$.

Let $C$ be a Weyl chamber and let $P$ be the positive system of roots associated to this chamber and $S$ the corresponding simple system of roots. We wish to prove that every element of $P$ is a non-negative integer linear combination of elements from $S$. Let$\alpha P$ and let $H \in C$. Suppose $\alpha \in S$. Then there is nothing to prove. Suppose then that $\alpha \notin S$. Then $\alpha = \beta + \gamma$ for some $\beta, \gamma \in P$. Then, $\alpha(H) = \beta(H) + \gamma(H)$ shows that $\beta(H), \gamma(H)$ are positive and strictly smaller than $\alpha(H)$ (Note that all roots in $P$ are strictly positive on $C$ and hence strictly positive when evaluated at $H$). Thus, if we choose $\alpha \in P$ so that $\alpha$ is not expressible as a non-negative integer linear combination of elements of $S$ and its value at $H$ is a minimum subject to this constraint, we get a contradiction since this hypothesis implies that $\beta$, and $\gamma$ are expressible as non-negative integer linear combinations of elements of $S$ and hence so also is $\alpha$. This forces us to conlcude that every element of $P$ is expressible as a non-negative integer linear combination of elements of $S$.

Cartan integers: Let $S = \{\alpha_1, ..., \alpha_l\}$ be the set of simple roots relative to a positive system. In order to develop the classification theory of simple Lie algebras over $\mathbb{C}$, we have to classify only connected Dynkin diagrams because a connected Dynkin diagram or a scheme is associated only to irreducible Cartan matrices. The Cartan integers are $a(i,j) = 2 < \alpha_i, \alpha_j > / < \alpha_j, \alpha_j >, i, j = 1, 2, ..., l$. These are negative integers. We say that the Cartan matrix $A = ((a(i,j))) \in \mathbb{Z}^{l\times l}$ is irreducible iff we cannot find non-empty disjoint sets $S_1, S_2$ such that $S = S_1 \cup S_2$ and such that $< \alpha, \beta >= 0$ for all $\alpha \in S_1, \beta \in S_2$, ie, $S_1$ and $S_2$ are orthogonal. The main result of this section is that a Cartan matrix is irreducible iff the Lie algebra (semisimple) is simple, ie, not expressible as a direct sum of non-trivial ideals. Equivalently, a semisimple Lie algebra is expressible as a direct sum of non-trivial ideals iff the associated Cartan matrix is irreducible, ie, relative to a permutation of the simple roots, the Cartan matrix appears as a direct sum of at least two non-trivial blocks, ie, it is block diagonal.

To see this first assume that $\mathfrak{g}$ is not simple. Then, we can write

$$\mathfrak{g} = \mathfrak{g}_1 \oplus \mathfrak{g}_2$$

where $\mathfrak{g}_k, k = 1, 2$ are both non-zero ideals in $\mathfrak{g}$, ie

$$[\mathfrak{g}, \mathfrak{g}_k] \subset \mathfrak{g}_k, k = 1, 2$$

In particular,

$$[\mathfrak{g}_1, \mathfrak{g}_2] \subset \mathfrak{g}_1 \cap \mathfrak{g}_2 = 0$$

ie $\mathfrak{g}_1$ and $\mathfrak{g}_2$ mutually commute. Now let $\alpha$ be a root. Choose $0 \neq X \in \mathfrak{g}_\alpha$. Then

$$X = X_1 + X_2, X_k \in \mathfrak{g}_k, k = 1, 2$$

Let $H \in \mathfrak{h}$. Since $\mathfrak{g}_k, k = 1, 2$, are ideals, they are in particular, invariant under $ad(H)$ and hence from the equation

$$ad(H)(X) = \alpha(H)X$$

we get

$$(\alpha(H)X_1 - [H, X_1]) + (\alpha(H)X_2 - [H, X_2]) = 0$$

with

$$\alpha(H)X_k - [H, X_k] \in \mathfrak{g}_k, k = 1, 2$$

and therefore,

$$[H, X_k] = \alpha(H)X_k, k = 1, 2, H \in \mathfrak{h}$$

Since however $\mathfrak{g}_\alpha$ is one dimensional and $X \neq 0$, it must necessarily follow that one of the $X_k' s, k = 1, 2$ vanishes. This proves exactly one of the alternatives

$$\mathfrak{g}_\alpha \subset \mathfrak{g}_k, k = 1, 2$$

occurs, ie, exactly one of the $X_k' s$ vanishes, or equivalently, $X = X_1$ or $X = X_2$. This shows that for any $\alpha \in \Delta$, either $g_\alpha \subset \mathfrak{g}_1$ or $\mathfrak{g}_\alpha \subset \mathfrak{g}_2$ but not both.

The next observation we make is that relative to the Cartan-Killing form, $\mathfrak{g}_1 \perp \mathfrak{g}_2$. To see this, let $X_k \in \mathfrak{g}_k, k = 1, 2$. Then, since $\mathfrak{g}_k, k = 1, 2$ are mutually commuting ideals, it follows that for $X \in \mathfrak{g}_1$ $ad(X_1)ad(X_2)X = ad(X_1)[X_2, X] = 0$ and for $X \in \mathfrak{g}_2$, $ad(X_1)ad(X_2)X = [X_1, [X_2, X]] = 0$ and therefore $ad(X_1)ad(X_2)X = 0$ for all $X \in \mathfrak{g}$ proving thereby that

$$< X_1, X_2 >= Tr(ad(X_1).ad(X_2)) = 0$$

ie $\mathfrak{g}_1 \perp \mathfrak{g}_2$. Now suppose $\alpha$ is a root and $\mathfrak{g}_\alpha \subset \mathfrak{g}_1$. Then since $\mathfrak{g}_1$ is an ideal, it follows that $\bar{H}_\alpha = [X_\alpha, X_{-\alpha}] \in \mathfrak{g}_1$ and hence also $X_{-\alpha} = (1/2)[X_{-\alpha}, \bar{H}_\alpha] \in \mathfrak{g}_1$. The same is valid for $\mathfrak{g}_2$. Thus, we can write

$$S = S_1 \cup S_2, S_1 \cap S_2 = \phi,$$

$$\{H_\alpha, X_\alpha, X_{-\alpha}\} \subset \mathfrak{g}_k, \forall \alpha \in \mathfrak{g}_k, k = 1, 2$$

and since $\mathfrak{g}_1 \perp \mathfrak{g}_2$, we get that

$$< \alpha, \beta >=< H_\alpha, H_\beta >= 0, \alpha \in S_1, \beta \in S_2$$

and this proves that the Cartan matrix is decomposable.

Conversely, suppose that the Cartan matrix is decomposable. Thus,

$$S = S_1 \cup S_2, S_1 \neq \phi, S_2 \neq \phi, S_1 \cap S_2 = \phi,$$

$$< \alpha, \beta >= 0, \alpha \in S_1, \beta \in S_2$$

Let $W_k$ be the group generated by $s_\alpha, \alpha \in S_k$ for each $k = 1, 2$. Let

$$\Delta_k = \bigcup_{\alpha \in S_k} W_1.\alpha, k = 1, 2$$

We first observe that if $\alpha \in S_1$ and $\beta \in S_2$, then since $< \alpha, \beta >= 0$, it follows that

$$s_\beta \alpha = \alpha - 2 < \beta, \alpha > \alpha / < \alpha, \alpha >= \alpha$$

and therefore,

$$s_\alpha = s_\beta s_\alpha s_\beta^{-1}$$

or equivalently,

$$s_\alpha s_\beta = s_\beta s_\alpha$$

ie, $W_1$ and $W_2$ commute. It follows that the Weyl group $W$, ie, the group generated by $s_\alpha, \alpha \in S$ is given by

$$W = W_1 W_2 = W_2 W_1$$

Further, we know that the set of all roots is given by

$$\Delta = \bigcup_{\alpha \in S} W.\alpha$$

and since we have seen that for any $\beta \in S_2, \alpha \in S_1$ $s_\beta \alpha = \alpha$ and $s_\alpha \beta = \beta$, it follows that

$$\Delta = (\bigcup_{\alpha \in S_1} W_1 W_2.\alpha) \cup (\bigcup_{\alpha \in S_2} W_2 W_1 \alpha)$$

$$= (\bigcup_{\alpha \in S_1} W_1 \alpha) \cup (\bigcup_{\alpha \in S_2} W_2 \alpha)$$

$$= \Delta_1 \cup \Delta_2$$

It is clear that $\Delta_1 \cap \Delta_2 = \phi$ for if not, then it will follow that a nontrivial linear combination of the elements of $S_1 \cup S_2 = S$ is zero which is false. Note that any $\alpha \in \Delta_1$ is an integer linear combination of the elements of $S_1$ with the integers all either non-negative or all non-positive and likewise for $\Delta_2$. Further, if $\alpha \in \Delta_1, \beta \in \Delta_2$, then $< \alpha, \beta >= 0$ since $\alpha$ is in the span of $S_1$ and $\beta$ is in the span of $S_2$ and by hypothesis $S_1 \perp S_2$. Further, if $\alpha \in \Delta_1$ then $-\alpha \in \Delta_1$ since otherwise $-\alpha \in \Delta_2$ will imply again that a non-trivial linear combination of elements of $S$ is zero. (Recall that if $\alpha$ is a root, then so also is $-\alpha$). It is also clear that if $\alpha \in \Delta_1, \beta \in \Delta_2$, then $\alpha + \beta$ cannot be a root because if it is a root then it is either in $\Delta_1$ or in $\Delta_2$ and in either case, it would follow that a nontrivial linear combination of the elements of $S$ is zero. For example if $\alpha+\beta \in \Delta_1$, then it would follow that $\beta$ is in the linear span of $\Delta_1$ which in turn is in the linear span of $S_1$ while $\beta$ itself is in the linear span of $S_2$. Thus, we have shown that if $\alpha \in \Delta_1$ and $\beta \in \Delta_2$, then $[X_{\pm\alpha}, X_{\pm\beta}] = 0$ and further, $[H_\alpha, X_\beta] = <\beta, \alpha> X_\beta = 0$. In other words, the set $\{X_\alpha, X_{-\alpha}, H_\alpha\}$ commutes with $\{X_\beta, X_{-\beta}, H_\beta\}$ for each $\alpha \in \Delta_1$ and each $\beta \in \Delta_2$.

Now, define $\mathfrak{g}_1$ to be the linear span of $\{X_\alpha, X_{-\alpha}, H_\alpha, \alpha \in \Delta_1\}$ and likewise $\mathfrak{g}_2$ to be the linear span of $\{X_\alpha, X_{-\alpha}, H_\alpha, \alpha \in \mathfrak{g}_2\}$. Then from the above observations, it is clear that $\mathfrak{g}_1$ and $\mathfrak{g}_2$ are mutually commuting Lie subalgebras (because if $\alpha, \beta \in \Delta_1$, and $\alpha+\beta$ is a root, then $\alpha+\beta \in \Delta_1$ and likewise for $\Delta_2$) and by the root space decomposition, their direct sum is $\mathfrak{g}$. Hence, $\mathfrak{g}_k, k = 1,2$ are ideals whose direct sum is $\mathfrak{g}$ proving thereby that $\mathfrak{g}$ is not simple.

## 1.32 Cartan's classification of the complex simple Lie algebras

Let $\mathfrak{g}$ be a complex simple Lie algebra. Let $S = \{\alpha_1, ..., \alpha_l\}$ be a simple system of roots w.r.t a positive system $P$. We have already seen that

$$a(i,j) = 2 < \alpha_i, \alpha_j > / < \alpha_j, \alpha_j >, i \neq j, i,j = 1,2,...,l$$

are non-positive integers and these are called the Cartan integers. Draw a diagram called a Dynkin diagram which has $l$ vertices labeled $\alpha_i, i = 1,2,...,l$. The weight of each vertex $\alpha_i$ is defined as a number $w_i$ proportional to $<\alpha_i, \alpha_i>$. Further, by the Cauchy-Schwarz inequality, it follows that for any two integers $i,j$, $n(i,j) = a(i,j)a(j,i)$ is a non-negative integer assuming only the values $0,1,2,3$. In the Dynkin diagram, we joint vertex $i$ with vertex $j$ using $n(i,j)$ links. The Dynkin diagram is clearly connected because the Cartan matrix $((a(i,j))) \in \mathbb{Z}_+^{l\times l}$ is irreducible because $\mathfrak{g}$ is assumed to be simple. Note that $a(i,i) = 2$. A Dynkin sub-diagram is a connected diagram obtained by retaining a subset of the vertices of the Dynkin diagram with the same number of links connecting any two of its vertices as in the original diagram.

Theorem 1: In a Dynkin subdiagram, $D$ having $m$ vertices, there cannot be more than $m-1$ links. In particular a Dynkin diagram cannot have more than $l$ links. Note that $l$ is the number of simple roots which is the rank of the simple Lie algebra which is the dimension of the Cartan sub-algebra.

Proof: Let $\alpha_1, ..., \alpha_m$ be the vertices of a Dynkin subdiagram and let $M$ denote the number of links in this subdiagram. Then since these are linearly independent, we have

$$0 < |\sum_{i=1}^m \alpha_i/|\alpha_i||^2 = m + 2 \sum_{1\leq i<j\leq m} < \alpha_i, \alpha_j > /|\alpha_i||\alpha_j|$$

$$= n - \sum_{1 \le i < j \le m} \sqrt{n(i,j)} \le n - M$$

since $n(i,j) \ge 1$ if $i$ and $j$ are connected and zero $n(i,j) = 0$ otherwise.

Theorem 2: In a subdiagram, there cannot be any cycle, ie a closed loop.

## 1.33 The cornerstone theorems of functional analysis

1.Baire's category theorem.

2.The principle of uniform boundedness.

3. The closed graph theorem.

4. The open mapping theorem and the bounded inverse theorem.

5.The Hahn-Banach theorem.

6. The Stone-Von-Neumann theorem.

[a] Closed operators.

[b] Adjoint of a densely defined operator in a Hilbert space.

[c] Symmetric operators.

[d] Self-adjoint operators.

[e] Closable operators.

[f] Essentially self-adjoint operators.

[g] Criteria based on the resolvent for an operator to be self-adjoint and essentially self-adjoint.

[h] Compact operators in a Banach space and the structure of the spectrum of a compact operator.

[i] The spectral theorem for normal compact operators in a Hilbert space.

[k] Some identities and facts about operators based on the contour integral of its resolvent.

[l] Reduced resolvent of a bounded operator in an infinite dimensional Banach space having an isolated eigenvalue.

## 1.34 Spectrum of a Compact operator

[2] Let $T$ be a compact operator in a Banach space $X$. We shall first prove that nonzero complex number cannot be a limit point of a sequence of eigenvalues of $T$. For suppose $c \ne 0$ and $Tx_n = c_n x_n, n = 1, 2, ...$ with $\| x_n \| = 1$ and $c_n \to c$. Then let

$$M_n = span(x_k : k \le n)$$

We can choose $y_{n+1} \in M_{n+1}$ such that $\| y_n \| = 1$ and $d(y_{n+1}, M_n) = 1$.

Note that $y_{n+1}$ is a linear combination of $x_k, k \le n+1$. Then for $m > n$,

$$c_m^{-1} T(y_m) - c_n^{-1} T(y_n) = c_m^{-1}(T - c_m) y_m + y_m - c_n^{-1} T(y_n)$$

Noting that

$$c_m^{-1}(T - c_m)y_m c_n^{-1}T(y_n) \in M_{m-1}$$

it follows that

$$\| c_m^{-1}T(y_m) - c_n^{-1}T(y_n) \| \geq 1 \forall m > n$$

and hence $\{c_n^{-1}T(y_n)\}$ cannot have any convergent subsequence. However since $c_n \to c \neq 0$ it follows that $y_n/c_n$ is a bounded sequence and hence by compactness of $T$, $T(y_n)/c_n$ has a convergent subsequence. This contradiction shows that a sequence of eigenvalues of $T$ cannot converge to a non-zero complex number.

Remark: Let $M$ be a closed subspace of a Banach space and let $x \notin M$. Define a linear functional $f$ on $< M, x >$ so that $f(x) = 1$ and $f|_M = 0$ and extend $f$ to the whole of $X$ using the Hahn-Banach theorem without increasing its norm. Then for all $\xi \in M$,

$$f(ax + \xi) = a$$

and hence,

$$|a| \leq \| f \| |a|.d(x, M)$$

so

$$d(x, M) \geq 1/ \| f \|$$

On the other hand,

$$|f(ax + \xi)|/ \| ax + \xi \| = |a|/ \| ax + \xi \| \leq 1/d(x, M)$$

Thus,

$$\| f \| = 1/d(x, M)$$

Again, let $M$ be a closed subspace of a Banach space $X$. Then, we claim that for each $\epsilon > 0$, there exists an $x \notin M$ such that $\| x \| = 1$ and $d(x, M) > 1 - \epsilon$ and in the special case when $M$ is finite dimensional, we can choose x such that $\| x \| = 1$ and $d(x, M) = 1$. To prove this, first choose any $y \notin M$. Since $M$ is closed, we have that $d(y, M) = \delta > 0$. Thus given any $0 < \epsilon < 1$, we can choose $u \in M$ such that $d(y, u) = \| y - u \| < \delta(1 + \epsilon)$. Define

$$x = (y - u)/ \| y - u \|$$

Then,

$$\| x \| = 1$$

and

$$d(x, M) = \| y-u \|^{-1} d(y-u, M) = \| y-u \|^{-1} d(y, M)$$

$$= \delta \| y-u \|^{-1} > (1+\epsilon)^{-1} > 1-\epsilon$$

Now consider the special case when $M$ is finite dimensional. Then let $y \notin M$. There exists a sequence $u_n \in M, n = 1, 2, ..$ such that $\| y - u_n \| < \delta(1 + 1/n)$ where $\delta = d(y, M)$. Then define

$$M_y = < M, y >, x_n = (y - u_n)/ \| y - u_n \|$$

Then $M_y$ is finite dimensional and $x_n \in M_y, \| x_n \|= 1$. By compactness of the unit ball in a finite dimensional Banach space, it follows that there is a subsequence $x_{n_k}$ of $x_n$ such that $x_{n_k} \to x \in M_y$ and by continuity of the norm

$$\| x \|= lim \| x_{n_k} \|= 1$$

Further,

$$d(x, M) = limd(x_{n_k}, M) = lim \| y - u_{n_k} \|^{-1} d(y - u_{n_k}, M)$$

$$= \geq lim\delta^{-1}(1 + 1/n_k)^{-1}d(y, M) = \delta^{-1}d(y, M) = 1$$

The proof is complete.

Remark: If $T$ is a compact operator and $c \neq 0$ is an eigenvalue of $T$, then $dimN(T - c) < \infty$. Indeed, suppose $dimN(T - c) = \infty$. Then, we can choose an infinite sequence $x_n \in X$ such that $\| x_n \|= 1$ $T(x_n) = cx_n$ and the $x_n's$ are linearly independent. Define $M_n =< x_1, ..., x_n >$. Choose $y_{n+1} \in M_{n+1}$ such that $\| y_{n+1} \|= 1$ and $d(y_{n+1}, M_n) = 1$. Then

$$c^{-1}T(y_m) - c^{-1}T(y_n) = c^{-1}(T - c)(y_m) + y_m - c^{-1}T(y_n), m > n$$

which shows that

$$\| T(y_m) - T(y_n) \| \geq |c|d(y_m, M_{m-1}) = |c|, m > n$$

and hence $T(y_n)$ cannot have any convergent subsequence, a contradiction to the compactness of $T$.

## 1.35 Hahn-Banach theorem

Let $X$ be a real normed linear space and let $M$ be a subspace of $X$. Let $f$ be a bounded linear functional on $M$. Then $f$ can be extended to the whole of $M$ without increasing its norm.

Proof: Assume without loss of generality that $\| f \|_M = 1$. let $x_1 \notin M$. For $x, y \in M$, we have

$$f(x) + f(y) = f(x + y) \leq \| x + y \| \leq \| x - x_1 \| + \| y + x_1 \|$$

Thus

$$sup_{x \in M}(f(x) - \| x - x_1 \|) \leq inf_{y \in M}(\| y + x_1 \| - f(y))$$

Define

$$\alpha = sup_{x \in M}(f(x) - \| x - x_1 \|)$$

Then extend $f$ from $M$ to $< M, x_1 >$ by setting

$$f(x + tx_1) = f(x) + t\alpha, t \in \mathbb{R}, x \in M$$

Then

$$f(-t^{-1}x) - \| -t^{-1}x - x_1 \| \leq \alpha, t\mathbb{R} - \{0\}, x \in M$$

gives

$$[f(x)+t\alpha] \geq -\parallel x+tx_1 \parallel, t>0$$

$$[f(x)+t\alpha] \leq \parallel x+tx_1 \parallel, t<0$$

Likewise, the inequality

$$\parallel t^{-1}x+x_1 \parallel -f(t^{-1}x) \geq \alpha, x \in M$$

gives us

$$\parallel x+tx_1 \parallel \geq f(x)+t\alpha, t>0$$

and

$$\parallel x+tx_1 \parallel +f(x) \geq -t\alpha, t<0$$

or equivalently,

$$f(x)+t\alpha \geq -\parallel x+tx_1 \parallel, t<0$$

Thus, we have proved

$$|f(x)+t\alpha| \leq \parallel x+tx_1 \parallel, x \in M, t \in \mathbb{R}$$

which shows that

$$|f(u)| \leq \parallel u \parallel, u \in < M, x_1 >$$

Then by transfinite induction, $f$ can be extended to the whole of $X$ without increasing its norm.

## 1.36 Polar decomposition in infinite dimensional Hilbert spaces

[1] Polar decomposition of a densely defined closed operator $T$ in a Hilbert space. $T$ closed implies $T^{**}=T$ since

$$Gr(T^{**}) = Gr(T^*)'^{\perp} = Gr(T)'^{\perp'\perp} = \bar{G}r(T) = Gr(T)$$

$|T| = \sqrt{T^*T}$ can be defined as the unique positive square root of $T^*T$ using the Dunford contour integral. It has the properties $D(|T|) = D(T)$. Define $U|T|x = Tx$ for $x \in D(|T|) = D(T)$. Then $U$ is a well defined isometry from $R(|T|)$ onto $R(T)$ and hence can be uniquely extended to an isometry from $\bar{R}(|T|)$ onto $\bar{R}(T)$. Further, we define $U$ to be zero on $R(|T|)^{\perp} = N(|T|)$. This then defines $U$ as the unique operator that satisfies $U|T|x = Tx, x \in D(T)$ and $U_{|}N(|T|) = 0$. This is called the polar decomposition of $T$.

Remark: Let $T = AB$ ($T$ densely defined). Then

$$B^*A^* \subset T^*$$

Note that the adjoint $T^*$ is uniquely defined because $T$ is densely defined. To see this, choose $x \in D(A^*) \cap A^{*-1}(D(B^*)$ and let $y \in D(T) = D(B) \cap B^{-1}(D(A))$. Then

$$< x, Ty >=< x, ABy >=< A^*x, By >=< B^*A^*x, y >$$

and therefore

$$x \in D(T^*), T^*x = B^*A^*x$$

This proves the claim

Remark: Let $A$ be bounded $T$ densely defined and let $T = AB$. Note that this means $D(T) = D(B) \cap B^{-1}(D(A)) = D(B)$ since $D(A) = \mathcal{H}$ because $A$ is bounded. Then, $B^*A^* = T^*$. We have already seen that $B^*A^* \subset T^*$. So suppose, $x \in D(T^*)$. Then, for $y \in D(T)$

$$< T * x, y >=< x, Ty >=< x, ABy >=< A^*x, By >$$

since $D(A) = \mathcal{H}$ because $A$ is bounded. Further the equation

$$< T^*x, y >=< A^*x, By >$$

implies that $A^*x \in D(B^*)$ and

$$< T^*x, y >=< B^*A^*x, y >$$

since this equation is true for all $y \in D(T)$ and $D(T)$ is dense in $H$, it follows that $T^* \subset B^*A^*$ and therefore, $T^* = B^*A^*$.

Let now $T$ be a closed operator as before and let

$$T = U|T|$$

be its polar decomposition. We know that $|T| = \sqrt{T^*T}$, the unique positive square root of $T^*T$ can be constructed using the Dunford-Taylor contour integral and

$$D(|T|) = D(T)$$

If in particular $T$ is bounded, then $|T|$ can be defined using an iterative process which expresses $|T|$ as the operator norm limit of a sequence of polynomials in $T^*T$. Here, we do not make any boundedness assumption. $|T|$ is self-adjoint and hence we get from the remark above that

$$|T|U^* \subset T^*$$

Now let $x \in D(T^*), y \in D(T)$. Then

$$< T^*x, y >=< x, Ty >=< x, U|T|y >=< U^*x, |T|y >$$

and therefore

$$< T^*x, y >=< |T|U^*x, y >$$

Since $D(T)$ is dense in $\mathcal{H}$, it follows then that

$$T^* = |T|U^*$$

Thus,

$$TT^* = U|T|^2U^*$$

Let

$$G = \sqrt{TT^*}$$

Then we get by uniqueness of the positive square root of a positive operator that

$$G = U|T|U^*$$

Note that

$$U|T|U^*U|T|U^* = U|T|^2U^*$$

because $U$ is isometric on $R(|T|)$ implies

$$< |T|x, U^*U|T|y >=< U|T|x, U|T|y >=< |T|x, |T|y >, x, y \in D(|T|) = D(T)$$

ie $U^*U$ is the identity on $R(|T|)$ and since $U$ is zero on $R(|T|)^\perp = N(|T|)$, it follows that $U^*U$ is zero on $N(|T|)$. Thus if $x, y \in \mathcal{H}$ and we define $P$ to be the projection onto $\bar{R}(|T|)$ so that $1 - P$ becomes the projection onto $N(|T|)$, then we get

$$< x, U^*Uy >=< Ux, Uy >=< UPx, UPy >=< Px, Py >=< x, Py >$$

or equivalently,

$$U^*U = P$$

This implies

$$G^2 = U|T|U^*U|T|U^* = U|T|P|T|U^* = U|T|^2U^*$$

and hence $G$ is the unique square root of $U|T|^2U^*$. We have

$$U^*G = |T|U^* = T^*$$

ie,

$$T^* = U^*G$$

is the polar decomposition of $T^*$, ie,

$$G = |T^*| = \sqrt{TT^*}$$

Now, let $H$ be a self-adjoint. Then, $H^* = H$ and so $|H^*| = |H|$ and if we denote the polar decomposition of $H$ by

$$H = U|H|$$

then

$$U^*|H| = U^*|H^*| = H^* = H = |H|U^*$$

In particular, by the uniqueness of the polar decomposition of $H$, it follows that

$$U = U^*$$

and hence

$$U^2 = U^*U = P$$

where $P$ is the projection onto $\bar{R}|H|) = \bar{R}(H)$.

Remark: $R(|T^*|) = R(T)$ can be seen as follows. We've already seen that

$$T = U|T|, T^* = |T|U^* = U^*|T^*|$$

The definition of $U$ implies that $R(U) = R(T)$ and likewise, from the polar decomposition of $T^*$, $R(U^*) = R(T^*)$. Thus,

$$R(U^*) = R(T^*) \subset R(T^*T) \subset R(|T|)$$

$$R(U) = R(T) \subset R(TT^*) \subset R(|T^*|)$$

Also,

$$U^*T = U^*U|T| = P|T| = |T|$$

and hence

$$R(|T|) \subset R(U^*) \subset R(|T|)$$

so that

$$R(|T|) = R(U^*)$$

and likewise, by the polar decomposition of $T^*$,

$$R(|T^*|) = R(U^{**}) = R(U) \subset R(|T^*|)$$

Thus

$$R(|T^*|) = R(U) = R(T)$$

and likewise

$$R(|T|) = R(|T^{**}|) = R(T^*)$$

Now, let $H$ be Hermitian with polar decomposition

$$H = U|H|$$

We've seen that $U = U^*$ and hence $U^2 = P$ where $P$ is the projection onto $R(H) = R(|H|)$. For $x \in R(P) = R(H) = R(|H|) = R$, define

$$x_+ = (u + Ux)/2, x_- = (x - Ux)/2$$

Then since $U^2 = 1$, we get

$$Ux_+ = x_+, Ux_- = -x_-$$

Let

$$M_- = \{x \in R : Ux = x\}, M_- = \{\in R : Ux = -x\}, M_0 = R^\perp$$

First we observe that

$$R = M_+ \oplus M_-$$

For let $x \in R$. Then

$$x = x_+ + x_-, x_+ \in M_+, x_- \in M_-$$

Further,

$$x_+ = Ux_+ \subset R(U) \subset R(H) = R,$$

$$x_- = Ux_- \subset R(U) = R(H) = R$$

and hence

$$R \subset M_+ \oplus M_-$$

Conversely, let $x \in M_+$. Then, $x \in R$ by definition of $M_+$ and likewise for $M_-$. Let $P_+$ denote the projection onto $M_+$, $P_-$ onto $M_-$ and $P_0$ onto $M_0$. Then we have the following orthogonal resolution of the identity:

$$1 = P_+ + P_- + P_0$$

Note that

$$P = P_+ + P_-$$

Now observe the following: $M_+, M_-, M_0$ are reduced by $H$. Note that a subspace $M$ is said to be reduced by $H$ if $P_M H \subset HP_M$ where $P_M$ is the orthogonal projection onto $M$. This is the same as saying that if $x \in D(H)$, then $P_M x \in D(H)$ and $P_M Hx = HP_M x$. This also clearly implies that if $x \in D(H) \cap M$ then $Hx \in M$, ie, $M$ is $H$-invariant in this restricted sense. Note that $U$ commutes with $H$ and in fact, we have $UH = HU$. To see this, we observe that

$$H = U|H| = H^* = |H^*|U^* = |H|U^*$$

and hence,

$$HU = |H|U^*U = |H|P$$

and further,

$$UH = |H|$$

Further, since $U$ is bounded, by a remark above,

$$|H| = |H|^* = (UH)^* = HU^* = HU = |H|P$$

since $U^* = U$. Thus,

$$HU = |H|$$

and hence we deduce that

$$UH = HU$$

proving the claim.

Remark: Let $A$ be a bounded operator that commutes with $H$, ie, $AH \subset HA$. Then $A$ commutes with $U$. In fact, we have that since $A$ commutes with $H$, $A$ also commutes with $|H|$ and hence

$$UA|H| \subset U|H|A = HA,$$

$$AU|H| = AH \subset HA$$

and hence $AU$ and $UA$ agree on $R = R(|H|)$. Further, if $x \in R^{\perp} = N(|H|)$, then $Ax \in N(|H|) = N(H)$ since $AHx = 0$ and therefore $HAx = 0$ because $AH \subset HA$. Thus, $UAx = 0$ for such an $x$ since $U$ is zero on $R^{\perp} = N(|H|) = N(H)$. On the other hand, for such an $x$, $AUx = 0$ since again $U$ is zero on $N(|H|)$. Thus, we have proved that $UA = AU$.

Remark: Since $UH = HU$, we have in particular that $P_{-}H = HP_{-}$. In fact, this follows from the fact that $R(P_{-})$ is the set of all $x \in R$ which have eigenvalue $-1$ for $U$. Thus, let $x \in H$ be arbitrary. Then, $P_{-}x \in R$ and $UP_{-}x = -P_{-}x$. Now the equation $UH = HU$ implies that $UH \subset HU$ which implies that $U(D(H)) \subset D(H)$ and hence $P_{-}x = -UP_{-}x \in D(H)$ and therefore,

$$HP_{-}x = -HUP_{-}x = -UHP_{-}x$$

ie, $HP_{-}x$ is an element of $R$ belonging to the eigenvalue $-1$ of $U$. In other words, $HP_{-}x \in R(P_{-})$ which implies that $P_{-}HP_{-}x = HP_{-}x$. Since $x \in \mathcal{H}$ was arbitrary, it follows that

$$HP_{-} = P_{-}HP_{-}$$

which gives on taking adjoints and using a remark above that

$$P_{-}H = P_{-}HP_{-}$$

ie,

$$HP_{-} = P_{-}H$$

Of course, this means that $x \in D(H)$ implies $P_{-}x \in D(H)$ and $HP_{-}x \in R(P_{-})$. Thus $x \in D(H) \cap R(P_{-})$ implies $P_{-}x = x \in D(H)$ implies $Hx = HP_{-}x \in R(P_{-})$. In this restricted sense, $R(P_{-})$ is $H$-invariant (This is the notion of an invariant subspace for unbounded operators).

Remark: We claim that $D(H) \bigcap M_{-}$ is dense in $M_{-}$. This is an elementary consequence of the density of $D(H)$ in $\mathcal{H}$ and the commutativity of $P_{-}$ with $H$. To see this, let $x \in M_{-}$ be arbitrary. Since $D(H)$ is dense in $\mathcal{H}$, it follows that there is a sequence $x_n \in D(H)$ that converges to $x$. However, by what we saw above, $P_{-}x_n \in D(H)$ and since $P_{-}$ is bounded, $P_{-}x_n \to P_{-}x = x$. Since $P_{-}x_n \in M_{-}$, it follows that $P_{-}x_n \in D(H) \cap M_{-}$ and this proves our claim.

Remark: Let $M$ be a subspace that reduces $H$, ie, if $P_M$ denotes the projection onto $M$, then $P_M H \subset HP_M$ which means that if $x \in D(H)$, then $P_M x \in D(H)$ and $HP_M x \in M$. This is also implies that if $x \in D(H) \cap M$, then $Hx \in M$. Now suppose $< Hx, x > \leq 0 \forall x \in M$. Then, we claim that $x \in M_{-}$.

Indeed, since $P_M$ commutes with $H$, it follows that $P_M$ commutes with $P_-$. In fact, by one of the above remarks, since $P_M$ is bounded and commutes with $H$, it also commutes with $U$. We now claim that $P_M$ also commutes with $P_-$. To prove this, we first show that $(P-U)/2 = P_-$. Note that $P$ is the projection onto $R$. To see this, suppose $x \in \mathcal{H}$. Then, $(P-U)x/2 \in R$ since the ranges of $P$ and $U$ are both $R$ and further $PU = U$. This latter equation follows because $R(U) = R$ and $P$ is the identity on $R$. Taking adjoints on both sides of this equation noting that $P$ and $U$ are bounded with domain the whole of $\mathcal{H}$, it follows that $U^* = U^*P$ and hence $U = UP$ since $U^* = U$. Thus,

$$(P-U)^2 = P + U^2 - 2UP = 2(P-U)$$

since $U^2 = U^*U = P$. Thus,

$$((P-U)/2)^2 = (P-U)/2$$

and

$$(P-U)^* = P - U^* = P - U$$

Thus, $(P-U)/2$ is an orthogonal projection whose range is contained in $R$. Now,

$$U(P-U)/2 = (UP - U^2)/2 = (U-P)/2$$

and hence since $R((P-U)/2) \subset R$, it follows that

$$R((P-U)/2) \subset R(P_-)$$

Now $UP_- = -P_-$ since $U$ equals $-1$ on $R(P_-)$ and further $PP_- = P_-$ since $R(P_-) \subset R = R(P)$. Therefore,

$$(P-U)P_-/2 = P_-$$

and therefore,

$$R(P_-) \subset R((P-U)/2)$$

Thus, we have proved that

$$(P-U)/2 = P_-$$

Now $P_M$ commutes with $U$ and also with $P$. This latter commutativity can be seen from the fact that $P_M$ commutes with $U$ and that $P = U^2$. Thus, $P_M$ also commutes with $P_-$. Likewise, $P_M$ also commutes with $P_+$ and with $P_0 = 1 - P_+ - P_-$.

Remark: $M_-$ reduces $H$, ie, $P_-H \subset HP_-$ and in fact, $P_-H = HP_-$. This follows from the fact that $HU = UH$ and $P_- = (P-U)/2 = (U^2 - U)/2$.

**References**

[1] Walter Rudin, "Functional Analysis".

[2] Tosio Kato, "Perturbation theory for linear operators".

## 1.37 Von-Neumann's theorem

Let $H$ be a Hilbert space and $T$ a closed operator in $H$. Then $T^*T$ is self-adjoint.

Proof:

$$Gr(T^*) = Gr(T)'^{\perp}$$

and since $Gr(T)$ is closed because $T$ is assumed to be closed, it follows that

$$H \times H = Gr(T^*) \oplus Gr(T)'$$

as an orthogonal direct sum of closed subspaces. Thus, for any $u, v \in H$, we have

$$(u, v) = (x, T^*x) + (-Ty, y)$$

for some $x \in D(T^*), y \in D(T)$. Thus,

$$u = x - Ty, v = T^*x + y$$

Taking $u = 0$ gives

$$x - Ty = 0, v = T^*x + y$$

for some $x \in D(T*), y \in D(T)$. Thus,

$$v = T^*Ty + y = (1 + T^*T)x$$

for some $x$. Since $v \in H$ was arbitrary, it follows that

$$R(1 + T^*T) = H$$

Further $1 + T^*T \geq 1 > 0$. Clearly $1 + T^*T$ is injective since $(1 + T^*T)x = 0$ for some $x \in D(T^*T) \subset D(T)$ implies $< x, (1 + T^*T)x >= 0$ implies $< x, x > + < x, T^*Tx >= 0$ for some $x \in D(T)$ implies

$$< x, x > + < Tx, Tx >= 0$$

implies

$$\| x \|^2 + \| Tx \|^2 = 0$$

implies $x = 0$. Thus, $1 + T^*T$ is invertible. Further, $1 + T^*T$ is symmetric because $T^*T$ is:

$$T^*T = T^*T^{**} \subset (T^*T)^*$$

since $T^{**} = T$ because $T$ is closed. Here, we have used the result established above that for any $T$ with dense domain, if $T = AB$, then, $B^*A^* \subset T^*$. Writing

$$S = 1 + T^*T$$

it follows that since $S$ is invertible and symmetric, so is $S^{-1}$:

$$Gr(S^{-1}) = PGr(S) \subset PGr(S^*) = Gr(S^{*-1}) = Gr(S^{-1*})$$

where

$$P(x, y) = (y, x)$$

is a bounded operator on $H \times H$. Note that here we have used the fact that if $A$ is an invertible operator (on its domain) with $A^{-1}$ bounded, then $A^{-1*}$ is bounded and hence

$$< A^{-1*}x, Ay >=< x, A^{-1}Ay >=< x, y >, x \in H, y \in D(A)$$

and therefore,

$$R(A^{-1*}) \subset D(A^*), A^*A^{-1*} = I$$

on the appropriate domain. On the other hand,

$$< A^*x, A^{-1}y >=< x, AA^{-1}y >=< x, y >, x \in D(A^*), y \in H$$

(Note that $A^{-1}y \in D(A) \forall y \in R(A) = H$) and therefore,

$$R(A^*) \subset D(A^{-1*}), A^{-1*}A^* = I$$

on the appropriate domain. From these two equations, it follows that $A^*$ is invertible and

$$A^{*-1} = A^{-1*}$$

Now clearly $S^{-1} = (1 + T^*T)^{-1}$ is a bounded symmetric operator (bounded by 1) and its domain is $R(S) = R(1 + T^*T) = \mathcal{H}$. Hence, $S^{-1}$ is a bounded self-adjoint operator.

Note: [1] Let $A, B$ be two operators in $\mathcal{H}$ with dense domain. Suppose $AB =\subset I$ and $BA \subset I$. Suppose further that $B$ is bounded so that its domain becomes the whole of $\mathcal{H}$ by a trivial extension. Then, can we say that $A^{-1}$ exists and equals $B$ ? First observe that $Ax = 0$ for some $x \in D(A)$ implies $BAx = 0$ implies $x = 0$ so that $A$ is one-one on $D(A)$. Secondly, let Since $A$ maps $D(A)$ onto $R(A)$, it follows that $A : D(A) \to R(A)$ is one-one onto and hence invertible with inverse $A^{-1} : R(A) \to D(A)$. Since $R(AB) \subset R(A)$, it follows then from the equation $AB \subset I$ that $A^{-1}AB = A^{-1}$ on $D(AB)$, ie, $B \subset A^{-1}$ on $D(AB)$. Note that the equation $AB \subset I$ implies that $D(AB) \subset R(A)$.

[2] Let $AB \subset I$ and let $D(AB)$ be dense in $\mathcal{H}$. Now $Bx = 0$ for $x \in D(AB)$ implies $x = ABx = 0$. Thus, $B$ is one-one on $D(AB)$. Since we assume that $D(AB)$ is dense in $\mathcal{H}$, then the equation $AB \subset I$ implies that $AB$ is bounded on a dense domain and hence it can be extended uniquely to a bounded operator on the whole of $\mathcal{H}$. Therefore, we may conclude that after such an extension has been made, $AB = I$. This implies that $B$ is one one on $D(B)$ for $Bx = 0$ implies $x = ABx = 0$. Thus, $B^{-1} : R(B) \to D(B) = \mathcal{H}$ is defined. Then, the equation $AB = I$ implies $ABB^{-1} = B^{-1}$ so that $A = B^{-1}$ (on $R(B)$).

## 1.38 An iterative method for constructing the positive square root of a bounded positive operator in a Hilbert space

## 1.39 Hilbert-Schmidt operators

[4] Let $T$ be a compact operator in a Hilbert space. Then show that $T^*$ is also compact.

Remark: Let $e_n, n = 1, 2, ...$ be an onb for the Hilbert space. the sequence of operators

$$T_N = \sum_{n,m=1}^{N} |e_n><e_n|T|e_m><e_m|, N \geq 1$$

If $T$ is Hilbert-Schmidt, then

$$\sum_{n,m} |<e_n|T|e_m>|^2 < \infty$$

and then it is easy to see that $T_N$ converges in Hilbert Schmidt norm to $T$:

$$\| T - T_N \|_{HS}^2 = \| T \|_{HS}^2 - \sum_{n,m=1}^{N} |<e_n|T|e_m>|^2$$

$$\rightarrow 0, N \rightarrow \infty$$

In other words, every HS operator can be expressed as an HS limit of a sequence of finite rank operators. Since convergence in HS norm implies convergence in operator norm, it follows that every HS operator is also compact. (Every finite rank operator is compact and the operator norm limit of a sequence of compact operators is compact).

Problem: Let $T$ be HS. Then

$$\| T \|_{HS} \geq \| T \|$$

In fact, for an onb $e_n$, we have

$$\| T \|_{HS}^2 = Tr(T^*T) = \sum_n <e_n|T^*T|e_n>$$

$$= \sum_{n,m} <e_n|T^*|e_m><e_m|T|e_n> = \sum_{n,m} |<e_m|T|e_n>|^2$$

on the one hand and hence

$$\| T \|^2 = sup_{\|x\|=1} \| Tx \|^2$$

we get

$$\| Tx \|^2 = \sum_n |<e_n|T|x>|^2 = \sum_{n,m} |<e_n|T|e_m>|^2 |<e_m|x>|^2$$

$$\leq \sum_{n,m} | < e_n|T|e_m > |^2 \parallel x \parallel^2 = \parallel T \parallel_{HS}^2 \parallel x \parallel^2$$

Now observe that if $T$ is any bounded operator, then $\parallel T^* \parallel = \parallel T \parallel$ because,

$$\parallel Tx \parallel^2 = < x, T^*Tx > \leq \parallel T^*T \parallel . \parallel x \parallel^2$$

and hence

$$\parallel T \parallel \leq \parallel T^*T \parallel \leq \parallel T^* \parallel . \parallel T \parallel$$

so that

$$\parallel T \parallel \leq \parallel T^* \parallel$$

and replacing $T$ by $T^*$, we get

$$\parallel T^* \parallel = \parallel T \parallel = \sqrt{\parallel T^*T \parallel} = \parallel TT^* \parallel$$

Let $u_n$ be a bounded sequence let $T$ be compact. Then, $T^*u_n$ is a bounded sequence since $T^*$ is bounded and hence by compactness of $T$, $TT^*u_n$ has a convergent subsequence say $TT^*v_k$ where $v_k = u_{n_k}$. Since $v_k$ is a bounded sequence, it follows that

$$< T^*(v_k - v_m), T^*(v_k - v_m) >$$

$$= | < v_k - v_m, TT^*(v_k - v_m) > | \leq \parallel v_k - v_m \parallel . \parallel TT^*(v_k - v_m) \parallel$$

$$\rightarrow 0, k, m \rightarrow \infty$$

and hence $T^*v_k$ is Cauchy and therefore converges. Thus, $T^*$ is compact.

Remark: If $T$ is compact then $T$ is bounded. If $T$ is bounded then $T^*$ is bounded.

Proof: Let Suppose $T$ is unbounded. Then, we can find a sequence $u_n$ such that $\parallel u_n \parallel = 1$ and $\parallel Tu_n \parallel \rightarrow \infty$. It is then clear that $Tu_n$ cannot have any Cauchy subsequence since any Cauchy sequence is necessarily bounded. Now for the second part. Let $T$ be bounded. Then, we have that

$$\parallel T^*x \parallel^2 = < x, TT^*x > \leq \parallel TT^*x \parallel . \parallel x \parallel$$

$$\leq \parallel T \parallel . \parallel T^*x \parallel . \parallel x \parallel$$

from which we deduce that

$$\parallel T^*x \parallel / \parallel x \parallel \leq \parallel T \parallel$$

so that $T^*$ is also bounded.

## 1.40 Spectral theorem for compact normal operators in a Hilbert space

First we prove the following: Let $T$ is compact, then its spectrum is countable with each non-zero value in the spectrum being an eigenvalue of $T$. Further, the set of eigenvalues of $T$ cannot accumulate at a non-zero complex number, in other words, the only possible accumulation point in the spectrum of $T$ is zero. Further, the generalized eigensubspace of any non-zero eigenvalue of $T$ is necessarily finite dimensional.

Let $T$ be compact. Let $c_1, c_2, ...$ be nonzero eigenvalues of $T$ with normalized eigenvectors $e_1, e_2, ..$ respectively. Then define

$$M_n = span(e_k : k \leq n)$$

For each $n$, we can choose a $x_{n+1} \in M_{n+1}$ having unit norm such that $d(x_{n+1}, M_n) = 1$ and then for all $m > n$,

$$c_m^{-1} T x_m - c_n^{-1} T x_n = c_m^{-1}(T - c_m) x_m + x_m - c_n^{-1} T x_n$$

and the rhs is of the form $x_m - u_m$ where $u_m \in M_{m-1}$. It follows then that

$$\| c_m^{-1} T x_m - c_n^{-1} T x_n \| \geq 1$$

This shows that $c_n^{-1} T x_n$ cannot have a convergent subsequence. On the other hand, since $T$ is compact, $Tx_n$ has a convergent subsequence, say $Tx_{n_k}$. if $c_n \to c \neq 0$, then it follows that $c_{n_k}^{-1} T x_{n_k}$ converges which is a contradiction. Hence the eigenvalues of $T$ cannot accumulate at a nonzero value.

Now let $c$ be any nonzero complex number that is not an eigenvalue of $T$. We shall show that $R(T - c)$ is closed. Indeed, suppose $(T - c)x_n \to y$ with $x_n$ a bounded sequence. By compactness of $T$, $Tx_{n_k}$ converges to some $z$ for some subsequence $x_{n_k}$ of $x_n$. Then, it follows that $x_{n_k}$ converges to $u = (z - y)/c$. Therefore, since $T$ is bounded, $(T - c)(x_{n_k})$ converges to $(T - c)u$ and therefore,

$$y = (T - c)u$$

which proves that $y \in R(T - c)$. Now suppose that $(T - c)x_n \to y$ with $x_n$ unbounded. Then defining $u_n = x_n / \| x_n \|$, it follows by compactness of $T$ that $Tu_{n_k}$ converges for some subsequence $u_{n_k}$ of $u_n$. Let $v = limTu_{n_k}$. Thus, the equation

$$lim(T - c)u_{n_k} = lim(T - c)x_{n_k} / \| x_{n_k} \| \to 0$$

since $(T - c)x_{n_k} \to y$ gives

$$limu_{n_k} = v/c$$

and hence by boundedness of $T$,

$$(T - c)v = 0$$

and further $\| v \|= |c| lim \| u_{n_k} \|= |c|$ shows that $v \neq 0$. Thus $c$ is an eigenvalue of $T$ which is a contradiction to our assumption. This completes the proof of the statement that $R(T - c)$ is closed if $c$ is not an eigenvalue of $T$.

Now if $c \neq 0$ is not an eigenvalue of $T$ and $\bar{c}$ is also not an eigenvalue of $T^*$, then $T - c$ is invertible on $R(T - c)$ which as we have seen is closed. Further, we have

$$R(T - c)^{\perp} = N(T^* - \bar{c}) = 0$$

and since $R(T - c)$ is closed, it must necessarily follow that

$$R(T - c) = \mathcal{H}$$

Thus, by $(T - c)^{-1}$ is a closed operator (because $T - c$ is bounded and hence closed) on the whole of $\mathcal{H}$ and hence by the closed graph theorem, $(T - c)^{-1}$ : $\mathcal{H} \to \mathcal{H}$ is bounded and therefore, $c$ belongs to $\rho(T) = \sigma(T)^c$ the resolvent of $T$ which is the complement of the spectrum of $T$. Now if $S$ is a bounded invertible operator then so is $S^*$ and

$$S^{*-1} = S^{-1*}$$

provided that $S^{-1}$ is also bounded. This can be proved by noting that

$$< x, S^* S^{-1*} y >=< Sx, S^{-1*} y >$$

for all $x \in \mathcal{H}, y \in D(S^{-1*})$ and if $S^{-1}$ is also a bounded operator, then

$$< Sx, S^{-1*} y >=< S^{-1} Sx, y >=< x, y >, x, y \in \mathcal{H}$$

so

$$S^* S^{-1*} = I$$

if both $S, S^{-1}$ are bounded. Interchanging $S$ and $S^{-1}$ then gives us

$$S^{-1*} S^* = I$$

thereby proving our claim. Suppose now that $c \in \rho(T)$ where $T$ is compact. Then, $T - c, (T - c)^{-1}$ are bounded operators and hence by the above remark,

$$(T^* - \bar{c})^{-1} = (T - c)^{-1*}$$

is also bounded and vice versa. Thus, $c \in \rho(T)$ iff $c \in \rho(T^*)$ and therefore, $c \in \sigma(T)$ iff $\bar{c} \in \sigma(T^*)$.

Remark: Suppose $0 \neq c \in \sigma(T)$ but $c$ is not an eigenvalue of $T$. Then $T - c$ is injective and we just saw that $R(T - c)$ closed. By the closed graph theorem therefore, $(T - c)^{-1}$ must exist and must be bounded on its range which contradicts the assumption that $c \in \sigma(T)$. Thus apart from zero, the spectrum of $T$ must consist of only a countable set of eigenvalues with no accumulation point at any non-zero complex number. Let now $c$ be any non-zero eigenvalue of $T$. The generalized eigen projection corresponding to this eigenvalue is given by

$$P(c) = -(2\pi i)^{-1} \int_{\Gamma} (T - z)^{-1} dz$$

with the contour $\Gamma$ encircling $c$ but excluding zero and all the other eigenvalues of $T$. In particular, $\Gamma$ lies entirely in $\rho(T)$ and hence $(T - z)^{-1}$ is bounded in norm as $z$ varies over $\Gamma$. It follows that. Now,

$$T(T - z)^{-1} = 1 + z.(T - z)^{-1}$$

and hence,

$$P(c) = (-2\pi i)^{-1} \int_{\Gamma} T(T - z)^{-1} dz$$

since

$$\int_{\Gamma} dz/z = 0$$

because $\Gamma$ excludes 0. Now, $T(T-z)^{-1}$ being the product of a compact operator and a bounded operator, is also compact for all $z \in \Gamma$ and hence its contour integral which is a limit in the operator norm of a finite linear combination of such compact operators must also be compact. Thus, $P(c)$ is a compact projection and hence it must be a finite dimensional projection. Indeed, if $P(c)$ is an infinite dimensional projection, then we can choose recursively an infinite sequence $x_n, n = 1, 2, ...$ of linearly independent elements in $R(P(c))$ each having unit norm and such that $d(x_{n+1}, < x_1, ..., x_n >) = 1 \forall n$ which would imply that $P(c)x_n = x_n$ has no convergent subsequence and this would contradict the compactness of $P(c)$.

Remark: Let $T$ be compact and normal. Then $T^*$ is also compact and normal. We now claim that $\| T^n \| = \| T \|^n, n = 1, 2, ....$ To see this, first observe that

$$\| T^2 \|^2 = \| (T^2)^* T^2 \|$$

$$= \| (T^*T)^2 \|$$

since $T$ commutes with $T^*$. Now put $H = T^*T$. $H$ is Hermitian and hence

$$\| H^2 \| = \| H^* H \| = \| H \|^2$$

Thus,

$$\| (T^*T)^2 \| = \| T * T \|^2 = \| T \|^4$$

and therefore,

$$\| T^2 \| = \| T \|^2$$

This is true for any normal $T$. Note that we have used the fact proved earlier that for any bounded operator $S$,

$$\| S \|^2 = \| S^* \|^2 = \| S^* S \|$$

Then, by iteration, we get

$$\| T^n \| = \| T \|^n, n = 2^m, m \geq 1$$

Now let $n$ be arbitrary. Choose positive integers $r, s$ so that $2^r = n + s$. Then,

$$\| T^{n+s} \| = \| T \|^{n+s}$$

by what we just saw. Hence,

$$\| T \|^{n+s} = \leq \| T^n \| . \| T \|^s$$

and therefore,

$$\| T \|^n \leq \| T^n \| \leq \| T \|^n$$

and this proves our claim.

# Chapter 2

# Antenna Theory

## 2.1 Course Outline

[1] Maxwell's equations in the frequency domain. Solution using retarded potentials, the far field approximation. Calculation of the total power radiated at a given frequency in the far field zone.

[2] Maxwell's equations in the frequency domain taking inhomogeneities, anisotropicity and field dependence (nonlinearity) of the permittivity and permeability tensors into consideration. Perturbative solution of the differential equations.

[3] Construction of the Green's function of the Helmholtz operator for boundaries of various shapes for the Dirichlet and Neumann boundary conditions:Applications to cavity resonator antennas.

[4] The reciprocity theorem between two electromagnetic fields driven by two different current densities.

[5] The radar equation, directivity and antenna aperture, reciprocity theorem between transmitter and receiver antenna.

[6a] The basic antenna parameters: Loss resistance, radiation resistance total resistance, effective aperture. Equivalent circuit of a transmitter and receiver antenna.

[6b] Poynting's theorem and its evaluation in the far field zone in terms of the current distribution in the source.

[7] The fields of an infinitesimal electric dipole in the near field zone, far field zone, intermediate zone.

[8] The fields produced by a finite straight wire of length $L$ carrying a spatial sinusoid current distribution vanishing at its ends.

[9] The fields produced by a circular loop carrying sinusoidal current in the far field zone.

[10] The far field pattern produced by an infinitesimal loop of wire in terms of the magnetic moment of the wire.

[11] The far field pattern produced by a volume carrying a sinusoidal current

density.

[12] The far field pattern produced by an infinitesimal dipole, a finite straight wire and a circular loop placed above an infinite ground plane. Calculation based on the method of images.

[13] The total power radiated by an antenna in the far field zone.

[14] The radiation resistance of an antenna in terms of the current distribution in it and the feeding current.

[15] A planar aperture on which an electromagnetic field is incident as an antenna. Computation of the far field Poynting vector.

[16] A waveguide terminated by a horn:The far field radiation pattern.

[17] Helical antennas as broad band antennas.

[18] A cavity resonator as a microstrip antenna.

[19] Waveguide feeding an aperture as an antenna.

[20] The method of stationary phase.

[21] The mutual impedance between two antennas, the self impedance of an antenna.

[22] Antenna arrays: The pattern multiplication theorem, application to binomial and Chebyshev arrays.

[23] Plotting of antenna pattern lobes.

[24a] Computing the current density induced on an antenna surface by an incident electromagnetic field using generalizations of the Pocklington and Hallen integral equations.

[24b] Application to the problem of determining the induced currents on the driver and driven elements of a Yagi-Uda array.

[24c] Solving Pocklington's integral equations by the method of moments.

[25] The effect of a gravitational field and inhomogeneous, anisotropic and field dependent permittivity and permeability on the pattern of an antenna:General relativistic considerations based on Covariant form of the Maxwell equations in a background curved space-time.

[26] A brief description of the finite element method for numerically solving antenna problems.

[25] An introduction to quantum antennas.

[a] Quantum current generated by electrons, positrons and photons within a cavity on the atomic scale. Description of currents on the cavity surface generated by the quantum magnetic field and the currents within the cavity generated by the electrons-positron Dirac field. Evaluation of the quantum statistical moments of the near and far field radiation patterns in a given state of the electrons, positron and photons. A description of coherent states for photons and electrons-positrons.

## 2.2 The far field Poynting vector

The far field magnetic vector potential is given by

$$\mathbf{A}(t,\mathbf{r}) = (\mu/4\pi)\int \mathbf{J}(\omega,\mathbf{r})exp(-jk|\mathbf{R}-\mathbf{r}|)d^3r/|\mathbf{R}-\mathbf{r}|$$

$$\approx \mathbf{P}(\omega,\hat{R})exp(-jkR)/R$$

where

$$\mathbf{P}(\omega,\hat{R}) = (\mu/4\pi)\int_S \mathbf{J}(\omega,\mathbf{r})exp(jk\hat{R}.\mathbf{r})d^3r$$

where $S$ denotes the source volume. Thus, the far field electric field (ie upto $O(1/R)$) is given by

$$\mathbf{E}(\omega,\mathbf{R}) = -\nabla\Phi - j\omega\mathbf{A}$$

$$= -\nabla((jc^2/\omega)div\mathbf{A})) - j\omega\mathbf{A}$$

$$= [-(jc^2/\omega)(-jk\hat{R}.\mathbf{P})(-jk\hat{R})/R - j\omega\mathbf{P}/R]exp(-jkR)/R$$

$$= [(jk^2c^2/\omega)P_r\hat{R}/R - j\omega(P_r\hat{R} + P_\theta\hat{\theta} + P_\phi\hat{\phi})]exp(-jkR)/R$$

$$= -j\omega(P_\theta\hat{\theta} + P_\phi\hat{\phi})exp(-jkR)/R$$

Note: The only $O(1/R)$ term in when we take the gradient of a function of the form $f(R,\theta,\phi)exp(-jkR)/R$ is given by

$$-jk\hat{R}.f(R,\theta,\phi)/R$$

All the other terms are of order $1/R^2$ and they do not contribute anything to the outward radiated power. In other words, the $O(1/R)$ terms come only by differentiation of the phase factor or equivalently from differentiation of the far field delay factor, not by differentiating the amplitude factor. We can likewise evaluate the far field magnetic field (ie, with neglect of $O(1/R^2)$ terms):

$$\mathbf{B}(\omega,\mathbf{R}) = \nabla\times\mathbf{A}(\omega,\mathbf{R}) =$$

$$-jk\hat{R}\times\mathbf{P}.exp(-jkR)/R$$

$$= -jk(P_\theta\hat{\phi} - P_\phi\hat{\theta}).exp(-jkR)/R$$

Therefore, the time averaged far field Poynting vector, ie, power flux is given by (upto $O(1/R^2)$ terms)

$$\mathbf{S}(\omega,R,\hat{R}) = \mathbf{S}(\omega,R,\theta,\phi) = Re[\mathbf{E}\times\mathbf{B}^*]/2\mu_0 =$$

$$= [\omega k(|P_\theta|^2 + |P_\phi|^2)/2\mu_0R^2]\hat{R}$$

$$= (\omega^2/2\mu_0cR^2)(|P_\theta(\omega,\hat{R})|^2 + P_\phi(\omega,\hat{R})|^2)\hat{R}$$

where we use the relations

$$\hat{\theta}\times\hat{\phi} = \hat{R}$$

The total power radiated out by the antenna when it operates at the frequency $\omega$ is then

$$W = \int_{sphereorradiusR} \mathbf{S}(\omega, R, \hat{R}).\hat{R}dS(\hat{R})$$

$$= \int_{unitsphere} \mathbf{S}(\omega, R, \hat{R}).\hat{R}.R^2 d\Omega(\hat{R})$$

$$= (\omega^2/2\mu_0 c)\int_0^{\pi}\int_0^{2\pi} |P_\theta(\omega,\theta,\phi)|^2 + |P_\phi(\omega,\theta,\phi)|^2)sin(\theta)d\theta.d\phi$$

Note that this result is independent of the radial distance $R$ as long as we are operating in the far field zone. This formula can be used to define the radiation resistance $R_r$ of the antenna at frequency $\omega$ as

$$|I_0(\omega)|^2 R_r/2 = W$$

where $I_0(\omega)$ is the input sinusoidal current phasor at frequency $\omega$.

## 2.3 Reciprocity theorem

Sources $(\rho_1, J_1, \rho_{m1}; M_1)$ generate fields $(E_1, H_1)$ while sources $(\rho_2, J_2, \rho_{m2}, M_2)$ generate fields $(E_2, H_2)$. Thus,

$$curlE_k = -j\omega\mu H_k - M_k, curlH_k = j\omega\epsilon E_k + J_k, k = 1, 2$$

Compute

$$div(E_1 \times H_2) = curl(E_1).H_2 - E_1.curl(H_2) =$$

$$= (-j\omega\mu H_1 - M_1).H_2 - E_1.(j\omega\epsilon E_2 + J_2),$$

$$div(E_2 \times H_1) = curl(E_2).H_1 - E_2.curl(H_1) =$$

$$= (-j\omega\mu H_2 - M_2).H_1 - E_2.(j\omega\epsilon E_1 + J_1),$$

Thus,

$$div(E_1 \times H_2 - E_2 \times H_1) = M_2.H_1 - M_1.H_2 + J_1.E_2 - J_2.E_1$$

Integrating this identity over the entire three dimensional space, making use of Gauss' integral theorem and the fact that the electromagnetic fields vanish at $\infty$ gives us the fundamental reciprocity relation

$$\int_{\mathbb{R}^3} (J_1.E_2 - J_2.E_1 + M_2.H_1 - M_1.H_2)d^3r = 0$$

## 2.4 Exercises

### 2.4.1 Equivalent circuit of a transmitter-receiver antenna system

### 2.4.2 Radiation resistance of an infinitesimal dipole

Start from the relation

$$\mathbf{A}(\omega, r) = (\mu dl/4\pi)\hat{z}.exp(-jkr)/r$$

While calculating the fields by differentiating the potentials, make use of the fact that spatial derivatives of only the phase term $exp(-jkr)$ need to be taken not of the other amplitude terms, for only the phase derivatives contribute to $O(1/r)$ terms in the field and hence to $O(1/r^2)$ terms in the Poynting vector in the far field zone and hence to non-zero total radiated power. Amplitude derivatives of the potentials contribute to $O(1/r^2)$ terms in the fields and hence to $O(1/r^3)$ terms in the Poynting vector which do not contribute anything to the radiated power in the far field zone.

Note that the electric field can be computed using

$$j\omega\epsilon E = curlB = curlcurlA = \nabla(divA) - \nabla^2 A = \nabla(divA) + k^2 A$$

and the magnetic field using

$$H = curlA/\mu$$

Thus, upto $O(1/r)$, we have

$$H = curlA/\mu = (-jk.dl/4\pi)\hat{r} \times \hat{z}.exp(-jkr)/r$$

$$= (jkdl/4\pi r)sin(\theta)\hat{\phi}.exp(-jkr)$$

and hence also

$$E = curlH/j\omega\epsilon = (-jk^2dl/(\omega\epsilon 4\pi r)sin(\theta)\hat{r} \times \hat{\phi}.exp(-jkr)$$

$$= (jk^2dl/\omega\epsilon.4\pi r)sin(\theta)\hat{\theta}.exp(-jkr)$$

where we have used

$$\hat{r} \times \hat{z} = -sin(\theta)\hat{\phi},$$

$$\hat{r} \times \hat{\phi} = -\hat{\theta}$$

Note that the above formulae imply

$$E_\theta/H_\phi = k/\omega\epsilon = 1/c\epsilon = \sqrt{\mu/\epsilon}$$

which is the characteristic impedance of the medium, as it should be.

### 2.4.3 Radiation from an infinitesimal current loop

[3] Show that for a small loop of wire carrying a current $I(\omega)$ located around the origin of the coordinate system with the parametric equation of the loop being given by $s \to \mathbf{R}(s), 0 \le s \le 1$, the far field magnetic vector potential is given by

$$\mathbf{A}(\omega, \mathbf{r}) = (\mu I(\omega)/4\pi) \int_0^1 d\mathbf{R}(s).exp(-jk|\mathbf{r} - \mathbf{R}(s)|)/|\mathbf{r} - \mathbf{R}(s)|$$

$$\approx (\mu I/4\pi r)exp(-jkr). \int_0^1 d\mathbf{R}(s)exp(j\hat{r}.\mathbf{R}(s))$$

Show that when this is a circular loop of radius $a$, the above formula simplifies to

$$\mathbf{A}(\omega, r, \theta, \phi) =$$

$$(\mu Ia/4\pi r)exp(-jkr) \int_0^{2\pi} (-\hat{x}.sin(\phi')+\hat{y}.cos(\phi')).exp(jka.sin(\theta)cos(\phi'-\phi))d\phi'$$

and then using the formulas

$$A_\rho = A_x.cos(\phi) + A_y.sin(\phi), A_\phi = -A_x.sin(\phi) + A_y.cos(\phi)$$

deduce that in the far field zone.

$$A_\rho = 0, A_z = 0,$$

$$A_\phi = (\mu Ia/4\pi r)exp(-jkr) \int_0^{2\pi} exp(jka.sin(\theta).cos(\phi'))cos(\phi')d\phi'$$

## 2.5 Appendix, B.E and M.Tech projects

### 2.5.1 A.1 Order of magnitudes in quantum antenna theory

Consider a cavity resonator of one Angstrom size, ie, a cube with each side of length $a = 10^{-10}m$. The Maxwell equations in such a cube have solutions of the from

$$A_r(t, x, y, z) = \sum_{mnp} c(mnp, t)u_{r,mnp}(x, y, z), r = 1, 2, 3$$

where $u_{r,mnp}$ are spatial functions obtained by integrating the electric field w.r.t time. These functions are of the form

$$\{cos(m\pi x/a), sin(m\pi x/a)\}\otimes\{cos(n\pi y/a), sin(n\pi y/a)\}\otimes\{cos(p\pi z/a), sin(p\pi z/a)\}$$

multiplied by some constants depending on the indices $(m,n,p)$. We may, without loss of generality, assume that the functions $u_{r,mnp}$ are normalized so that

$$\int_C u_{r,mnp}(\mathbf{r})\bar{u}_{s,m'n'p'}(\mathbf{r})d^3r = \delta_{rs}\delta_{mm'}\delta_{nn'}\delta_{pp'}$$

The dependence of $c(mnp,t)$ on $t$ is $exp(i\omega(mnp)t)$ where $\omega(mnp)$ are the characteristic frequencies of oscillation:

$$\omega(mnp) = (\pi c/a)\sqrt{m^2+n^2+p^2}, m,n,p = 1,2,...$$

which are of the order of magnitude

$$\omega = \pi c/a$$

The electric field is

$$E_r = \partial_t A_r = \sum_{mnp} c(mnp,t)i\omega(mnp)u_{r,mnp}(\mathbf{r})$$

The magnetic field is

$$B = curl A$$

which is of the order of magnitude $|c(mnp,t)|/a$ where by $c(mnp,t)$ we actuall mean its average in a coherent state. The total electric field energy within the cavity $C$ is

$$U_E = (\epsilon_0/2)\int_C |E|^2 d^3r$$

which has components of the order of magnitude

$$\epsilon_0|\omega(mnp)c(mnp,t)|^2a^3 = \epsilon_0\omega(mnp)^2a^3|c(mnp,t)|^2$$

The total magnetic field energy within the cavity is

$$U_B = (2\mu_0)^{-1}\int_C |B|^2 d^3r$$

which is has components of the order of magnitude

$$|c(mnp,t)/a|^2a^3/\mu_0 = |c(mnp,t)|^2a/\mu_0$$

The ration of the orders of magnitude of the electric field energy and the magnetic field energy within the cavity therefore has the order of magnitude

$$U_E/U_B \approx \mu_0\epsilon_0\omega(mnp)^2a^2 \approx \omega^2a^2/c^2 \approx 1$$

as expected. The canonical commutation relations are

$$[A_r(t,\mathbf{r}), \partial_t A_s(t,\mathbf{r}')] = (ih/2\pi)\delta^3(\mathbf{r}-\mathbf{r}')$$

These yield,

$$[c_r(mnp,t), \omega(mnp)c_s(m'n'p',t)^*] = (h/2\pi)\delta_{rs}\delta_{mm'}\delta_{nn'}\delta_{pp'}$$

so that the eigenvalues of $c_r(mnp,t)^*c_r(mnp,t)$ are positive integer multiples of $h/2\pi\omega(mnp)$. This means that the field energy within the cavity when a finite number of modes are excited assumes eigenvalues that are of the same order of magnitude as positive integer multiples of $h\omega/2\pi$ as expected by Planck's quantum theory of radiation. This fact also yields the result that $|c(mnp,t)|$ is of the order of magnitude of $\sqrt{h/(2\pi\omega)}$.

Now we come to the question of computing the order of magnitude of the Poynting vector power flux at a given radial distance $R$ from the quantum cavity antenna caused by the surface current density induced by the magnetic field on on the antenna surface. The magnetic field on the surface and hence the corresponding induced surface current density both have the order of magnitudes of $|c(mnp,t)|/a$ which is of the order $a^{-1}\sqrt{h/\omega}$. Therefore, the far field magnetic vector potential at a distance $R$ from the cavity is of the order of magnitude (use the retarded potential formula) $(a/R)\sqrt{h/\omega}$ and hence the corresponding far field radiated magnetic field is of the order of magnitude $(\omega/c)(a/R)\sqrt{h/\omega}$ while the near field magnetic field is of the order of magnitude $(a/R^2)\sqrt{h/\omega}$. Actually, these expressions for the magnetic field must be multiplied by $\sqrt{N}$ where $N$ is a positive integer corresponding to the largest modal eigenvalue of the operators $(2\pi\omega(mnp)/h)c(mnp,t)^*c(mnp,t)$.

The far field Poynting vector has the order of magnitude of $B^2c/2\mu_0$ which is of the order

$$(c/2\mu_0)(\sqrt{N}.(\omega/c)(a/R).\sqrt{h/\omega})^2$$
$$= (h/2\mu_0)N.(\omega/c)(a^2/R^2)$$

and the total power radiated outward by this quantum antenna in the far field zone is thus of the order of magnitude

$$P = N(h/2\mu_0)(a^2\omega/c)$$

Now we look at the order of magnitude of the power radiated in the far field zone by the Dirac field of electrons and positrons within the cavity. The Dirac equation is

$$[i\gamma^\mu\partial_\mu - m]\psi(x) = 0$$

or more precisely in arbitrary units,

$$[(ih/2\pi)\partial_t - c(\alpha, (-ih/2\pi)\nabla) - \beta mc^2]\psi(x) = 0$$

Here, the appearance of the constants $h, m, c$ is explicitly shown. Now the $|\psi(x)|^2$ is the probability density of the electron which must integrate to unity over the cavity volume. Thus $\psi(x)$ is of the order of magnitude $a^{-3/2}$. The Dirac current density $J^\mu = e\psi^*\gamma^0\gamma^\mu\psi$ has the same order of magnitude as $e|\psi(x)|^2c$ which is $ec/a^{3/2}$. Therefore the far field magnetic vector potential at a radial distance of $R$ from the cavity is, in accordance with the retarded potential theory of the order

$$(ec/a^{3/2}).(a^3/R) = eca^{3/2}/R$$

The electric field in the far field zone is then of the order

$$E \approx \omega.eca^{3/2}/R$$

where $\omega$ is the characteristic oscilation frequency of the Dirac current. The magnetic field is of the order

$$B \approx a^{-1}.eca^{3/2}/R = ec\sqrt{a}/R$$

If $P$ is the characteristic momentum of the electrons and positrons in a given state, for example $P$ may be the average momentum of an electron in a given state, then according to De-Broglie, $P$ is of the order $h/a$ since $a$ is the order of the electron wavelength. Then the electron energy is of the order

$$E_e = c\sqrt{m^2c^2 + P^2} \approx c\sqrt{m^2c^2 + h^2/a^2}$$

and the characteristic frequency of oscillation of the Dirac wave field is then

$$\omega = E_e/h$$

The Poynting vector corresponding to the power radiated by the Dirac field in the far field zone then has the order of magnitude

$$S \approx c(\epsilon_0 E^2 + B^2/\mu_0) = c^3\epsilon_0\omega^2 ea^3/R^2 + e^2c^3a/\mu_0 R^2$$

and the total power radiated in the far field zone is of the order

$$W = SR^2 = c^3\epsilon_0\omega^2 ea^3 + e^2c^3a/\mu_0$$

## 2.5.2 A.2 The notion of a Fermionic coherent state and its application to the computation of the quantum statistical moments of the quantum electromagentic field generated by electrons and positrons within a quantum antenna

Aim: The aim of this section is to present a calculation involving the computation of the quantum statistical moments of the electromagnetic field produced by an ensemble of electrons and positrons whose state is specified by a mixed state superposition of Fermionic coherent states. Fermionic coherent states are parameterized by Grassmannian/Fermionic numbers and in order to attach physical signficance to the final results, we must use the Berezin integral for Fermionic variables to determine the above mentioned superposition of Fermionic coherent states. We can incorporate some unknown real parameters into the Berezin linear combination of coherent states and estimate these parameters by minimizing the distance between the average value of the electromagnetic field generated by the Fermions and the desired electromagnetic field pattern. If need be, we may modify this cost function to be minimized by constraining the higher order quantum statistical moments of the generated quantum electromagnetic field to be specified. An example of an application of this circle of ideas is to use a quantum antenna to generate a set of desired spatial patterns at a given set of frequencies.

First consider just a single Fermion specified by the annihilation operator $a$ and the creation operator $a^*$. Thus,

$$a^2 = a^{*2} = 0, aa^* + a^*a = 1$$

Let $\gamma$ be a Grassmannian variable that will be used to specify the coherent state of this Fermion just as a complex number $z$ is used to specify a the coherent state of a single Boson. $\gamma$ anticommutes with itself, with $\gamma^*$ and with $a, a^*$, just as in the Bosonic situation, the complex number $z$ that specifies the coherent state commutes with itself, with $\bar{z}$ and with the Boson creation and annihilation operators:

$$\gamma^2 = 0, \gamma\gamma^* + \gamma^*\gamma - 0, \gamma.a + a.\gamma = 0,$$

$$\gamma^*a + a\gamma^* = 0, \gamma^*a^* + a^*\gamma^* = 0$$

Define now the Fermionic Weyl operator

$$D(\gamma) = exp(\gamma.a^* - a\gamma^*)$$

Clearly, $D(\gamma)$ is a unitary operator since it is the exponential of a skew Hermitian operator. Now,

$$D(\gamma) = 1 + \gamma a^* - a\gamma^* + (1/2)(\gamma a^* - a\gamma^*)^2$$

$$= 1 + \gamma a^* - a\gamma^* - (1/2)(\gamma a^* a\gamma^* + a\gamma^*\gamma a^*)$$

$$= 1 + \gamma a^* - a\gamma^* + (1/2)(\gamma^*\gamma a^*a - \gamma^*\gamma(1 - a^*a))$$

$$= 1 + \gamma a^* - a\gamma^* + \gamma^*\gamma(a^*a - 1/2)$$

Then,

$$aD(\gamma) = a - \gamma aa^* + \gamma^*\gamma a/2 = (1 + \gamma^*\gamma/2)a - \gamma aa^*$$

$$D(\gamma)a = a + \gamma a^*a - \gamma^*\gamma a/2 = (1 - \gamma^*\gamma/2)a + \gamma a^*a$$

Thus,

$$aD(\gamma) - D(\gamma)a = \gamma^*\gamma a - \gamma$$

However,

$$D(\gamma)\gamma = \gamma - \gamma^*\gamma a$$

$$\gamma.D(\gamma) = \gamma - \gamma^*\gamma a$$

ie,

$$[\gamma, D(\gamma)] = 0$$

Thus,

$$aD(\gamma) - D(\gamma)a = -D(\gamma)\gamma = -\gamma.D(\gamma)$$

These equations can be rearranged as

$$D(\gamma)aD(\gamma)^{-1} = a + \gamma,$$

$$D(\gamma)^{-1}.aD(\gamma) = a - \gamma$$

We define the Fermionic single particle coherent state as

$$|\gamma >= D(-\gamma)|0 >= D(\gamma)^{-1}|0 >$$

where $|0>$, is the vacuum, ie, zero particle state. Then

$$a|\gamma >= aD(\gamma)|0 >= D(\gamma)^{-1}.D(\gamma)a.D(\gamma)^{-1}|0 >= D(\gamma)^{-1}(a+\gamma)|0 >$$

$$= \gamma.D(\gamma)^{-1}|0 >= \gamma|\gamma >$$

This proves the desired property of a coherent state, namely that it should be an eigenvector of the annihilation operator. We observe that

$$D(\gamma)^{-1} = 1 - \gamma a^* + a\gamma^* + \gamma^*\gamma(a^*a - 1/2)$$

and hence

$$|\gamma >= |0 > -\gamma|1 > -(1/2)\gamma^*\gamma|0 >= (1 - \gamma^*\gamma/2)|0 > -\gamma|1 > ---(1)$$

From this expression, we can directly verify the coherent state property:

$$a|\gamma >= \gamma.a|1 >= \gamma|0 >,$$

while

$$\gamma|\gamma >= \gamma|0 >$$

since

$$\gamma^2 = 0, \gamma\gamma^*\gamma = -\gamma^*\gamma^2 = 0$$

proving thereby the coherent state property for the state (1).

Now we are in a position to discuss physical implications for Fermionic coherent states. The first observation is that a coherent state is not parametrized by a complex number, it is parametrized by a Fermionic/Grassmannian parameter or a set of anticommuting Grassmannian parameters. Then, if we compute average values of quantities like for example the Dirac four current density in such a state, we will get a Grassmannian number. What physical significance does this have when our averages are not real or complex numbers ? The answer to this question is provided by the Berezin integral: Let $\phi(\gamma,\gamma^*)$ be a function of the Grassmannian parameters $\gamma,\gamma^*$ so that the Berezin integral

$$\rho=\int \phi(\gamma,\gamma^*)|\gamma><\gamma|d\gamma.d\gamma^*$$

defines a mixed state. Then, the average value of a function $F(a,a^*)$ of the Fermionic operators $a,a^*$ in the state $\rho$ becomes a complex number to which we can attach physical meaning:

$$Tr(\rho.F(a,a^*))=\int \phi(\gamma,\gamma^*)<\gamma|F(a,a^*)|\gamma>d\gamma.d\gamma^*$$

Another example involving computing average values of the electromagnetic field emitted by a field of electrons and positrons in a given coherent state of the electron-positron field. Let $a_k, k=1,2,...$ denote the annihilation operators of the electrons and positrons after discretizing in momentum space. They satisfy the CAR

$$[a_k,a_m]_+=0,[a_k,a_m^*]=\delta_{km}$$

The current density field generated by this field is according to Dirac's theory, a quadratic function of these operators and hence the electromagnetic field generated by this current density according to the retarded potential formula, is also a quadratic function of these operators. We can express this electromagnetic field as

$$F_{\mu\nu}(x)=\sum_{k,m=1}^{N}[G_{\mu\nu}(x,k,m,1)a_k a_m+\bar{G}_{\mu\nu}(x,k,m,1)a_m^* a_k^*$$
$$+G_{\mu\nu}(x,k,m,2)a_k^* a_m],x\in\mathbb{R}^4$$

This should be a Hermitian operator field and hence

$$\bar{G}_{\mu\nu}(x,k,m,2)=G_{\mu\nu}(x,m,k,2)$$

The coherent state of the electrons and positrons is given by

$$|\gamma>=D(\gamma)|0>,D(\gamma)=\Pi_{k=1}^{N}exp(\gamma(k)a_k^*-a_k\gamma(k)^*)$$

where $\gamma=((\gamma(k)))_{k=1}^{N}$ are Fermionic/Grassmannian parameters and $\gamma(k)$ and $\gamma(k)^*$ anticommute with $\gamma(l),\gamma(l)^*,a_l,a_l^*$ for all $l$. We can write

$$D(\gamma)=\Pi_{k=1}^{N}(1+\gamma(k)a_k^*-a_k\gamma(k)^*+\gamma(k)^*\gamma(k)(a_k^*a_k-1/2))$$

The state of the electrons and positrons is assumed to be given by a Berezin integral based superposition of the coherent states:

$$\rho(\theta)=\int \phi(\gamma,\gamma^*|\theta)|\gamma><\gamma|d\gamma.d\gamma^*$$

and hence the average electromagnetic field in this state is

$$<F_{\mu\nu}(x)>(\theta)=Tr(\rho(\theta)F_{\mu\nu}(x))=$$

$$\int \phi(\gamma,\gamma^*|\theta) < \gamma|F_{\mu\nu}(x)|\gamma > d\gamma.d\gamma^*$$

where

$$< \gamma|F_{\mu\nu}(x)|\gamma >=$$

$$\sum_{k,m=1}^{N} [G_{\mu\nu}(x,k,m,1) < \gamma|a_k a_m|\gamma > +\bar{G}_{\mu\nu}(x,k,m,1) < \gamma|a_m^* a_k^*|\gamma >$$

$$+G_{\mu\nu}(x,k,m,2) < \gamma|a_k^* a_m|\gamma >], x \in \mathbb{R}^4$$

$$= \sum_{k,m=1}^{N} [G_{\mu\nu}(x,k,m,1)\gamma(k)\gamma(m)+\bar{G}_{\mu\nu}(x,k,m,1)\gamma(m)^*\gamma(k)^*$$

$$+G_{\mu\nu}(x,k,m,2)\gamma(k)^*\gamma(m)], x \in \mathbb{R}^4$$

We can now control the parameter vector $\theta$ so that this average electromagnetic field is as close as possible to a desired electromagnetic field $F_{d\mu\nu}(x)$ over a given space-time region $x \in D$ by minimizing

$$E(\theta) = \int_D | < F_{\mu\nu}(x) > (\theta) - F_{\mu\nu}(x)|^2 d\mu(x)$$

where $\mu(.)$ is a measure on $D$.

Remark 1: More generally, we can compute all the statistical moments of the radiation field

$$Tr(\rho(\theta)F_{\mu_1\nu_1}(x_1)...F_{\mu_k \ni_k}(x_k) >$$

in the superposed coherent state $\rho(\theta)$. This computation will involve determining coherent state expectations such as

$$< \gamma|a_{k_1}^*...a_{k_r}^* a_{s_1}...a_{s_m}|\gamma >$$

and noting that this evaluates to

$$\gamma(k_r)^*...\gamma(k_1)^*\gamma(s_1)...\gamma(s_m)$$

The reference for Fermionic coherent state for us has been the master's thesis by Greplova, title "Fermionic Gaussian States".

Remark 2: From Steven Weinberg's book, "The quantum theory of fields, vol.1", it is known that the free Dirac field can be expanded in terms of momentum-spin space electron annihilation operators $a(P,\sigma)$ and positron creation operators $b(P,\sigma)^*$ which satisfy the CAR (canonical anticommutation relations)

$$[a(P,\sigma), a(P',\sigma')^*]_+ = \delta^3(P-P')\delta_{\sigma,\sigma'},$$

$$[b(P,\sigma), b(P',\sigma')^*]_+ = \delta^3(P-P')\delta_{\sigma,\sigma'}$$

and all the other anticommutators evaluating to zero. The second quantized Dirac wave field is then the solution to Dirac's relativistic wave equation and is given by

$$\psi(x) = \psi(t,r) = \int [a(P,\sigma)u(P,\sigma)exp(-ip.x) + b(P,\sigma)^* v(P,\sigma)exp(ip.x)]d^3P$$

where

$$p^0 = E(P) = \sqrt{m^2 + P^2}$$

The Dirac current density operator field is then

$$J^\mu(x) = -e\psi(x)^*\gamma^0\gamma^\mu\psi(x)$$

and it is evident that this can be expressed as a linear combination of the quadratic operators

$$a(P,\sigma)^*a(P,\sigma'), a(P,\sigma)^*b(P,\sigma')^*, b(P,\sigma)a(P,\sigma')^*, b(P,\sigma)b(P',\sigma')^*$$

Thus, using the retarded potential formula for the Maxwell equations in the form

$$A^\mu(x) = \int G(x - x')J^\mu(x')d^4x'$$

it is evident that once again $A^\mu(x)$ is expressible as a linear combination of the above quadratic operators. After discretizing the integrals in 3-momentum space, we then club all the electron and positron annihilation operators into one set $\{a_k\}$ and their adjoints into $\{a_k^*\}$ and then use the above coherent state formalism of Greplova to determine the quantum averages of the electromagnetic field.

### 2.5.3 The optimization problem in quantum antenna design

The ultimate aim of all these computations can be formulated in very simple terms as an optimization problem: Design the control parameters $\theta$ or the control classical fields to a quantum antenna so that the error energy between the average value of the quantum electromagnetic field produced by the quantum antenna and the desired classical electromagnetic field pattern is a minimum subject to the constraint that the second order central moments of the quantum electromagnetic field (ie variance of fluctuations) is smaller than a given threshold.

Remark:More generally, we can control the wave function operator of the Dirac field of electrons and positrons as well as the Maxwell photon field operators within the cavity resonator antenna by introducing classical control current and electromagnetic field sources into the cavity. The quantum cavity photon and electron-positron fields will then be expressible in terms of the free quantum fields plus additional perturbation terms involving the classical current and field sources. Once this is done, we can in principle calculate the far field antenna pattern produced by the cavity surface currents induced by the tangential components of the quantum magnetic field operators as well as that produced by the Dirac field of electrons and positrons and then design these classical control fields so that the far field quantum Poynting radiation pattern has a mean value and correlations in a given quantum coherent state of the photons and electrons-positrons within the cavity as close as possible to specified values.

To formulate this optimization problem in abstract terms, let

$$X(t,r) = X(x) = \sum_k [f_k(\theta)X_k(t,r) + \bar{f}_k(\theta)X_k(t,r)^*]$$

be the quantum field radiated out by the quantum antenna where $\theta$ is a control parameter vector or a classical field. This form arises typically by perturbatively solving the Maxwell-Dirac field equations upto linear orders in the perturbing classical current and electromagnetic field. $f_k(\theta)$ is a complex valued function of $\theta$ and $X_k(t,r)$ are quantum operator fields. The average value of this radiated field in a state $|\Phi>$ is given by

$$< X(t,r) >=< \Phi|X(t,r)|\Phi >=$$

$$\sum_k [f_k(\theta) < \Phi|X_k(t,r)|\Phi > +\bar{f}_k(\theta) < \Phi|X_k(t,r)^*|\Phi >]$$

and the central correlations in the field are

$$C(t,r|t',r') =< X(t,r).X(t',r')^* > - < X(t,r) >< X(t',r')^* >$$

$$= \sum_{k,m} f_k(\theta)\bar{f}_m(\theta) < \Phi|(X_k(t,r)- < X_k(t,r) >).(X_m(t',r')^* - < X_m(t',r') >^* |\Phi >$$

$$+\sum_{k,m} f_k(\theta)f_m(\theta) < \Phi|(X_k(t,r)- < X_k(t,r) >).(X_m(t',r')- < X_m(t',r') >)|\Phi >$$

$$+c.c$$

where $c.c$ denotes complex conjugate of the previous terms. It is then easy to see that if the $f_k's$ are linear functions of $\theta$, then the problem of minimizing $\int_D (< X(t,r) > -X_d(t,r))^2 dtd^3r$ subject to the constraint that $\int_{D\times D} W(t,r|t',r')C(t,r|t',r')dtd^3rdt'd^3r'$ is fixed is equivalent to finding the minimum of the ratio of two quadratic forms:

$$E(\theta) = \frac{\theta^T Q_1 \theta}{\theta^T Q_2 \theta}$$

where $Q_1, Q_2$ are two Hermitian positive definite matrices and the optimal equations for $\theta$ then result in the generalized eigenvalue problem

$$(Q_1 - cQ_2)\theta = 0$$

The minimum value of this ratio is the minimum of all the generalized eigenvalues $c$, ie, the minimum of all the $c's$ for which

$$det(Q_1 - cQ_2) = 0$$

and the the optimal value of $\theta$ is a generalized eigenvector corresponding to the minimum of $c$ with the normalization condition

$$\theta^T Q_2 \theta = E$$

where $E$ is the prescribed value of the energy of the quantum fluctuations $\int_{D\times D} W(t,r|t',r')C(t,r|t',r')dtd^3rdt'd^3r'$.

### 2.5.4 Approximate analysis of a rectangular quantum antenna

The quantum antenna is assumed to be the cuboid region $[0,a]\times[0,b]\times[0,d]$. This rectangular cavity is assumed to comprise of photons, electrons and positrons. The exact equations governing the quantum fields corresponding to these particles are (a) The Maxwell equations for the four vector potential driven by the Dirac field current and (b) The Dirac field equations driven by an interaction between the Dirac field and the Maxwell field four vector potential. These exact field equations are:

$$\Box A_\mu(x) = \mu_0 e\psi^*(x)\alpha^\mu\psi(x), x = (t,r) --(1)$$

$$((\alpha,-i\nabla)+\beta m_0)\psi(t,r) + eA_\mu(x)\alpha^\mu\psi(x) = i\partial_t\psi(t,r) ---(2)$$

where

$$\alpha^\mu = \gamma^0\gamma^\mu, \beta = \gamma^0$$

and $\gamma^\mu$ are the Dirac Gamma matrices. Note that $(\gamma^0)^2 = I_4$ and hence $\alpha^0 = I_4$. The boundary conditions under which we need to solve these Maxwell-Dirac equations are that the Dirac operator wave field $\psi(x)$, the tangential components of the electric field $F_{0r} = A_{r,0} - A_{),r}, r = 1,2,3$ and the normal components of the magnetic field $F_{rs} = A_{s,r} - A_{r,s}, 1 \le r < s \le 3$ must vanish on the boundaries of the cavity. In particular, the freed Dirac field must have an expansion

$$\psi^{(0)}(t,r) = \sum_{mnp} \mathbf{c}(mnp,t)u_{mnp}(r)$$

where $m,n,p$ run over positive integers and

$$u_{mnp}(r) = (2\sqrt{2}/\sqrt{abd})sin(m\pi x/a)sin(n\pi y/b)sin(p\pi z/d)$$

Substituting this into the free Dirac equation, ie, without any electromagnetic interactions, we get

$$\sum_{mnp}(i\partial_t c(mnp,t))u_{mnp}(r) =$$

$$((\alpha,-i\nabla)+\beta m_0).\sum_{mnp}\mathbf{c}(mnp,t)u_{mnp}(r)$$

from which we derive on taking the inner products on both sides with $u_{kls}(r)$ and using the orthonormality of this set of functions over the cavity volume, ie,

$$< u_{kls}, u_{mnp} >= \int_B u_{kls}(r)u_{mnp}(r)d^3r = \delta_{km}\delta_{ln}\delta_{sp}$$

where $B$ is the cavity volume

$$B = [0,a] \times [0,b] \times [0,d],$$

the following sequence of differential equations

$$i\partial_t c(kls,t) = \sum_{mnp}[< u_{kls}, -i\partial_x u_{mnp} > \alpha_1 c(mnp,t)$$

$$+ < u_{kls}, -i\partial_y u_{mnp} > \alpha_2 c(mnp,t) + < u_{kls}, -i\partial_z u_{mnp} > \alpha_3 c(mnp,t)] + m_0\beta c(kls,t)$$

Now we evaluate

$$< u_{kls}, -i\partial_x u_{mnp} >= -i\delta_{ln}\delta_{sp}(m\pi/a)(\int_0^a (2/a)sin(k\pi x/a)cos(m\pi x/a)dx$$
$$= a_1(k,m)\delta_{ln}\delta_{sp}$$

where

$$a_1(k,m) = (-2im\pi/a^2)\int_0^a sin(k\pi x/a).cos(m\pi x/a)dx$$

Likewise,

$$< u_{kls}, -i\partial_y u_{mnp} >= a_2(l,n)\delta_{km}\delta_{sp},$$

and

$$< u_{kls}, -i\partial_z u_{mnp} >= a_3(s,p)\delta_{km}\delta_{ln}$$

Combining all these equations gives us finally,

$$i\partial_t c(kls,t) = \sum_m a_1(k,m)\alpha_1 c(mls,t)$$

$$+\sum_n a_2(l,n)\alpha_2 c(kns,t) + \sum_p a_3(s,p)\alpha_3 c(klp,t)] + m_0\beta c(kls,t)$$

Note that $\alpha_1, \alpha_2, \alpha_3, \beta$ are $4\times 4$ Hermitian matrices while $c(mnp,t)$ is a $4\times 1$ complex vector. Arranging the $4 \times 1$ vectors $c(mnp,t), m,n,p \geq 1$ in lexicographic order to give an infinite vector $\mathbf{c}(t)$ and likewise defining a block structured infinite dimensional Dirac Hamiltonian matrix $\mathbf{H}_0$ by

$$\mathbf{H}_0 = \sum_{klsm} a_1(k,m)(\mathbf{I}_4 \otimes \mathbf{e}(kls))\alpha_1(\mathbf{I}_4 \otimes \mathbf{e}(mls)^T)$$

$$+\sum_{lksn} a_2(l,n)(\mathbf{I}_4 \otimes \mathbf{e}(kls))\alpha_2(\mathbf{I}_4 \otimes \mathbf{e}(kns)^T)$$

$$+\sum_{klsp} a_3(s,p)(\mathbf{I}_4 \otimes \mathbf{e}(kls))\alpha_3(\mathbf{I}_4 \otimes \mathbf{e}(klp)^T)$$

$$+m_0\beta \otimes \mathbf{I}$$

where we may choose $\mathbf{e}(mnp), m, n, p \geq 1$ as any orthonormal basis for $l^2(\mathbb{Z}_+)$, the Hilbert space of all one sided square summable infinite sequences and define

$$\mathbf{c}(t) = \sum_{mnp} c(mnp, t)\mathbf{e}(mnp)$$

By orthonormal, we mean that

$$\mathbf{e}(kls)^T \mathbf{e}(mnp) = \delta_{km}\delta_{ln}\delta_{sp}$$

Thus the free Dirac equation in the RDRA has been put in "Standard" block matrix form:

$$i\frac{d\mathbf{c}(t)}{dt} = \mathbf{H}_0\mathbf{c}(t)$$

the general solution to which can be expressed as

$$\mathbf{c}(t) = \sum_{n} d(n).\mathbf{c}_n exp(-iE(n)t)$$

where $\mathbf{c}_n, n \geq 1$ form an orthonormal basis for $l^2(\mathbb{Z}_+)$ and the $d(n)'s$ are arbitrary complex numbers such that

$$\sum_{n} |d(n)|^2 = 1$$

$E(n)'s$ are the (energy) eigenvalues of the infinite dimensional Hermitian $H_0$:

$$det(\mathbf{H}_0 - E(n)\mathbf{I}) = 0$$

The average energy of the free Dirac field of electrons and positrons within the cavity is then

$$< \mathbf{c}(t), \mathbf{H}_0\mathbf{c}(t) >= \sum_{n} E(n)d(n)^*d(n)$$

It is easy to see as in the case of the Dirac equation in free space that if $E(n)$ is an eigenvalue of $\mathbf{H}_0$ then so is $-E(n)$ where the $E(n)'s$ may be taken as positive, Hence if $\mathbf{c}_{en}$ is an eigenvector of $\mathbf{H}_0$ corresponding to the eigenvalue $E(n)$ and $\mathbf{c}_{pn}$ is an eigenvector corresponding to the eigenvalue $-E(n)$, then the solution can be expressed as

$$\mathbf{c}(t) = \sum_{n}[d_e(n)\mathbf{c}_{en} exp(-iE(n)t) + d_p(n)^*\mathbf{c}_{pn} exp(iE(n)t)]$$

Therefore, it is plausible in the second quantized theory, to look upon the $d_e(n)'s$ as annihilation operators of the electrons and the $d_p(n)^*$'s as the creation operators of the positrons. The actual Dirac wave function $\psi(t, r)$ in the absence of electromagnetic interactions is then

$$\psi(t,r) = \sum_{kmnp} [d_e(k)c_{ek}(mnp)u_{mnp}(r)exp(-iE(k)t) + d_p(k)^* c_{pk}(mnp)u_{mnp}(r)exp(iE(k)t)] --- (3)$$

A simple calculation then shows that the second quantized Hamiltonian of the free Dirac field of electrons and positrons within the cavity is given by

$$H_{D0} = \int_B \psi(t,r)^*((\alpha, -i\nabla) + \beta m)\psi(t,r)d^3r$$

$$= \int_B \psi(t,r)^* i\partial_t \psi(t,r)d^3r$$

$$= \sum_k E(k)(d_e(k)^* d_e(k) - d_p(k)d_p(k)^*)$$

Now from the basic anticommuation relations for the Dirac field, we have

$$\{\psi(t,r), \psi(t,r')^*\} = \delta^3(r-r')I$$

and this immediately implies the following anticommutation relations for the electron and positron creation and annihilation operators:

$$\{d_e(k), d_e(m)^*\} = \delta_{km}, \{d_p(k), d_p(m)^*\} = \delta_{km}$$

with all the other anticommutators vanishing. This completes our description of the free Dirac field of electrons and positrons within the RDRA. Using these anticommutation relations, we immediately get that the total second quantized Hamiltonian of the free Dirac field in the cavity can equivalently be expressed as

$$H_{D0} = \sum_k E(k)(d_e(k)^* d_e(k) + d_p(k)^* d_p(k))$$

namely, the sum of the total electron and positron energies. Likewise, when we solve the free Maxwell equations within the cavity after incorporating the appropriate boundary conditions, we get that the scalar potential is zero since there are no charges while the magnetic vector potential admits an expansion obtained from

$$\mathbf{E} = -\partial_t \mathbf{A}$$

as

$$\mathbf{A}(t,r) = \sum_k [b(k)\mathbf{w}_k(r)exp(-i\omega(k)t) + b(k)^* \mathbf{w}_k(r)^* exp(i\omega(k)t)] - - - (4)$$

where now $\mathbf{w}_k(r)$ has three components that are calculated from the expansion of $E_z$ and the relationship between the transverse and longitudinal components of the electric field within the cavity. Note that the electromagnetic field is being computed in the Coulomb gauge which implies that the electric scalar potential becomes zero in view of the fact that in the Coulomb gauge, the scalar potential satisfies Poisson's equation and is therefore a matter field which evaluates to zero since there is no unperturbed charge density.

We also note that the third, ie, $z$ component of $\mathbf{w}_k(r)$ where the index $k$ is identified with the modal triplet $(mnp)$ is proportional to $sin(m\pi x/a)sin(n\pi y/b)cos(p\pi z/d)$

in view of the boundary conditions on the electric field an the fact that each mode of the magnetic vector potential is proportional to the electric field ($-j\omega\mathbf{A} = \mathbf{E}$). $b(k) = b(mnp)$ is identified with a photon annihilation operator while $b(k)^*$ with a photon creation operator. They satisfy the canonical commutation relations

$$[b(k), b(m)^*] = \delta_{km}$$

Formally, we can compute both the free Dirac current density $\psi(t,r)^*\alpha^\mu\psi(t,r)$ of electrons and positrons within the cavity as well as the surface current density on the RDRA walls induced by the tangential components of the quantum magnetic field $\mathbf{B} = curl\mathbf{A}$ and obtain the far field radiation pattern generated by both of these cavity current components. Obviously, this far field radiation pattern will have its first component being a quadratic form in the electron-positron creation and annihilation operators $d_e(k), d_p(k), d_e(k)^*, d_p(k)^*$ while the second component will be linear in the photon creation-annihilation operators $b(k), b(k)^*$ and therefore, in principle, we can compute all the statistical moments of the radiation field in a joint coherent state of the photons, electrons and positrons. However, this picture of the far field quantum radiation pattern is incomplete because it does not take into account the cavity current density terms caused by perturbation in the Dirac wave field due to interaction with the photons and it does not also take into account the cavity surface current density terms caused by perturbation in the Maxwell field caused by its interaction with the Dirac field. We shall now indicate an approximate first order calculation by which these extra correction terms may be obtained due interactions between the Maxwell field and the Dirac field.

We denote the free Dirac field within the cavity derived above by $\psi^{(0)}(t,r)$ and the corresponding momentum space wave function $c(mnp,t)$ by $c^{(0)}(mnp,t)$. Likewise, we denote the free Maxwell field within the cavity by $\mathbf{A}^{(0)}$. Let $\delta\mathbf{A}$ denote the perturbation to the Maxwell field caused by the Dirac current and $\delta\psi, \delta c(mnp,t)$ the perturbation to the Dirac field caused by the Maxwell current. Then, clearly if $S(x-y)$ denotes the electron propagator and $D(x-y)$ the photon propagator, we have using (1) and (2), approximately,

$$\delta A^\mu(t,r) = \mu_0 e \int D(t-t', r-r')\psi^{(0)*}(t',r')\alpha^\mu\psi^{(0)}(t',r')dt'd^3r'$$

$$\delta\psi(t,r) = e\int S(t-t', r-r')A_\mu^{(0)}(t',r')\alpha^\mu\psi^{(0)}(t',r')dt'd^3r'$$

where we substitute for $\psi^{(0)}$ and $A_\mu^{(0)}$ the expressions given in (3) and (4). Then,

$$\psi^{(0)*}\alpha^\mu\psi^{(0)}(t,r) =$$

$$\sum_{kmnpk'm'n'p'} [d_e(k)^*\bar{c}_{ek}(mnp)exp(iE(k)t) + d_p(k)\bar{c}_{pk}(mnp)exp(-iE(k)t)].\alpha^\mu.$$

$$.[d_e(k')c_{ek'}(m'n'p')exp(-iE(k')t) + d_p(k')^*c_{pk'}(m'n'p')exp(iE(k')t)]u_{mnp}(r)u_{m'n'p'}(r)$$

$$= \sum [d_e(k)^* d_e(k') \bar{c}_{ek}(mnp) \alpha^\mu c_{ek'}(m'n'p') exp(i(E(k)-E(k'))t) u_{mnp}(r) u_{m'n'p'}(r)]$$

$$+ \sum [d_e(k)^* d_p(k')^* \bar{c}_{ek}(mnp) \alpha^\mu c_{pk'}(m'n'p') exp(i(E(k)+E(k'))t) u_{mnp}(r) u_{m'n'p'}(r)]$$

$$+ \sum [d_p(k) d_e(k') \bar{c}_{pk}(mnp) \alpha^\mu c_{ek'}(m'n'p') exp(-i(E(k)+E(k'))t) u_{mnp}(r) u_{m'n'p'}(r)]$$

$$+ \sum [d_p(k) d_p(k')^* \bar{c}_{pk}(mnp) \alpha^\mu c_{ek'}(m'n'p') exp(-i(E(k)-E(k'))t) u_{mnp}(r) u_{m'n'p'}(r)]$$

We see that the frequencies of the Dirac current that generate the perturbation to the quantum electromagnetic field are $E(k) \pm E(k'), k, k' = 1, 2, ...$ or more precisely, these divided by Planck's constant. Here $E(k)$ was obtained by solving the free Dirac eigenvalue equation inside the rectangular cavity with zero boundary conditions. The $E(k)'s$ were obtained as the eigenvalues of the Dirac Hamiltonian. From basic principles of special relativity, it is easy to see that these $E(k)'s$ are of the order

$$c\sqrt{m_0^2 c^2 + P^2}$$

where

$$P^2 = (h/2\pi)^2((m\pi/a)^2 + (n\pi/b)^2 + (p\pi/d)^2)$$

with $m, n, p$ being positive integers determined by the mode of oscillation of the field within the cavity. Now this current is of the general form

$$-e\psi^{(0)*}\alpha^\mu \psi^{(0)}(t, r)$$

$$= \sum_{k,k'} [d(k)^* d(k') f^\mu_{kk'}(t, r) + d(k) d(k') g^\mu_{kk'}(t, r) + d(k)^* d(k')^* \bar{g}^\mu_{kk'}(t, r)]$$

where the $d(k)'s$ are the annihilation operators of the electrons and positrons and their adjoints $d(k)^*$ are the corresponding creation operators. The functions $f_{kk'}, g_{kk'}$ are constructed by superposing $exp(\pm i(E(k) \pm E(k')) u_{mnp}(r) u_{m'n'p'}(r)$ and these components are easily seen to be expressible as superpositions of space-time sinusoids with the temporal frequencies being $E(k) \pm E(k')$ or their negatives and the spatial frequencies, ie, wave-numbers being

$$m\pi/a, n\pi/b, p\pi/d, m'\pi/a, n'\pi/b, p'\pi/d.$$

Note that because the current density is a Hermitian operator field, it follows that

$$\bar{f}^\mu_{kk'}(t, r) = f^\mu_{k'k}(t, r)$$

Then, the perturbation in the electromagnetic potentials can be expressed as

$$\delta A^\mu(t, r) =$$

$$\sum_{k,k'} [d(k)^* d(k') \int D(t-t', r-r') f^\mu_{kk'}(t', r') dt' d^3r' + d(k) d(k') \int D(t-t', r-r') g^\mu_{kk'}(t', r') dt' d^3r' + d(k)^* d(k')^* \int D(t-t', r-r') \bar{g}^\mu_{kk'}(t', r') dt' d^3r']$$

Let

$$J^\mu(k) = -e \int \psi^{(0)*}(x) \alpha^\mu \psi^{(0)}(x) exp(-ik.x) d^4x$$

$$= \int \psi^{(0)*}(t,r)\alpha^\mu \psi^{(0)}(t,r)exp(-i(k^0 t - K.r))dtd^3r$$

where

$$k = (k^\mu) = (k^0, K)$$

denote the space-time four dimensional Fourier transform of the unperturbed Dirac four current density. Then, we can write down the space-time Fourier transform of the correction $\delta A^\mu(x), x = (t,r)$ to the electromagnetic four potential caused by this Dirac current as

$$\delta A^\mu(k) = \int \delta A^\mu(x).exp(-ik.x)d^4x$$

$$= \mu_0 D(k)J^\mu(k) = \mu_0 J^\mu(k)/k^2, k^2 = k_\mu k^\mu = (k^0)^2 - |K|^2$$

in units where $c = 1$. It should be noted that by the convolution theorem for Fourier transforms, if $\psi^{(0)}(k)$ denotes the space-time Fourier transform of $\psi^{(0)}(x)$, then

$$J^\mu(k) = (2\pi)^{-4} \int \psi^{(0)*}(k' - k)\alpha^\mu \psi^{(0)}(k')d^4k'$$

and hence, the perturbation to the electromagnetic four potential in the space-time Fourier domain, ie, in four momentum space of the photon can be expressed as

$$\delta A^\mu(k) = (\mu_0/(2\pi)^4 k^2) \int \psi^{(0)*}(k' - k)\alpha^\mu \psi^{(0)}(k')d^4k'$$

Remark: The unperturbed electromagnetic field is in the Coulomb gauge, ie, $div\mathbf{A}^{(0)} = 0$ and also since there is no charge/current for the unperturbed field, the unperturbed electric scalar potential is a matter field which is identically zero, ie, $A^{(0)0} = 0$. Hence, we are guaranteed that the unperturbed electromagnetic potentials also satisfy the Lorentz gauge conditions, ie, $div\mathbf{A}^{(0)} + \partial_t A^{(0)0} = 0$. This means that while computing the perturbations to the electromagnetic potentials caused by currents coming from the Dirac field, we can safely work in the Loretnz gauge.

Likewise, the change in the Dirac field caused by interaction with the electromagnetic field within the cavity is given upto first order perturbation theory by

$$\delta\psi(x) = \delta\psi(x) = e \int S(x - x')A_\mu^{(0)}(x')\alpha^\mu \psi^{(0)}(x')d^4x'$$

$$= e \int S(x - x')A_r^{(0)}(x')\alpha^r \psi^{(0)}(x')d^4x'$$

$$= -e \int S(x - x')(\alpha, \mathbf{A}^{(0)}(x'))\psi^{(0)}(x')d^4x'$$

$$= -e \int S(t-t', r-r') \sum_k [b(k)(\alpha, \mathbf{w}_k(r')) exp(-i\omega(k)t') + b(k)^*(\alpha, \mathbf{w}_k(r')^*) exp(i\omega(k)t')].$$

$$.[\sum_{kmnp} [d_e(k) c_{ek}(mnp) u_{mnp}(r') exp(-iE(k)t') + d_p(k)^* c_{pk}(mnp) u_{mnp}(r') exp(iE(k)t')] dt' d^3r'$$

$$= -e \sum_{kk'mnp} b(k) d_e(k') \int S(t-t', r-r')(\alpha, \mathbf{w}_k(r')) c_{ek'}(mnp) u_{mnp}(r') exp(-i(\omega(k)+E(k'))t') dt' d^3r'$$

$$-e \sum_{kk'mnp} b(k) d_p(k')^* \int S(t-t', r-r')(\alpha, \mathbf{w}_k(r')^*) c_{pk'}(mnp) u_{mnp}(r') exp(-i(\omega(k)-E(k'))t') dt' d^3r'$$

$$-e \sum_{kk'mnp} b(k)^* d_e(k') \int S(t-t', r-r')(\alpha, \mathbf{w}_k(r')) c_{ek'}(mnp) u_{mnp}(r') exp(i(\omega(k)-E(k'))t') dt' d^3r'$$

$$-e \sum_{kk'mnp} b(k)^* d_p(k')^* \int S(t-t', r-r')(\alpha, \mathbf{w}_k(r')^*) c_{pk'}(mnp) u_{mnp}(r') exp(i(\omega(k)+E(k'))t') dt' d^3r'$$

From this expression, it is clear that the characteristic frequencies of the interaction term between the electromagnetic potentials and the Dirac field and hence the characteristic frequencies of the perturbation in the Dirac field caused by electromagnetic interaction are $\pm\omega(k) \pm E(k')$. In terms of the compact notation introduced above, namely using the same symbol $d(k)$ for both electron and positron annihilation operators and likewise $d(k)^*$ for both electron and positron creation operators, we can write

$$\delta\psi(x) = \int S(t-t', r-r')[\sum b(k) d(k') h_{1kk'}(t', r') + b(k) d(k')^* h_{2kk'}(t', r') +$$

$$+ b(k)^* d(k') h_{3kk'}(t', r') + b(k)^* d(k')^* h_{4kk'}(t', r')] dt' d^3r'$$

where the functions $h_{mkk'}(t, r)$ are built by superposing the functions $exp(i \pm (\omega(k) \pm E(k'))t)(\alpha, w_k(r)) c_{k'}(mnp) u_{mnp}(r)$ and the same expression with $w_k(r)$ replaced by its complex conjugate $w_k(r)^*$. Here, the symbol $c_{k'}(mnp)$ stands for either $c_{ek'}(mnp)$ or $c_{pk'}(mnp)$.

In particular this expression shows that the perturbation to the Dirac field caused by electromagnetic interactions have frequencies $\pm\omega(k) \pm E(k')$, namely linear combinations of the unperturbed electromagnetic characteristic frequencies and the unperturbed Dirac characteristic frequencies. This represents a new feature of our model.

Before proceeding further, observe that we can write in the four dimensional momentum/space-time frequency domain,

$$\delta\psi(k) = S(k) \mathcal{F}(e A_\mu^{(0)} \alpha^\mu \psi^{(0)})(k)$$

where

$$S(k) = (k^0 - (\alpha, \mathbf{K}) - \beta m_0 + i0)^{-1}$$

is the electron propagator in the four momentum domain $k = (k^\mu) = (k^0, \mathbf{K})$ and

Control of the quantum electromagnetic field and the Dirac field of electrons and positrons within the rectangular cavity by means of a classical electromagnetic field coming from a laser source connected to the cavity plus a classical current source coming from a probe inserted into the cavity: Let $A^c_\mu(x)$ denote the classical electromagnetic four potential from the laser and $J^c_\mu(x)$ the classical current density coming from the probe insertion. The relevant equations are

$$\Box \mathbf{A}_\mu = -e\mu_0\psi^*\alpha_\mu\psi + \mu_0\mathbf{J}^c_\mu,$$

$$((\alpha, -i\nabla) + \beta m)\psi = [-e(\alpha, \mathbf{A}) - e(\alpha, \mathbf{A}^c)]\psi$$

The first order perturbative solution to these equations is with $x = (t, \mathbf{r})$,

$$\psi(x) = \psi^{(0)}(x) + \delta\psi(x),$$

$$A^r(x) = A^{r(0)}(x) + \delta A^r(x), r = 1, 2, 3$$

where

$$\psi^{(0)}(x) =$$

$$\sum_{kmnp} [d_e(k)c_{ek}(mnp)u_{mnp}(r)exp(-iE(k)t) + d_p(k)^*c_{pk}(mnp)u_{mnp}(r)exp(iE(k)t)]$$

$$A^{r(0)}(x) = \sum_k [b(k)w^r_k(\mathbf{r})exp(-i\omega(k)t) + b(k)^*\bar{w}^r_k(\mathbf{r})exp(i\omega(k)t)]$$

$$\delta\psi(x) =$$

$$-e\int S_e(x-y)[(\alpha, \mathbf{A}^{(0)}(y)) + (\alpha, \mathbf{A}^c(y))]\psi^{(0)}(y)d^4y$$

$$= -e\int S_e(x-y)[\alpha^r A^{(0)}_r(y) + \alpha^r A^c_r(y)]\psi^{(0)}(y)d^4y$$

$$= \delta\psi_1(x) + \delta\psi_{ctr}(x),$$

$$\delta A_r(x) = -e\mu_0\int D(x-y)(\psi^*\alpha_r\psi)(y)d^4y + \mu_0\int D(x-y)J^c_r(y)d^4y$$

$$\approx -e\mu_0\int D(x-y)(\psi^{(0)*}\alpha_r\psi^{(0)})(y)d^4y + \mu_0\int D(x-y)J^c_r(y)d^4y$$

where the classically controllable part of the Dirac field is

$$\delta\psi_{ctr}(x) = -e\int S_e(x-y)\alpha^r A^c_r(y)\psi^{(0)}(y)d^4y$$

and this component contains a classical field component $A^c_r$ and a quantum field component $\psi^{(0)}$, while the part of the Dirac field perturbation that is not controllable is

$$\delta\psi_1(x) = -e\int S_e(x-y)\alpha^r A^{(0)}_r(y)\psi^{(0)}(y)d^4y$$

On the other hand, the controllable part of the electromagnetic field is purely classical:

$$\delta A_{r,ctr}(x) = \mu_0 \int D(x-y) J_r^c(y) d^4y$$

If we go one step further in the perturbation series, then we get an additional term in the controllable part of the electromagnetic field so that the above equation gets modified to:

$$\delta A_{r,ctr}(x) = \mu_0 \int D(x-y) J_r^c(y) d^4y$$

$$-e\mu_0 \int D(x-y)(\delta\psi_{ctr}(y)^* \alpha_r(\psi^{(0)} + \delta\psi_1)(y) d^4y$$

$$-e\mu_0 \int D(x-y)(\psi^{(0)*} + \delta\psi_1^*)(y) \alpha_r \delta\psi_{ctr}(y) d^4y$$

Note that in this analysis, the perturbation parameter is the electron charge $e$ and if we neglect $O(e^2)$ terms, then the above expression for the controllable part of the electromagnetic field simplifies to

$$\delta A_{r,ctr}(x) = \mu_0 \int D(x-y) J_r^c(y) d^4y +$$

$$-e\mu_0 \int D(x-y) \delta\psi_{ctr}(y)^* \alpha_r \psi^{(0)}(y) d^4y$$

$$-e\mu_0 \int D(x-y) \psi^{(0)*}(y) \alpha_r \delta\psi_{ctr}(y) d^4y$$

In the particular case of the rdra considered here, we find that the controllable part of the Dirac field has the expansion

$$\delta\psi_{ctr}(x) = -e \int S_e(x-y) \alpha^r A_r^c(y) \psi^{(0)}(y) d^4y$$

$$= -e \int S_e(t-t', r-r') \alpha^r A_r^c(t', r') [\sum_{kmnp} [d_e(k) c_{ek}(mnp) u_{mnp}(r') exp(-iE(k)t') +$$

$$d_p(k)^* c_{pk}(mnp) u_{mnp}(r') exp(iE(k)t')] dt' d^3r'$$

Now define the following Fourier components of the control classical laser generated electromagnetic field w.r.t the cavity boundary conditions and the energy spectrum of the free Dirac field in the cavity after:

$$\int A_r^c(t', r') u_{mnp}(r') exp(-iK.r') exp(i\omega t') d^3r' dt' = C_{A,r}(\omega, K|m, n, p)$$

Then, we can express the above controllabe part of the Dirac field in the following form in the spatio-temporal Fourier domain:

$$\int \delta\psi_{ctr}(t, r).exp(i(\omega t - K.r) dt d^3r =$$

$$\sum_{mnpk}[-ed_e(k)S(\omega,K)c_{ek}(mnp)\alpha^r[C_r(\omega-E(k),K|mnp)$$

$$-ed_p(k)^*S(\omega,K)c_{pk}(mnp)\alpha^r C_r(\omega+E(k),K|mnp)]$$

$$=-eS(\omega,K)\sum_{mnpk}[d_e(k)c_{ek}(mnp)\alpha^r C_r(\omega-E(k),K|mnp)+d_p(k)^*c_{pk}(mnp)\alpha^r C_r(\omega+E(k)|mnp)]$$

The controllable part of the Dirac four current density is then given upto first order perturbation terms by ($x=(t,r)$)

$$\delta J^\mu(t,r)=-e\psi^{(0)*}(x)\alpha^\mu\delta\psi_{ctr}(x)-e\delta\psi_{ctr}(x)^*\alpha^\mu\psi^{(0)}(x)$$

and it is immediately clear from the above expression that the far field radiated electromagnetic potential generated by this controllable current field can be expressed in the form

$$\delta A_R^\mu(t,r)=\int D(t-t',r-r')\delta J^\mu(t',r')dt'd^3r'$$

$$=\sum_{mnpkrm'n'k's} d_e(k)^*d_e(k')\int C_r(\omega-E(k),K|mnp)\bar{C}_s(-\omega'-E(k'),K'|m'n'p')$$

$$.F^{\mu rs}(t,r|\omega,K,\omega',K',mnpk,m'n'p'k')d\omega d^3K d\omega' d^3K'$$

plus three other similar terms involving $d_e(k)^*d_p(k')^*, d_p(k)d_e(k'), d_p(k)d_p(k')^*$. In compact notation, the expected value of this controllable far field pattern can be expressed as a Hermitian quadratic form in the complex numbers $C_r(\omega,K|mnp),\omega\in\mathbb{R},K\in\mathbb{R}^3,m,n,p\in\mathbb{Z}_+$. These complex numbers are controllable since they represent in some sense the spatio-temporal components of the Fourier components of the classical control electromagnetic field $A_\mu^c$.

### 2.5.5 Quantum Antennas constructed using supersymmetric field theories

Reference: Steven Weinberg, "The quantum theory of fields, vol.III, Supersymmetry", Cambridge University Press.

[1] Let $\Phi$ be a left Chiral field, ie, it is a function of only $\theta_L=(1+\gamma_5)\theta/2$ and

$$x_+^\mu=x^\mu+(1/2)\theta_R^T\epsilon\gamma^\mu\theta_L$$

Let $V^A$ be gauge super-fields for each Yang-Mills gauge group index $A$. Expand $V^A$ as

$$V^A(x,\theta)=\theta^T\epsilon\gamma^\mu\theta.V_\mu^A(x)+\theta^T\epsilon\theta.\theta^T\gamma_5\epsilon\lambda^A(x)+(\theta^T\epsilon\theta)^2D^A(x)$$

$V_\mu^A$ is called the gauge field, $\lambda^A$ is called the gaugino field and $D^A$ is called the auxiliary field. The transformation law of the gauge superfield under extended gauge transformations defined by an arbitrary left Chiral superfield $\Omega$ is given by

$$\Gamma \to exp(i\Omega)\Gamma.exp(-i\Omega^*)$$

Now define a left Chiral spinor supefield

$$W_L = D_R^T\epsilon D_R exp(t.V)D_L(exp(-t.V))$$

where $t.V = t_A V^A$. Note that the gauge superfield transformation law implies

$$exp(t.V) \to exp(i\Omega)exp(t.V).exp(-i\Omega^*)$$

$$exp(-t.V) \to exp(i\Omega^*)exp(-t.V)exp(-i\Omega)$$

Then since $\Omega$ is left Chiral, $\Omega^*$ becomes right Chiral and therefore under the gauge transformation,

$$W_L \to exp(i\Omega)D_R^T\epsilon D_R exp(t.V)D_L(exp(-t.V)exp(-i\Omega))$$

$$= exp(i\Omega)D_R^T\epsilon D_R exp(t.V)(D_L exp(-t.V))exp(-i\Omega))$$

$$+exp(i\Omega)D_R^T\epsilon D_R D_L(exp(-i\Omega))$$

The second term is zero since $D_R^T\epsilon D_R.exp(-i\Omega) = 0$ because $D_R.exp(-i\Omega) = 0$ and $[D_R^T\epsilon D_R, D_L]$ is proportional to $\gamma^\mu\partial_\mu D_R$. Thus under gauge transformations, $W_L$ transforms as

$$W_L \to exp(i\Omega)W_L.exp(-i\Omega)$$

which is consistent with the fact that since $W_L$ is left Chiral, its gauge transform should also be left Chiral. It is then easy to see that the quantity

$$Tr[W_L^T\epsilon W_L]_F = [W_L^{AT}\epsilon W_L^A]_F$$

is gauge invariant, Lorentz invariant and supersymmetry invariant where the subscript $F$ denotes the coefficient of $\theta_L^T\epsilon\theta_L$ in the expansion of a Chiral superfield. The matter superfield $\Phi$ is left Chiral and has a scalar field component, a left handed Dirac field component and an auxiliary $F$ component. The field $L_1 = [\Phi^* exp(-t.V)\Phi]_D$ is a supersymmetric Lagrangian that is also gauge invariant since under a gauge transformation,

$$exp(-t.V) \to exp(-i\Omega^*)exp(t.V)exp(i\Omega),$$

while on the other hand,

$$\Phi \to exp(-i\Omega)\Phi, \Phi^* \to \Phi^* exp(i\Omega^*)$$

$L_1$ describes the scalar field, the Dirac field and their interactions with gauge field in a way that generalizes the interaction of the Dirac field with the gauge fields like the electromagnetic field and more generally, with non-Abelian gauge fields. If the left Chiral superpotential field $f(\Phi)$ is also taken into account by using its $F$-term, then we obtain a Lagrangian for matter interacting with gauge fields that is supersymmetry, Lorentz and gauge invariant. The component superfield equations derived from such a Lagrangian will containing interaction terms. If we first make all the interactions in these field equations zero, we then then a set of free field equations for the bosonic and fermionic components whose solution can be expressed in terms of the boson and fermion creation and annihilation operator fields in momentum space. We then express the interaction terms in terms of these creation and annihilation operators and calculate the radiation fields for the gauge fields generated by the current densities of these fields obtained by using the component matter fields (comprising the Dirac field, the Yang-Mills matter fields and the scalar Klein-Gordon field) with their corrections due to the interactions taken into account. Thus, the radiation fields will be in a perturbation approximation, polynomial functionals of the boson and fermion creation and annihilation operator field. The quantum statistical moments of the radiated gauge fields which include the Abelian electromagnetic field, the non-Abelian Yang Mills gauge fields (ie, the W and Z bosonic fields in the electroweak theory) can then be computed in any given state of the bosons and fermions. It should be noted that the gaugino field which is the fermionic superpartner of the Yang-Mills gauge field is a matter field like the Dirac field and there will also be a conserved current associated with this field which will cause a radiation of Yang-Mills gauge bosons. Likewise the supergravity Lagrangian will have a classical bosonic Einstein graviton metric field component with the Riemann curvature as the associated Lagrangian and it will correspondingly have a superpartner field, namely a gravitino fermionic field having spin 3/2 and there will also be a current associated with this gravitino field which will will act as a source of gravitational radiation in the form of gravitons. All these bosonic radiation fields will be described by operators whose statistical moments in any state of the bosons and their fermionic superpartners can in principle be evaluated and controlled using classical fields.

### 2.5.6 Quantization of the Maxwell and Dirac field in a background curved metric of space-time

Let $\Gamma_\mu$ denote the spinor connection of the gravitational field. If $V_a^\mu$ is the tetrad field of the metric, ie,

$$g^{\mu\nu} = \eta^{ab} V_a^\mu V_b^\nu$$

where $\eta$ is the Minkowski metric, then it is known (see for example, [a] Steven Weinberg, "Gravitation and Cosmology: Principles and Applications of the General Theory of Relativity", Wiley, or [b] Steven Weinberg, "The quantum theory of fields, vol.III, Supersymmetry", Cambridge University Press) that

$$\Gamma_\mu = (1/2) V_{a\nu:\mu} V_b^\nu [\gamma^a, \gamma^b]$$

and that Dirac's equation that is diffeomorphic as well as locally Lorentz invariant is given by

$$[\gamma^a V_a^\mu(i\partial_\mu + eA_\mu + i\Gamma_\mu) - m_0]\psi = 0$$

Local Lorentz invariance is checked by noting that if $\Lambda(x)$ is a local Loretnz transformation (ie, a space-time dependent Lorentz transformation matrix w.r.t. the Minkowski metric $\eta$, ie, $\Lambda(x)^T\eta\Lambda(x) = \eta$), and if $D(.)$ denotes Dirac's spinor representation of the Lorentz group, then

$$V_a^\mu(x)D(\Lambda(x))\gamma^a(\partial_\mu + \Gamma_\mu(x))D(\Lambda(x))^{-1} =$$

$$= V_a^\mu D(\Lambda)\gamma^a D(\Lambda)^{-1}D(\Lambda)(\partial_\mu + \Gamma_\mu(x))D(\Lambda(x))^{-1}$$

$$= V_a^\mu \Lambda_b^a \gamma^b D(\Lambda)(\partial_\mu + \Gamma_\mu(x))D(\Lambda(x))^{-1}$$

$$= V_a^{\mu'}(x)\gamma^a(\partial_\mu + \Gamma'_\mu(x))$$

where $\Gamma'_\mu(x)$ is obtained from $\Gamma_\mu(x)$ by replacing $V_a^\mu(x)$ by $V_a^{\mu'}(x)$ with

$$V_a^{\mu'}(x) = \Lambda_a^b(x)V_b^\mu(x)$$

This is proved by assuming $\Lambda(x) = I + \omega(x)$ to be an infinitesimal Local Lorentz transformation. Note that $i\Gamma_\mu(x)$ is not a Hermitian matrix since $[\gamma^a, \gamma^b]$ is not skew-Hermitian for all $a, b = 0, 1, 2, 3$.

Remark: $\gamma^{0*} = \gamma^0, \gamma^{a*} = -\gamma^a, a = 1, 2, 3$ and hence $[\gamma^0, \gamma^a]$ is Hermitian for $a = 1, 2, 3$ while $[\gamma^a, \gamma^b]$ is skew-Hermitian for $a, b = 1, 2, 3$.

However, it can be shown using integration by parts that the action functional for the Dirac field in curved space-time defined by

$$S[\psi, \psi^*] = i\int \psi^*(x)\gamma^0[\gamma^a V_a^\mu(x)(\partial_\mu + \Gamma_\mu(x)) - m_0]\psi(x)\sqrt{-g(x)}d^4x$$

is real and apart from being locally Lorentz invariant, it is also diffeomorphic invariant.

The electron propagator in a curved background metric: The electron propagator is clearly given by the formal operator theoretic expression

$$S_e = (V_a^\mu\gamma^a(i\partial_\mu + i\Gamma_\mu))^{-1}$$

More specifically, if $S_e(x, y)$ is the position space kernel representation of the electron propagator, then

$$[i\gamma^a V_a^\mu(x)(\partial_\mu + \Gamma_\mu(x)) - m_0]S_e(x, y) = \delta^4(x - y)$$

We shall obtain an approximate solution to the electron propagator using perturbation theory. Let $S_{e0}(x - y)$ denote the unperturbed electron propagator. Then,

$$[i\gamma^a\partial_a - m_0]S_{e0}(x - y) = \delta^4(x - y)$$

with solution

$$S_{e0}(x - y) = [i\gamma^a\partial_a - m_0]^{-1}\delta^4(x - y)$$

or equivalently, using Fourier transforms,

$$S_{e0}(x-y) = (2\pi)^{-4}\int(i\gamma^a p_a - m + i0)^{-1}exp(ip.(x-y))d^4p$$

Let $\delta S_e(x,y)$ denote the correction to the electron propagator upto first order perturbation theory caused by the gravitational terms. We then have upto first order in the gravitational metric perturbations from the flat space-time Minkowski metric,

$$V_a^\mu = \delta_a^\mu + \delta V_a^\mu(x)$$

and hence upto the first order,

$$\Gamma_\mu = \delta\Gamma_\mu = (1/2)[\gamma^a,\gamma^b]\delta_a^\nu[\delta V_{b\nu,\mu} - \Gamma^\rho_{\nu\mu}\eta_{b\rho}]$$

$$= (1/2)[\gamma^a,\gamma^b][\delta V_{ba,\mu} - \Gamma_{ba\mu}]$$

Note that upto first order,

$$g_{\mu\nu} = \eta_{ab}V_\mu^a V_\nu^b = \eta_{ab}(\delta_\mu^a + \delta V_\mu^a)(\delta_\nu^b + \delta V_\nu^b)$$

$$= \eta_{\mu\nu} + \eta_{ab}\delta_\mu^a\delta V_\nu^b + \eta_{ab}\delta_\nu^b\delta V_\mu^a$$

$$= \eta_{\mu\nu} + 2\delta V_{\mu\nu}$$

where the tetrad $V_{a\mu}$ has been chosen to be symmetric in its two indices. In fact, if the metric is expressed as

$$g_{\mu\nu} = \eta_{\mu\nu} + \delta g_{\mu\nu}$$

then we can choose

$$V_{\mu\nu} = \delta g_{\mu\nu}/2$$

Now writing the differential equation satisfied by the propagator as

$$[i\gamma^a(\delta_a^\mu + \delta V_a^\mu(x))(\partial_\mu + \delta\Gamma_\mu(x)) - m_0][S_{e0}(x-y) + \delta S_e(x,y)) = \delta^4(x-y)$$

we get on equating first order terms,

$$[i\gamma^\mu\partial_\mu - m_0]\delta S_e(x,y) + i\gamma^a\delta V_a^\mu(x)\partial_\mu S_{e0}(x-y) + i\gamma^\mu\delta\Gamma_\mu(x)S_{e0}(x-y) = 0$$

from which we deduce the following formula for the first order propagator correction:

$$\delta S_e(x,y) = -i\int S_{e0}(x-z)[\gamma^a\delta V_a^\mu(z)\partial_\mu S_{e0}(z-y) + \gamma^\mu\delta\Gamma_\mu(z)S_{e0}(z-y)]d^4z$$

$$= -i\int S_{e0}(x-z)\gamma^a[V_a^\mu(z)\partial_\mu S_{e0}(z-y) + \delta\Gamma_a(z)S_{e0}(z-y)]d^4z$$

In order to relate all this to quantum antennas, we must also calculate the Dirac four current density in curved space time. Consider

$$\psi^*\gamma^0[(V_a^\mu\gamma^a(i\partial_\mu + i\Gamma_\mu) - m_0]\psi = 0$$

or equivalently,

$$\psi^*[V_a^\mu\alpha^a(i\partial_\mu + i\Gamma_\mu) - m_0\beta]\psi = 0$$

or equivalently,

$$i\partial_\mu[\psi^* V_a^\mu\alpha^a\psi] - i\partial_\mu[\psi^* V_a^\mu]\alpha^a\psi$$
$$+\psi^*[iV_a^\mu\alpha^a\Gamma_\mu - m_0\beta]\psi = 0$$

Taking the conjugate of this equation gives

$$-i\partial_\mu[\psi^* V_a^\mu\alpha^a\psi] + i\psi^*\alpha^a\partial_\mu[V_a^\mu\psi]$$
$$+\psi^*[-iV_a^\mu\Gamma_\mu^*\alpha^a - m_0\beta]\psi = 0$$

### 2.5.7 Relationship between the electron self energy and the electron propagator

Consider an electron bound to its nucleus. Let $E_n$ denote its energy eigenvalue corresponding to the eigenfunction $u_n(\mathbf{x})$. Then, the propagator of the electron in the second quantized picture can be expressed as

$$S(x,y) = S(t,\mathbf{x}|(t',\mathbf{y}) = \theta(t-t')\sum_n u_n(\mathbf{x})\bar{u}_n(\mathbf{y})exp(-iE_n(t-t'))$$

To check this, we prove that $S$ satisfies the propagator differential equation

$$(i\partial_t - H)S(t,\mathbf{x}|t',\mathbf{y}) = i\delta^4(x-y), x = (t,\mathbf{x}), y = (t',\mathbf{y})$$

This is proved using the identities

$$\theta'(t-t') = \delta(t-t'), \sum_n u_n(\mathbf{x})\bar{u}_n(\mathbf{y}) = \delta^3(\mathbf{x}-\mathbf{y})$$

In the frequency/energy domain, the propagator is given by

$$S(\mathbf{x},\mathbf{y}|E) = \int_{\mathbb{R}} S(t,\mathbf{x}|0,\mathbf{y})exp(iE(t-t'))dt =$$
$$i\sum_n u_n(\mathbf{x})\bar{u}_n(\mathbf{y})/(E-E_n)$$

or equivalently, in operator theoretic notation,

$$S(E) = i\sum_n |u_n><u_n|/(E-E_n), <u_n|u_m>= \delta_{n,m}$$

In particular,

$$\int \bar{u}_n(\mathbf{x})S(\mathbf{x},\mathbf{y}|E)d^3x =$$

$$i\bar{u}_n(y)/(E-E_n)$$

Then the change in the propagator caused by radiative effects in which the energy levels get perturbed by $\delta E_n$ and correspondingly, the stationary state eigenfunctions get perturbed by $\delta u_n(\mathbf{x})$ is given by

$$-i\delta S(\mathbf{x},\mathbf{y}|E) = \sum_n [\delta u_n(\mathbf{x})\bar{u}_n(\mathbf{y})+u_n(\mathbf{x})\delta u_n(\mathbf{y})]/(E-E_n)+\sum_n u_n(\mathbf{x})\bar{u}_n(\mathbf{y})\delta E_n/(E-E_n)^2$$

It follows then that on writing the one loop radiative correction to the electron propagator as

$$\delta S = S.\Sigma.S$$

that

$$-i < u_n|\delta S|u_n >= -i < u_n|S\Sigma.S|u_n >= -i < u_n|\Sigma|u_n > /(E-E_n)^2$$

on the one hand while on the other,

$$-i < u_n|\delta S|u_n >= \delta E_n/(E-E_n)^2$$

where we have used the orthogonality relation

$$< \delta u_n|u_n >= 0$$

since both $u_n$ and $u_n + \delta u_n$ are normalized, ie have unit norm. This gives us the fundamental relation between the change in the electron propagator caused by one loop radiative corrections and the shift in the electron energy as

$$\delta E_n = -i < u_n|\Sigma|u_n >$$

This is the extra energy gained by the electron due to propagator corrections coming from radiative as well as gravitational effects.

Electron self energy corrections induced by quantum gravitational effects. For the free gravitational field, let

$$< 0|T(\delta\Gamma_\mu(x)\delta\Gamma_\nu(y))|0 >= D_\Gamma(x,y)$$

This can be viewed as some sort of propagator for the free quantum gravitational field. The wave equation satisfied by the Dirac field in the presence of quantum gravitational effects is given by

$$[i\gamma^a(\delta_a^\mu + \delta V_a^\mu(x))(\partial_\mu + \delta\Gamma_\mu(x)) - m_0][\psi^{(0)}(x) + \delta\psi(x)] = 0$$

$\psi^{(0)}$ is the free electron-positron wave operator field. It satisfies the zeroth order perturbation equation:

$$[i\gamma^\mu\partial_\mu - m_0]\psi^{(0)} = 0$$

and its solution is expressible as a superposition of the electron annihilation operators and the positron creation operators in momentum space. $\delta\psi$ is the correction to the free Dirac field caused by gravitational effects upto first order. It satisfies the first order perturbation equation:

$$[i\gamma^{\mu}\partial_{\mu} - m_0]\delta\psi(x)$$

$$+i\gamma^{a}\delta V_a^{\mu}(x)\partial_{\mu}\psi^{(0)}(x) + i\gamma^{\mu}\delta\Gamma_{\mu}(x)\psi^{(0)}(x) = 0$$

ie $\delta\psi(x)$ satisfies the same differential equation as the first order perturbation $\delta S_e(x,y)$ in the electron propagator and its solution is given by

$$\delta\psi(x) = -i\int S_{e0}(x-y)[\gamma^{a}\delta V_a^{\mu}(y)\partial_{\mu}\psi^{(0)}(y) + \gamma^{\mu}\delta\Gamma_{\mu}(y)\psi^{(0)}(y)]d^4y$$

and hence the approximate corrected electron propagator upto linear orders in the graviton propagator is given by

$$S_e(x,y) =< 0|T\{(\psi^{(0)}(x) + \delta\psi(x)).(\psi^{(0)}(y) + \delta\psi(y))^*\}|0 >$$

$$= S_{e0}(x-y)+ < 0|T\{\psi^{(0)}(x).\delta\psi(y)^*\}|0 > + < 0|T\{\delta\psi(x).\psi^{(0)*}(y)\}|0 >$$

$$+ < 0|T\{\delta\psi(x).\delta\psi(y)^*\}|0 >$$

$$= S_{e0}(x-y)+ < 0|T\{\delta\psi(x).\delta\psi(y)^*\}|0 >$$

with

$$< 0|T\{\delta\psi(x).\delta\psi(y)^*\}|0 >=$$

$$\int S_{e0}(x-u)[\gamma^{a}\delta V_a^{\mu}(u)\partial_{\mu}\psi^{(0)}(u)+\gamma^{\mu}\delta\Gamma_{\mu}(u)\psi^{(0)}(u)].[(\delta V_b^{\nu}(v)\partial_{\nu}\psi^{(0)}(v)^*\gamma^{b*}$$

$$+\psi^{(0)}(v)^*\delta\Gamma_{\nu}(v)^*\gamma^{\nu*}]S_{e0}(y-v)^*d^4ud^4v$$

### 2.5.8 Antennas designed using robotic links carrying current

Let $\mathbf{R}_k(t), k = 1,2,...,N$ denote the positions of moving point objects which have to be picked up by a robot. More generally, instead of point objects, we can assume extended rigid bodies moving on the ground or in space which have to be picked up by the robot. We have a camera in synchronization with the robot at the location $\mathbf{R}(t)$ (ie the position of its centre of mass). Now the camera takes pictures of the point objects and also of the robot and a digital computer calculates the distances and bearings of the images of the objects with that of the robot and accordingly generates control torques that are used to manipulate the robot so that it moves closer to one of the objects, say the $m^{th}$ one in succession, ie, the robot uses the error in the images of the position of the $m^{th}$ object and that of the robot to generate a control torque signal that eventually enables the robot to track this object and finally reduce the error in its position relative to the robot to zero and finally pick the object up. This series of jobs is performed successively on the different objects so that finally all the objects are picked up. The mathematical details of formulating an algorithm are based on the gradient descent algorithm and could be described as follows: Let $I(\mathbf{R}_k(t), x, y)$ denote the image field on the camera screen generated by the $k^{th}$ object at time $t$ and let $I(\mathbf{R}(t), \mathbf{q}(t), x, y)$ be the image field on the camera screen generated by the robot at time $t$ whose centre of mass is located at $\mathbf{R}(t)$ and whose link angles relative to a given direction are denoted by $\mathbf{q}(t)$. The computer calculates the error energy

$$E_k(t, \mathbf{R}(t), \mathbf{q}(t)) = \int_{screen} (I(\mathbf{R}_k(t), x, y) - I(\mathbf{R}(t), \mathbf{q}(t), x, y))^2 dxdy$$

and then the computer generates the following algorithm for moving the robot using a force and torque that causes the robot's location and link angles respectively to change after a small time $\delta t$ to

$$\mathbf{R}(t) + \delta\mathbf{R}(t), \mathbf{q}(t) + \delta\mathbf{q}(t)$$

where

$$\delta\mathbf{R}(t) = -\mu.\delta t \nabla_{\mathbf{R}(t)} E_k(t, \mathbf{R}(t), \mathbf{q}(t)),$$

$$\delta\mathbf{q}(t) = -\mu.\delta t \nabla_{\mathbf{q}(t)} E_k(t, \mathbf{R}(t), \mathbf{q}(t))$$

Our aim will be to generalize this model to the tracking and picking up rigid bodies and even non-rigid extended bodies. The force and torque generation mechanism are based on Newtonian mechanics:

$$\mathbf{F}(t) = M\mathbf{R}''(t), \tau(t) = J(\mathbf{q}(t))\mathbf{q}''(t) + N(\mathbf{q}(t), \mathbf{q}'(t))$$

where $M, J$ are respectively the robot mass and its mass moment of inertia matrix and $N(\mathbf{q}(t), \mathbf{q}'(t))$ consists of centrifugal, coriolis, frictional and gravitational potential contributions to the computed torques. In practice, these control forces and torques are generated by discretizing the time derivatives with a time step of $\delta t$:

$$\delta\mathbf{q}''(t) \approx (\delta\mathbf{q}(t) - 2\delta\mathbf{q}(t - \delta t) + \delta\mathbf{q}(t - 2\delta t))/(\delta t)^2$$

$$\mathbf{R}''(t) = (\mathbf{R}(t) - 2\mathbf{R}(t - \delta t) + \mathbf{R}(t - 2\delta t))/(\delta t)^2$$

We propose in our project to do a noise analysis of this algorithm based on Varadhan's large deviation theory. We also propose to do a robustness analysis of this problem based on how sensitive is the tracking error energy to errors induced in camera imaging and also to errors induced in the digital computer due to finite register effects while computing the control forces and torques. This work is to be regarded as an extension of a paper [1] based on the gradient search algorithm. [1] does not consider a mathematical analysis of noise effects on the algorithm. We propose to do such an analysis by adding WGN to the rhs of the gradient algorithm thereby resulting in a nonlinear stochastic difference equation and we shall use the standard techniques based on mean and variance propagation to analyze these effects [6]. The gradient algorithm for developing the computed force and torques are based on the gradient algorithm which take into account only the instantaneous error. In our project, we shall also be considering generalizations of this based on past error history.

### 2.5.9 A project proposal for developing an experimental setup for transmitting quantum states over a channel in the presence of an eavesdropper

In quantum computation and information theory, it is by now a well established fact that a qubit state and more generally $d$-qubit state (ie a pure state in $\mathbb{C}^{2^d}$) can be transmitted over a channel from $A$ to $B$ by transmitting just $2d$ classical bits provided that $A$ and $B$ share a maximally entangled state, ie, a state of the form $d^{-1/2}\sum_{k=0}^{d-1}|k,k>$. The idea is simply to append this state to the maximally entangled state at $A's$ end, then perform a unitary transformation on the total $2d$-qubit state of $A$, perform a measurement at $A's$ end thereby causing $B's$ state to collapse to one of $2d$ possible $d$-qubit states. When $A$ then reports to $B$ about his measurement outcome via $2d$ classical bits, $B$ is able to apply an appropriate unitary gate at his end to recover the original state that $A$ had intended to transmit. In quantum information theory, another important problem is the Cq problem in which $A$ wishes to transmit classical information over a quantum channel by encoding his classical bits in the form of quantum states. Thus, if $A's$ classical information source is the alphabet $A = \{1,2,...,a\}$, with the alphabet $k$ occurring with probability $p(k)$, then the total information contained in this source that $A$ wishes to transmit is $H(A) = H(p) = -\sum_{x=1}^{a} p(x).log(p(x))$. $A$ encodes the alphabet $x \in A$ in the form of a density matrix $\rho(x)$ (ie, a mixed state in a finite dimensional Hilbert space $\mathcal{H}$), and transmits this state over the channel assumed to be noiseless. The state received by $B$ is then $\rho(x)$ and the average state received by $B$ is $\bar{\rho} = \sum_{x\in A} p(x)\rho(x)$. It is natural to expect that the total information that $A$ has transmitted to $B$ must be given by

$$I_p(A,B) = H(B) - H(B|A) = H(\bar{\rho}) - \sum_{x\in A} p(x)H(\rho(x))$$

where

$$H(W) = -Tr(W.log(W))$$

is the Von-Neumann entropy of the state $W$. In order for this to be a meaningful measure of the information transmitted, we can ask the following question: Suppose $A$ encodes a string of his source alphabets $\mathbf{x} = (x(1),...,x(n)) \in A^n$ into the state $\rho(\mathbf{x}) = \rho(x(1)) \otimes ... \otimes \rho(x(n))$ in the tensor product Hilbert space $\mathcal{H}^{\otimes n}$, then can he choose $M_n$ such distinct sequences $\mathbf{x}_1,...,\mathbf{x}_M$ such that (a) these sequences are all typical for $A'$ source w.r.t to the probability distribution $p^n(\mathbf{x}) = p(x(1))...p(x(n))$, (b) There exist positive "Detection Operators"

$\mathbf{D}_1, ..., \mathbf{D}_{\mathbf{M_n}}$ for $B$ in the Hilbert space $\mathcal{H}^{\otimes n}$ such that for any $\epsilon > 0$ with $n$ sufficiently large, one has

$$\mathbf{D}_1 + ... + \mathbf{D}_{M_n} \leq I,$$

$$Tr(\rho(\mathbf{x}_k)\mathbf{D}_k) > 1 - \epsilon, k = 1, 2, ..., M_n$$

and

$$Tr(D_k) \leq Tr(E(\mathbf{x}_k, n, \delta)), k = 1, 2, ..., M_n$$

where $E(\mathbf{x}, n, \delta)$ is a $\delta$-typical projection on $\mathcal{H}^{\otimes n}$ corresponding to the situation when $\mathbf{x}$ is a $d$-typical sequence for $A's$ source. These requirements amount to saying that the detection operator $\mathbf{D}_k$ of $B$ does not have too large a dimension so as to "leak" into another sequence $\mathbf{x}_j, j \neq k$ when $\mathbf{x}_k$ is transmitted, and further, that with a large probability, $B's$ decision on what sequence $A$ had transmitted is correct when he uses his detection operators. Then the question is that if $M_n$ is maximal subject to these requirements, so that the rate of reliable transmission of information (ie, with error probability smaller than $\epsilon$), is $\frac{log(M_n)}{n}$, then $limlog(M_n)/n = sup_p I_p(A, B)$. In other words, the maximum rate of reliable transmission of information on a Cq channel is precisely the Cq channel capacity defined by

$$C = sup_p I_p(A, B) = sup_p(H(\sum_x p(x)\rho(x)) - \sum_x p(x)H(\rho(x)))$$

Our project will involve verifying this capacity formula by preparing the states $\rho(x), x \in A$ using lasers and ions, so that by shining the laser on an ion, we can generate excited ion states used for transmission. In other words, one of the primary objectives of our experimental setup will be to prepare a large class of quantum states using the quantum electromagnetic field generated by a laser interacting with ions. If the ion and the laser field start in an initial state $|k, \phi(u)>$ where $|k>$ represents a stationary state of the ion and $|\phi(u)>$ a coherent state of the laser, then after interacting with each other for a time duration $T$, the final state of the ion and the laser field will be the pure state $U(T)|k, \phi(u)>$ and by partially tracing this out over the laser field state, the ion state becomes the mixed state

$$\rho_{ion}(k, u, T) = Tr_2(U(T)|k, \phi(u)><k, \phi(u)|U(T)^*)$$

By varying $k, u, T$, namely, the intial state of the ion, the coherent state of the laser field and the time duration of interaction of these two, we can thus generate a host of mixed states on the Hilbert space of the ion. These mixed states can be used for transmission.

Another application of our experimental setup will be to create entangled states between three people $A, B, E$ or more generally a mixed state $\rho_{ABE}$ in the tensor product Hilbert space $\mathcal{H}_{ABE} = \mathcal{H}_A \otimes \mathcal{H}_B \otimes \mathcal{H}_E$ and to transmit maximal information from $A$ to $B$ while restricting the information transmitted from $A$ to $E$ to be a minimum. Let $A$ have a classical source with alphabet $A$

and source probability distribution $p(x), x \in A$ and let for each $x \in A$ $\rho_{BE}(x)$ be a state in $\mathcal{H}_{BE} = \mathcal{H}_B \otimes \mathcal{H}_E$. Then the information transmitted from $A$ to $B$ is given by

$$I(A,B) = H(\sum_x p(x)\rho_B(x)) - \sum_x p(x)H(\rho_B(x))$$

while the information transmitted from $A$ to $E$ is given by

$$I(A,E) = H(\sum_x p(x)\rho_E(x)) - \sum_x p(x)H(\rho_E(x))$$

where

$$\rho_B(x) = Tr_E(\rho_{BE}(x)), \rho_E(x) = Tr_B(\rho_{BE}(x))$$

The problem is to select the probability distribution $p(x), x \in A$ for $A's$ source and the states $\rho_{BE}(x), x \in A$ so that

$$I(A,B) - I(A,E)$$

is a maximum, ie, to transmit maximum information across the Cq channel from $A$ to $B$ while keeping the Cq information transmitted to $E$ a minimum. Such a setup can be arranged using lasers and ion trap experiments as follows: Let $A$ generate a current $I(t,x)$ dependent upon his source alphabet $x \in A$ and let him connect this current to a classical antenna that transmits electromagnetic waves to both $B$ and $E$. The magnetic vector potential at $B$ is then given by

$$\mathbf{A}_B(t,\xi) = \int_{S_A} G(t-s, \mathbf{R}_B + \xi, \mathbf{u})\mathbf{J}(s,\mathbf{u},x)dsdu$$

while the magnetic vector potential at $E's$ end is given by

$$\mathbf{A}_E(t,\xi) = \int_{S_A} G(t-s, \mathbf{R}_C + \xi, \mathbf{u})\mathbf{J}(s,\mathbf{u},x)dsdu$$

where $G(t-s, \mathbf{R}, u)$ is the standard retarded potential Green's function between the point $\mathbf{u}$ on $A's$ antenna surface the point $\mathbf{R}$ of reception of the field. $\mathbf{R}_B$ is the location of the nucleus of $B's$ atomic receiver while $\mathbf{R}_E$ is the location of the nucleus of $E's$ atomic receiver. $\mathbf{J}(t,\mathbf{u},x)$ is the surface current density on $A's$ antenna surface $S_A$ and it depends upon the alphabet $x$ that $A$ wishes to transmit and hence from basic antenna theory, it can be expressed as

$$\mathbf{J}(t,\mathbf{u},x) = \int \mathbf{F}(t-s,\mathbf{u})I(s,x)ds, x \in A$$

where the function $\mathbf{F}(t,\mathbf{u}), t \in \mathbb{R}, \mathbf{u} \in S_A$ depends only upon the antenna surface geometry and the point on this surface where the current source $I(t,x)$ us fed in. There is an interaction potential $V_{BE}$ between the systems used by $B$ and $E$ and hence, the Schrodinger equation for the joint state $\rho_{BE}$ when $A$ transmits the symbol $x$ has the form

$$i\rho'_{BE}(t) = [H_{BE}(t), \rho_{BE}(t)]$$

where

$$H_{BE}(t) = H_B(t) + H_E(t) + V_{BE}$$

with

$$H_B(t) = (\mathbf{p}_B + e\mathbf{A}_B(t,\xi_B)^2/2m - Ze^2/|\xi_B| - e\Phi_B(t,\xi_B),$$
$$H_E(t) = (\mathbf{p}_E + \mathbf{A}_E(t,\xi_E))^2/2m - Ze^2/|\xi_E| - e\Phi_E(t,\xi_E)$$

and

$$\mathbf{p}_B = -i\nabla_{\xi_B}, \mathbf{p}_E = -i\nabla_{\xi_E}$$

When $A$ uses a quantum antenna source to transmit a quantum electromagnetic field, then the fields $\mathbf{A}_B, \mathbf{A}_E$ also become quantum fields and then the above Schrodinger equation for $\rho_{BE}(t)$ must be partially traced out over the coherent state of the bath field to obtain the "system part" of $\rho_{BE}(t)$. Note that since the surface current density $\mathbf{J}(t, \mathbf{u}, x)$ in $A's$ antenna depends upon the symbol $x$ that $A$ wishes to transmit, it follows that $\rho_{BE}(t) = \rho_{BE}(t, x)$ will also depend upon the symbol $x$ and then the Cq approach mentioned above can be applied to design $A's$ antenna and his source probability distribution $p(x)$ for maximal transmission of information from $A$ to $B$ while keeping the information that has been leaked into $E's$ receiver at a minimum.

## 2.6 Problems in Antenna Theory

[5] **Aperture antenna pattern fluctuations**

Consider a surface antenna with the surface equation $z = f(x,y)$. Let $\mathbf{E}_i(x,y,z)$ be an electric field at fixed frequency $\omega$ that is incident upon this surface.

[a] Justify that the surface magnetic current density on the antenna surface is given by

$$\mathbf{M}_s(x,y) = -\hat{n} \times \mathbf{E}_i(x,y,f(x,y))$$

where

$$\hat{n} = (-f_{,x}(x,y)\hat{x} - f_{,y}(x,y)\hat{y} + \hat{z})/\sqrt{1 + f_{,x}^2 + f_{,y}^2}$$

where

$$f_{,x} = \frac{\partial f}{\partial x}, f_{,y} = \frac{\partial f}{\partial y}$$

[b] Show that the differential surface area element on the antenna surface is given by

$$dS(x,y) = \sqrt{1 + f_{,x}^2 + f_{,y}^2}dxdy$$

[c] Show that the far field electric vector potential radiated by the antenna surface aperture is given by

$$\mathbf{F}(\mathbf{r}) = (\epsilon/4\pi)(exp(-jkr)/r)\int \mathbf{M}_s(x',y')exp(jK\hat{r}.(x'\hat{x}+y'\hat{y}+f(x',y')\hat{z}))dS(x',y')$$

[d] Hence, if the surface fluctuates by a small amount so that its new equation is $z = f(x,y)+\delta f(x,y)$, then evaluate $\delta \mathbf{F}(\mathbf{r})$ and hence $\delta \mathbf{E}(\mathbf{r})$, the radiated fields in the far field zone as a linear functional of $\delta f$.

hint:

$$\delta\sqrt{1+f_{,x}^2+f_{,y}^2} = (f_{,x}\delta f_{,x} + f_{,y}\delta f_{,y})/\sqrt{1+f_{,x}^2+f_{,y}^2}$$

Hence, evaluate $\delta\hat{n}(x,y), \delta dS(x,y)$ in terms of $\delta f(x,y)$ and its partial derivatives.

[6] If $\delta f(x,y)$ in the previous problem is a random function with mean zero and correlations

$$\mathbb{E}(\delta f(x,y).\delta f(x',y'))R_{ff}(x,y|x',y')$$

then evaluate the correlations in the far field pattern fluctuations.

[7] In problem [5], evaluate the total power radiated out by the aperture surface antenna in the far field zone.

hint: In the far field zone, the electric field upto $O(1/r)$ is

$$\mathbf{E} = -\nabla \times \mathbf{F}/\epsilon = jk\hat{r} \times \mathbf{F}/\epsilon$$

and the corresponding far field zone magnetic field is given by

$$-j\omega\mu\mathbf{H} = \nabla \times \mathbf{E} = -jk\hat{r} \times \mathbf{E}$$

Now calculate the far field Poynting vector field $(1/2)Re(\mathbf{E} \times \mathbf{H}^*)$ in the far field zone upto $O(1/r^2)$, take the radial component, multiply it by the surface element $dS = r^2 d\Omega$ where $d\Omega$ is the solid angle differential element and integrate the result over all solid angles to get the total radiated power.

## 2.7 Design of a quantum unitary gate using superstring theory with noise analysis based on the Hudson-Parthasarathy quantum stochastic calculus

The superstring action is given by

$$S[X,\psi] = \int((1/2)\partial_\alpha X^\mu \partial^\alpha X_\mu - i\psi^{\mu T}\sigma^2\rho^\alpha\partial_\alpha\psi_\mu)d^2\sigma + \int B_{\mu\nu}(X)dX^\mu \wedge dX^\nu$$

Note that

$$\begin{aligned} dX^\mu \wedge dX^\nu &= (X^\mu_{,1}X^\nu_{,2} - X^\mu_{,2}X^\nu_{,1})d^2\sigma \\ &= \epsilon^{\alpha\beta}\partial_\alpha X^\mu.\partial_\beta X^\nu \end{aligned}$$

where

$$\epsilon^{12} = 1, \epsilon^{21} = -1, \epsilon^{11} = \epsilon^{22} = 0$$

The supersymmetry transformations under which this action is invariant are

$$\delta X^\mu = c_1 k^T \sigma^2 \psi^\mu,$$

$$\delta \psi^\mu = c_2 \rho^\alpha k \partial_\alpha X^\mu$$

Here,

$$\rho^0 = \sigma^1, \rho^1 = \sigma^3$$

where $k$ is an infinitesimal Fermionic parameter. This supersymmetric action can also be derived using basic superfield theory by defining our superfield on the space of two dimensional Bosonic and two dimensional Fermionic space as

$$\Phi^\mu(\sigma, \theta) = X^\mu(\sigma) + \theta^T \epsilon \psi^\mu(\sigma) + \theta^T \epsilon \theta . Y$$

where $Y$ is a scalar Bosonic field and $\epsilon = i\sigma^2$. The infinitesimal supersymmetry transformations are defined by the super-vector field

$$L = k^T(\epsilon \rho^\alpha \theta . \partial_\alpha + \partial / \partial \theta)$$

with $k$ being an infinitesimal Fermionic parameter. It is clear that under such an infinitesimal transformation of the superfield $\Phi$, we have

$$\delta X^\mu = k^T \epsilon \psi^\mu,$$

$$\theta^T \epsilon \delta \psi^\mu = k^T \epsilon \rho^\alpha \theta . \partial_\alpha X^\mu$$

or equivalently,

$$\delta \psi^\mu = \rho^\alpha k \partial_\alpha X^\mu$$

## 2.8 Quantum Boltzmann equation for a system of particles interacting with a quantum electromagnetic field

Let $\rho(t) = \rho_{123..N}(t)$ denote the state of the system of $N$ identical particles. This state is an operator on $\mathcal{H}^{\otimes N}$ and it satisfies Schrodinger's equation

$$i\rho'(t) = [H(t), \rho(t)]$$

where

$$H(t) = \sum_{j=1}^N ((p_j + eA(t, r_j))^2/2m - e\Phi(t, r_j)) + H_F(t)$$

where $H_F(t)$ is the electromagnetic field Hamiltonian in Boson Fock space. It is given by

$$H_F(t) = (\epsilon/2) \int |E(t, r)|^2 d^3r + (1/2\mu) \int |B(t, r)|^2 d^3r$$

where

$$E(t,r) = -\nabla\Phi(t,r) - \partial_t A(t,r), B(t,r) = curlA(t,r)/\mu$$

The Maxwell equations for $E, B$ are written down taking into account the quantum current density and charge density associated with the charges of the $N$ particles and their joint density operator $\rho(t)$. If $\rho_1(t,r,r')$ denotes the position space representation of the marginal density for one particle, then we know by analogy with the expression for the quantum current and charge density in a pure state $\psi(t,r)$,

$$\mathbf{J}(t,r) = (i/2m)(\psi(t,r)^*\nabla\psi(t,r) - \psi(t,r)\nabla\psi(t,r)^*),$$

$$\sigma(t,r) = \psi(t,r)^*\psi(t,r)$$

that the same quantities in the mixed state $\rho_1$ are given by

$$\mathbf{J}(t,r) = (i/2m)[\nabla_1\rho(t,r,r') - \nabla_2\rho(t,r,r'))]|_{r'=r}$$

$$\sigma(t,r) = \rho(t,r,r)$$

Note that

$$\rho_1(t) = Tr_{23...N}\rho_{123...N}(t)$$

These expressions for the current and charge densities are to be substituted into the Maxwell equations

$$curlE(t,r) = -\partial_t B(t,r), curlB(t,r) = \mu J(t,r) + \mu\epsilon\partial_t E(t,r),$$

$$divB(t,r) = 0, divE(t,r) = \sigma(t,r)/\epsilon$$

This model can be used to describe a quantum plasma within a quantum cavity resonator having a quantum electromagnetic field within it. The solutions for the quantum electromagnetic field will be given by a sum of two terms:The first term is the free field solutions as conventionally described in quantum electrodynamics in terms of the photon creation and annihilation operators. This part is the solution to the homogeneous (ie, source free) part of the Maxwell equations. The second term is the particular solution of the Maxwell equations that is linear in the current and charge densities. This expression for the electromagnetic field operators is to be substituted into the quantum Boltzmann equation for $\rho_1(t,r,r')$ in order to get an appropriate description of the plasma.

## 2.9 Device physics in a semiconductor using the classical Boltzmann kinetic transport equation

## 2.10 The quantum Boltzmann equation for a plasma

Suppose that the joint density matrix of $N$ particles is $\rho(123...N)$. It satisfies the Schrodinger equation

$$i\partial_t\rho_t(12...N) = [\sum_{a=1}^{N} H_a + \sum_{1\leq a<b\leq N} V_{ab}, \rho_t(12..N)]$$

In this equation, if we take a partial traced over $2, 3, ..., N$, we get

$$'i\partial_t\rho_{1t} = [H_1, \rho_{1t}] + (N-1)Tr_2[V_{12}, \rho_{12}]$$

and if we take the trace of the same over $3, 4, ..., N$, we get

$$i\partial_t\rho_{12t} = [H_1 + H_2 + V_{12}, \rho_{12t}] + (N-2)Tr_3[V_{13} + V_{23}, \rho_{123t}]$$

We write

$$\rho_{123} = (1/3)(\rho_{12} \otimes \rho_3 + \rho_{13} \otimes \rho_2 + \rho_1 \otimes \rho_{23}) + g_{123}$$

where $g_{123}$ is small. Then, neglecting second order of smallness terms like $V$ multiplied with $g_{123}$ gives us the approximate equation

$$i\partial_t\rho_{12t} = [H_1+H_2+V_{12}, \rho_{12}]+((N-2)/3)Tr_3[V_{13}+V_{23}, \rho_{12}\otimes\rho_3+\rho_{13}\otimes\rho_2+\rho_1\otimes\rho_{23}]$$

This is a bit hard to handle. So we content ourselves with the approximation

$$\rho_{12} = \rho_1 \otimes \rho_1 + g_{12}$$

where $g_{12}$ is small. We then get approximately,

$$i\partial_t\rho_{1t} = [H_1, \rho_{1t}] + (N-1)Tr_2[V_{12}, \rho_1 \otimes \rho_1]$$

Writing

$$V_{12} = \sum_a W_{1a} \otimes W_{2a}$$

gives us

$$Tr_2[V_{12}, \rho_1 \otimes \rho_2] = \sum_a Tr(\rho_1 W_{2a})[W_{1a}, \rho_1]$$

and the our Boltzmann equation becomes

$$i\partial_t\rho_1 = [H_1, \rho_1] + (N-1)\sum_a Tr(\rho_1 W_{2a})[W_{1a}, \rho_1]$$

Suppose we make the approximation

$$\rho_{123} = \rho_1 \otimes \rho_1 \otimes \rho_1 + g_{123}$$

where $g_{123}$ is small. Then we get

$$i\partial_t \rho_{12t} = [H_1 + H_2 + V_{12}, \rho_{12}] + (N-2)Tr_3[V_{13} + V_{23}, \rho_1 \otimes \rho_1 \otimes \rho_1]$$

Even this equation is hard to manipulate further without assuming some specific form of the interaction potential $V_{12}$. We consider

$$\rho_{12} = \rho_1 \otimes \rho_1 + g_{12},$$

$$\rho_{123} = (1/3)(\rho_{12} \otimes \rho_3 + \rho_{13} \otimes \rho_2 + \rho_1 \otimes \rho_{23}) + g_{123}$$

$$= \rho_1 \otimes \rho_1 \otimes \rho_1 + (1/3(g_{12} \otimes \rho_3 + g_{13} \otimes \rho_2 + \rho_1 \otimes g_{23}) + g_{123}$$

We first derive a differential equation for $g_{12}$ after neglecting second order of smallness terms:

$$i\partial_t \rho_{12} = i\partial_t \rho_1 \otimes \rho_1 + i\rho_1 \otimes \partial_t \rho_1$$

$$+ i\partial_t g_{12}$$

$$= [H_1, \rho_1] \otimes \rho_1 + (N-1)Tr_2[V_{12}, \rho_1 \otimes \rho_1] \otimes \rho_1 + \rho_1 \otimes [H_1, \rho_1]$$

$$+ (N-1)\rho_1 \otimes Tr_2[V_{12}, \rho_1 \otimes \rho_1] + i\partial_t g_{12}$$

$$= [H_1 + H_2 + V_{12}, \rho_{12}] + (N-2)Tr_3[V_{12} + V_{13}, \rho_{123}]$$

$$= [H_1 + H_2, \rho_1 \otimes \rho_1] + [V_{12}, \rho_1 \otimes \rho_1]$$

$$+ (N-2)Tr_3[V_{13} + V_{23}, \rho_1 \otimes \rho_1 \otimes \rho_1]$$

After making the appropriate cancellations, we get

$$i\partial_t g_{12} =$$

$$= [V_{12}, \rho_1 \otimes \rho_1] + (N-2)Tr_3[V_{13} + V_{23}, \rho_1 \otimes \rho_1 \otimes \rho_1]$$

$$-(N-1)Tr_2[V_{12}, \rho_1 \otimes \rho_1] \otimes \rho_1$$

$$-(N-1)\rho_1 \times Tr_2[V_{12}, \rho_1 \otimes \rho_1]$$

Note that on writing

$$V_{12} = \sum_a W_{1a} \otimes W_{2a}$$

and using the fact that the $V_{jk}'s$ are identical copies of each other acting on different copies of the tensor product of two identical copies a Hilbert space just as the $H_k's$ are identical copies of each other acting on different copies of the same Hilbert space, we get

$$Tr_3[V_{13} + V_{23}, \rho_1 \otimes \rho_1 \otimes \rho_1]$$

$$= \sum_a [Tr(\rho_1 W_{2a})([W_{1a}, \rho_1] \otimes \rho_1 + \rho_1 \otimes [W_{1a}, \rho_1])]$$

A better approximation to the quantum Boltzmann equation can then be obtained by solving this equation for $g_{12}(t)$ and substituting it into the equation

$$i\partial_t \rho_1 = [H_1, \rho_1] + (N-1)Tr_2[V_{12}, \rho_{12}]$$

$$= [H_1, \rho_1] + (N-1)Tr_2[V_{12}, \rho_1 \otimes \rho_1 + g_{12}]$$

Formally this equation has the form

$$i\partial_t \rho_1(t) = [H_1, \rho_1(t)] + \delta.F_1(\rho_1(s), s \leq t)$$

where $F$ is an operator valued nonlinear functional of $\rho_1(s), s \leq t$. This equation can be solved upto $O(\delta)$ using first order perturbation theory:

$$\rho_1(t) = U(t)\rho_1(0)U(t)^* + \delta. \int_0^t U(t-\tau)F_\tau(\rho_1(s), s \leq \tau)U(t-\tau)^* d\tau$$

where

$$U(t) = exp(-itH_1)$$

If we consider the Hamiltonian to comprise of an interaction between the particles and an electromagnetic field, then we can write

$$H_1 = (p_1 + eA(t,r))^2/2m - e\Phi(t,r) \approx p_1^2/2m - e\Phi(t,r) + (e/2m)((p_1, A) + (A, p_1))$$

and the particle interaction potential as

$$V_{12} = V(|r_1 - r_2|)$$

In that case, in the position representation, we have

$$[p_1^2, \rho_1] = [p_1, \rho_1].p_1 + p_1.[p_1, \rho_1]$$

and noting that $p_1$ is represented by the kernel

$$p_1(r, r') = -i\nabla_r \delta^3(r - r') = i\nabla'_r \delta^3(r - r')$$

we get

$$[p_1, \rho_1](r, r') = -i\nabla_r \rho_1(r, r') + i\nabla'_r \rho_1(r, r')$$

$$[p_1, \rho_1].p_1(r, r') = \nabla_r.\nabla'_r \rho_1(r, r') - \nabla'^2_r \rho_1(r, r')$$

and likewise,

$$p_1.[p_1, \rho_1](r, r') = \nabla_r.\nabla'_r \rho_1(r, r') - \nabla^2_r \rho_1(r, r')$$

$$[(A, p_1), \rho_1](r, r') = [A_k p_{1k}, \rho_1](r, r') =$$

$$A_k(t,r)p_{1k}\rho_1(r,r') - \int \rho_1(r,r'')A_k(t,r'')p_{1k}(r'',r')dr''$$

$$= -i(A(t,r),\nabla_r)\rho_1(r,r') - i(\nabla'_r, A(t,r'))\rho_1(r,r'))$$

Therefore with neglect of nonlinear terms in the electromagnetic field, our position space representation of the quantum Boltzmann dynamics of the single particle density operator is given by

$$i\partial_t\rho_1(t,r,r') = (2m)^{-1}(2\nabla_r.\nabla'_r\rho_1(t,r,r') - (\nabla_r^2 + \nabla_r'^2)\rho_1(t,r,r'))$$
$$-(i/m)(A(t,r),\nabla_r)\rho_1(t,r,r') - i(\nabla'_r, A(t,r'))\rho_1(t,r,r')$$
$$-e\Phi(t,r)\rho_1(t,r,r') + e\Phi(t,r')\rho_1(t,r,r')$$
$$+nonlinearterms.$$

Assume that in principle, we have solved this equation for $\rho_1(t,r,r')$

Remark:

$$Tr_2[V_{12}\rho_1 \otimes \rho_1](r_1,r_1') =$$
$$V(r_1,r_2)\delta(r_1 - r_1'')\delta(r_2 - r_2')\rho_1(t,r_1'',r_1')\rho_1(t,r_2',r_2)d^3r_2'd^3r_2d$$
$$= \int V(r_1,r_2)\rho_1(t,r_1,r_1')\rho_1(t,r_2,r_2)d^3r_2$$

and likewise,

$$Tr_2[(\rho_1 \otimes \rho_1)V_{12}](r_1,r_1') =$$
$$\int \rho_1(t,r_1,r_1')\rho_1(t,r_2,r_2)V(r_1',r_2)d^3r_2$$

Combining these two equations, we get

$$Tr_2[V_{12},\rho_1 \otimes \rho_1](r,r') =$$
$$\int (V(r_1,r_2) - V(r_1',r_2))\rho_1(t,r_1,r_1')\rho_1(t,r_2,r_2)d^3r_2$$

which means that our quantum Boltzmann equation assumes the following form for the single particle density kernel evolution:

$$i\partial_t\rho_1(t,r,r') =$$
$$(2m)^{-1}(2\nabla_r.\nabla'_r\rho_1(t,r,r') - (\nabla_r^2 + \nabla_r'^2)\rho_1(t,r,r'))$$
$$-(i/m)(A(t,r),\nabla_r)\rho_1(t,r,r') - i(\nabla'_r, A(t,r'))\rho_1(t,r,r')$$
$$-e\Phi(t,r)\rho_1(t,r,r') + e\Phi(t,r')\rho_1(t,r,r')$$
$$+(N-1)\int (V(r,r_2) - V(r',r_2))\rho_1(t,r,r')\rho_1(t,r_2,r_2)d^3r_2$$

and we may replace $V(r,r')$ by $V(|r-r'|)$ in the case when the interaction potential between two particles is a only a function of the distance between them.

In principle, using perturbation theory, this quantum Boltzmann equation can be solved for to obtain the single particle density operator kernel $\rho_1(t, r, r')$ as a function of the electromagnetic field and then the average dipole moment of the electron in this electromagnetic field will be given by

$$\mathbf{p}(t) = \int (-e\mathbf{r})\rho_1(t, \mathbf{r}, \mathbf{r})d^3r$$

and its average magnetic moment by

$$\mathbf{m}(t) = \int (-e\mathbf{L}(r, r')/2m)\rho_1(t, \mathbf{r}', \mathbf{r})d^3rd^3r'$$

where $\mathbf{L}(r, r')$ is the kernel of the orbital angular momentum operator in the position representation. This would then solve the problem of explaining the origin of permittivity and permeability of the plasma from the quantum statistical mechanical point of view. The angular momentum kernel is obtained as follows:

$$L_x(r, r') = (yp_z - zp_y)(r, r') = -iy\delta'(z-z')\delta(x-x')\delta(y-y') + iz\delta(x-x')\delta'(y-y')\delta(z-z')$$

$$L_y(r, r') = (zp_x - xp_z)(r, r') = -iz\delta'(x-x')\delta(y-y')\delta(z-z') + ix\delta(x-x')\delta(y-y')\delta'(z-z')$$

$$L_z(r, r') = (xp_y - yp_x)(r, r') = -ix\delta(x-x')\delta'(y-y')\delta(z-z') + iy\delta'(x-x')\delta(y-y')\delta(z-z')$$

where

$$r = (x, y, z), r' = (x', y', z')$$

Then the components of the average magnetic moment can be expressed as

$$m_x(t) = \int (-eL_x/2m)(r, r')\rho_1(t, r', r)d^3r'd^3r$$

$$= (ie/2m)\int (y\delta'(z-z')\delta(x-x')\delta(y-y') - z\delta(x-x')\delta'(y-y')\delta(z-z'))\rho_1(t, r', r)d^3r'd^3r$$

$$= (ie/2m)[\int [-y\partial_z\rho_1(t, r', r)|_{r'=r} + z\partial_y\rho_1(t, r', r)|_{r'=r}]dxdydz$$

and likewise for the other components. We could arrange these calculations in vectorial notation:

$$\mathbf{L}(r, r') = -i(r \times \nabla)(r, r') = -ir \times \nabla_r\delta(r - r')$$

and hence

$$\int \mathbf{L}(r, r')\rho_1(t, r', r)d^3rd^3r' = -i\int r \times (\nabla_r\delta(r - r'))\rho_1(t, r', r)d^3rd^3r'$$

$$= -i\int r \times \nabla_1\rho_1(t, r, r)d^3r$$

where $\nabla_1$ stands for the gradient w.r.t the first argument:

$$\nabla_1\rho_1(t,r,r) = (\nabla_{r_1}\rho_1(t,r_1,r))|_{r_1=r}$$

Perturbative solution of the Boltzmann equation :

$$i\partial_t\rho_1(t) = [H_1,\rho_1(t)] + \delta(N-1)Tr_2[V_{12},\rho_1(t)\otimes\rho_1(t)]$$

where $\delta$ is a perturbation parameter. We've also observed that an additional term can be added to this equation to improve its accuracy, namely,

$$\delta^2(N-1)Tr_2[V_{12},g_{12}]$$

where

$$g_{12}(t) =$$

$$= -i\int_0^t [[V_{12},\rho_1\otimes\rho_1] + (N-2)Tr_3[V_{13}+V_{23},\rho_1\otimes\rho_1\otimes\rho_1]$$

$$-(N-1)Tr_2[V_{12},\rho_1\otimes\rho_1]\otimes\rho_1$$

$$-(N-1)\rho_1\times Tr_2[V_{12},\rho_1\otimes\rho_1]](s)ds$$

$$= \int_0^t F_2(\rho_1(s)\otimes\rho_1(s),\rho_1(s)\otimes\rho_1(s)\otimes\rho_1(s))ds$$

so our perturbative solution would have the form

$$\rho_1(t) = U(t)\rho_1(0)U(t)^* + \delta.\int_0^t U(t-s)F_1(\rho_1(s)\otimes\rho_1(s))U(t-s)^*ds$$

$$+\delta^2(N-1)\int_0^t U(t-s)[\int_0^s F_2(\rho_1(\tau)\otimes\rho_1(\tau),\rho_1(\tau)\otimes\rho_1(\tau)\otimes\rho_1(\tau))d\tau]U(t-s)^*ds$$

$$= \rho_1(t) = U(t)\rho_1(0)U(t)^* + \delta.\int_0^t U(t-s)F_1(\rho_1(s)\otimes\rho_1(s))U(t-s)^*ds$$

$$+\delta^2(N-1)\int_{0<\tau<s<t} [U(t-s)F_2(\rho_1(\tau)\otimes\rho_1(\tau),\rho_1(\tau)\otimes\rho_1(\tau)\otimes\rho_1(\tau))d\tau]U(t-s)^*]dsd\tau$$

where

$$F_1(\rho_1\otimes\rho_1) = (N-1)Tr_2[V_{12},\rho_1\otimes\rho_1]$$

**The quantum Boltzmann equation derived from the Dirac relativistic wave equation**

$$H(t) = (\alpha,-i\nabla+eA) + \beta m - e\Phi = H_0 + H_I(t)$$

where

$$H_0 = (\alpha, -i\nabla) + \beta m, H_I(t) = e(\alpha, A) - e\Phi$$

Let $V_{12} = V(r_1, r_2)$ be the interaction potential between two particles. Our aim is to formulate the Quantum Boltzmann Equation

$$i\partial_t \rho(t) = [H(t), \rho(t)] + (N-1)Tr_2[V_{12}, \rho(t) \otimes \rho(t)]$$

in the position space representation. We first note that $\rho(t, r, r')$ for each $t, r, r'$ is a $4 \times 4$ matrix and

$$[H_0, \rho] = [\alpha_k p_k + \beta m, \rho](r, r')$$

$$= \alpha_k(-i\partial_k \rho(t, r, r')) - i\partial'_k \rho(t, r, r')\alpha_k$$

and

$$[\beta, \rho](r, r') = \beta.\rho(t, r, r') - \rho(t, r, r')\beta$$

Further,

$$((\alpha, A)\rho)(r, r') = A_k(t, r)\alpha_k \rho(t, r, r') = (\alpha, A(t, r))\rho(t, r, r')$$

$$(\rho(\alpha, A))(r, r') = \rho(t, r, r')(\alpha, A(t, r'))$$

Thus,

$$[(\alpha, A), \rho](r, r') = (\alpha, A(t, r))\rho(t, r, r') - \rho(t, r, r')(\alpha, A(t, r'))$$

and likewise,

$$[\Phi, \rho](r, r') = (\Phi(t, r) - \Phi(t, r'))\rho(t, r, r')$$

## 2.11 Study project on quantum antennas

Consider a cavity resonator $B$ with boundary $\partial B$ inside which we have confined both an electromagnetic field with vector potential expansion

$$\mathbf{A}(t, \mathbf{r}) = \sum_n a(t, n)\mathbf{u}_n(\mathbf{r})$$

and a corresponding Dirac field expansion

$$\psi(t, \mathbf{r}) = \sum_n \mathbf{c}(t, n)v_n(\mathbf{r})$$

where the 3-vector valued basis functions $\mathbf{u}_n(\mathbf{r})$ are chosen so that the corresponding electric and magnetic fields satisfy the appropriate boundary conditions and the scalar basis functions $v_n(\mathbf{r})$ also satisfy the appropriate boundary conditions. Owing to the Hermitianity of the Laplacian and the Dirac Hamiltonian, these basis functions can be chosen to be orthonormal:

$$< \mathbf{u}_n, \mathbf{u}_m >= \delta_{nm}, < v_n, v_m >= \delta_{nm}$$

Since $divE = 0$ because there is no volume charge density and $E = -\nabla\Phi - \partial_t A$, if we adopt the Coulomb gauge, $\nabla^2\Phi = 0, divA = 0$ and hence $\Phi = 0$. Thus in the Coulomb gauge $E = -\partial_t A$ and $B = curlA$. The coulomb gauge condition $divE = 0$ implies $div\mathbf{u}_n = 0$. The total energy density in the electromagnetic field within the cavity is then

$$U_F = (1/2)\int_B [(\partial_t A)^2 + (curlA)^2]d^3x$$

and by using the orthonormality of the basis functions, we get

$$U_F = (1/2)\sum_n (\partial_t a(t,n))^2 + (1/2)\sum_n a(t,n)^2 \int_B (curl\mathbf{u}_n(\mathbf{r}))^2 d^3x$$

Remark: Within the cavity,

$$(\nabla^2 + \omega(n)^2)\mathbf{u}_n(\mathbf{r}) = 0$$

where the $\omega(n)'s$ are the characteristic frequencies of oscillation. Then

$$\int_B (curl\mathbf{u}_n(r), curl\mathbf{u}_m(r))d^3x$$

$$= -\int (\mathbf{u}_n, \nabla^2\mathbf{u}_m)(r)d^3x$$

$$= \omega(n)^2\delta_{n,m}$$

Thus,

$$U_F = (1/2)\sum_n ((\partial_t a(t,n))^2 + \omega(n)^2 a(t,n)^2)$$

Likewise, we can quantize the Dirac field and its energy (Hamiltonian) within the cavity is given by

$$U_D = \int_B \psi(t,r)^*(\alpha, -i\nabla) + \beta m)\psi(t,r)d^3x$$

Now, the Dirac equation gives

$$\sum_n i\partial_t \mathbf{c}(t,n)v_n(r) = H_D \sum_n \mathbf{c}(t,n)v_n(r)$$

where

$$H_D = (\alpha, -i\nabla) + \beta m$$

This gives

$$i\partial_t c_l(t,n) = \sum_{m,k} < v_n(r), H_{Dlk}v_m > c_k(t,m)$$

and then, the second quantized Dirac Hamiltonian is given by

$$U_D = \sum_{lknm} c_l(t,n)^* c_k(t,m) < v_n, H_{Dlk}v_m >$$

In order to interpret this formula in terms of electron-positron creation and annihilation operators, we require to first diagonalize the quadratic form $U_D$.

## 2.12 Test:Antennas and Wave Propagation

Question 3 is compulsory. Attempt any three questions from the remaining. Each question carries ten marks.

[1] Explain how using the principle of pattern multiplication, you will calculate the far field radiation pattern produced by a microstrip antenna designed as a cuboidal cavity of dimensions $a, b, d$ with the only non-vanishing component of the magnetic vector potential being $A_z(x, y, z)$ satisfying the Helmholtz equation

$$(\nabla^2 + k^2)A_z = 0$$

with the boundary condition that the tangential components of the electric field and normal component of the magnetic field vanish on the boundaries. Derive an explicit formula for the far field radiation pattern.

[2] A planar Archimedian spiral antenna has the equation of its curve given by

$$\rho = A.exp(b\phi), z = 0$$

where $(\rho, \phi, z)$ are cylindrical coordinates. This antenna carries a current $I$ at frequency $\omega$. Calculate the radiation resistance of this antenna assuming that $N$ complete spirals are made. It is known that if all the dimensions are generated by simply rotating an antenna, then the antenna is broadband. Justify this for the Archimedian spiral.

[3] A horn antenna consists of a rectangular waveguide with transverse dimensions $a, b$ feeding into a horn having a spherical aperture of radius $R$ with centre at the centre of the mouth of the waveguide assumed to be on the xy plane and extending from azimuthal angle$\theta = 0$ to $\theta = \alpha$. Determine the following:

[a] The distance $\delta(x, y, \theta)$ between a point $(x, y, 0)$ at the mouth of the guide and a point $(R, \theta, \phi)$ on the spherical horn surface. Approximate this distance upto quadratic terms in $x, y$ assuming that $a, b << R$.

[b] Derive expressions for the fields $E_x(x, y), E_y(x, y), H_x(x, y), H_z(x, y)$ at the waveguide mouth assuming a $TE_{m,n}$ mode.

[c] Calculate using the phase shift principle, the fields at the horn aperture using the approximations made in [a]. In other words you must note that these field are given by

$$\int_0^a \int_0^b \mathbf{E}(x, y)exp(-ik\delta(x, y, \theta))dxdy,$$

and

$$\int_0^a \int_0^b \mathbf{H}(x, y)exp(-ik\delta(x, y, \theta))dxdy$$

[d] Using the formulas for the electromagnetic fields on the horn surface derived in [c], calculate the Poynting vector pattern in the far field zone by using the fact that the surface electric current density on the horn surface is

$\mathbf{J}_s = \hat{r} \times \mathbf{H}$ and the surface magnetic current density on the same is $\mathbf{M}_s = -\hat{r} \times \mathbf{E}$.

[4] Let $E(x, y)$ denote the electric field on a planar aperture $D$ in the xy plane. justify the statement based on the equation of wave propagation that the field in space at the point $(X, Y, Z)$ produced by this aperture must be of the form

$$F(X, Y, Z) = \int f(k_x, k_y) exp(i(k_x X + k_y Y + \sqrt{k^2 - k_x^2 - k_y^2 Z})) dk_x dk_y$$

where

$$k = \omega / c$$

with $\omega$ being the frequency of operation. Explain how you would evaluate the function $f(k_x, k_y)$ so that the boundary condition

$$F(X, Y, 0) = E(X, Y), (X, Y) \in D$$

is satisfied. Describe an approximate method for evaluating $F$ based on the method of stationary phase, ie, by selecting a contribution to the radiated field pattern from only a single $(k_x, k_y)$ at which the above integrand is stationary w.r.t small variations. Explain why you can make such an approximation.

[5] Consider a parabolic reflector antenna whose surface is defined by the equation

$$\rho^2 = 4az, \rho^2 = x^2 + y^2$$

[a] Prove that its focus is at $(0, 0, a)$ and directrix is the plane $z = -a$ by showing that if $(x, y, z)$ is any point on the parabolic surface, then the distance of this point from the focus equals its distance from the directrix.

[b] Prove that if a ray of light starting from the focus is incident upon the parabolic reflector, then the reflected ray is parallel to the $z$ axis. Do this problem by applying Snell's law of reflection to the light ray, namely [i] the anglse made by the reflected ray and the incident ray with the normal to the surface at the point of incidence are equal and [ii] the incident and reflected rays and the normal fall in one plane.

[c] If an plane electromagnetic wave propagates from the focus to the parabolic surface, then by applying the boundary conditions at the reflected surface assuming no transmitted ray, calculate the electromagnetic field in the reflected ray.

## 2.13 Some additional remarks about refractive index computation using classical statistical mechanics

Let the fluid consist of electric and magnetic dipoles and let $f(t,r,v,p,p',m)d^3rd^3vd^3pd^3p'd^3m$ denote the number of particles of the fluid that have positions in the volume $d^3r$, velocity in the volume $d^3v$, electric dipole moment in the volume $d^3p$, dipole rate of change in the volume $d^3p'$ and magnetic dipole moment in the volume $d^3m$. The equations of motion for these variables are

$$dr/dt = v, dv/dt = -\gamma v + Q(E(t,r) + v \times B(t,r)),$$
$$dL/dt = p \times E + m \times B,$$
$$m = QL/2m_0,$$
$$d^2\xi/dt^2 = -\gamma d\xi/dt - U'(\xi) + Q(E(t,r) + v \times B(t,r))$$
$$p = Q\xi$$

Note that $L$ is the orbital angular momentum of the charge $Q$ which is assumed to be an electron ($Q = -e$) moving in the nuclear electrostatic field of an atom and $\xi$ is the position of the electron w.r.t the nucleus of the atom of which it is a part. Thus, we can write

$$dm/dt = (Q/2m_0)(p \times E(t,r) + m \times B(t,r)),$$
$$d^2p/dt^2 = -\gamma.dp/dt - U'(p/Q) + Q^2(E(t,r) + v \times B(t,r))$$

Formally, we can formulate a classical Boltzmann equation for the rate of change of $f$ taking into account a collision term and then by solving it along with the Maxwell equations

$$divE = \rho/\epsilon_0, divB = 0,$$
$$curlE = -\partial_t B, curlB = \mu_0 J + \mu_0\epsilon_0\partial_t E$$

where

$$J(t,r) = \partial_t P(t,r) + \nabla \times M(t,r)$$
$$\rho(t,r) = -divP(t,r)$$

with

$$P(t,r) = \int pf(t,r,v,p,p',m)d^3vd^3pd^3p'd^3m,$$
$$M(t,r) = \int m.f(t,r,v,p,p',m)d^3vd^3pd^3p'd^3m$$

Note that since an atom including its electrons is neutral, we are assuming no free charges in the fluid. The charge density of the fluid comes only from polarized atoms and as is well known in classical electrodynamics, it is determined from the polarization in accordance with the above equation. Likewise, the only current density in the fluid comes from the polarization and magnetization currents of each atom, no contribution from free charges. Let us now see how these equations can be formulated from the quantum mechanical viewpoint.

## 2.14 Lecture on Antenna Theory

[1] Calculating the path of a light ray in a medium having an inhomogeneous refractive index.

$n(x, y, z)$ is the refractive index. $c$ is the speed of light in vacuum. Total time taken by a light ray in going from point $P_1 = (x_1, y_1, z_1)$ to the point $P_2 = (x_2, y_2, z_2)$ is given by

$$cT(1,2) = \int_1^2 nds = \int_{s_1}^{s_2} n(x(s), y(s), z(s))\sqrt{x'(s)^2 + y'(s)^2 + z'(s)^2}ds$$

Minimize this time (Fermat's principle of least time) with the boundary conditions

$$(x(s_k), y(s_k), z(s_k)) = (x_k, y_k, z_k), k = 1, 2$$

using the calculus of variations, ie, the Euler-Lagrange equations.

[2] Horn antenna: Distance of a point $(x, y, 0)$ on the waveguide mouth to the point $(R.cos(\phi)sin(\theta), R.sin(\phi)cos(\theta), R.cos(\theta))$ on the horn surface is given by

$$d^2 = (R.cos(\phi)sin(\theta) - x)^2 + (R.sin(\phi).sin(\theta) - y)^2 + (R.cos(\theta))^2$$

$$= R^2 + (x^2 + y^2) - 2R(x.cos(\phi) + y.sin(\phi))sin(\theta)$$

Using the Binomial approximation with $R >> a^2 + b^2$ gives us

$$d = d(\theta, \phi, x, y) =\approx R + (x^2+y^2)/2R - (x.cos(\phi)+y.sin(\phi)) + (x.cos(\phi)+y.sin(\phi))^2 sin^2(\theta)/2R$$

Derive the fields on the horn surface as follows: Let $\mathbf{E}(x, y), \mathbf{H}(x, y)$ be the fields on the waveguide mouth surface. These are calculated using the formula

$$(\nabla_\perp^2 + h(m,n)^2)E_{zm,n}(x, y) = 0$$

$$(\nabla_\perp^2 + h(m,n)^2)H_{zm,n}(x, y) = 0$$

$$h(m,n)^2 = \pi^2(m^2/a^2 + n^2/b^2)$$

with the boundary condition

$$E_z(x, y) = 0, x = 0, a, y = 0, b, \partial H_z(x, y)/\partial x = 0, x = 0, a, \partial H_z(x, y)/\partial y = 0, y = 0, b$$

so that

$$E_{z,m,n}(x, y) = C(m,n)sin(m\pi x/a)sin(n\pi y/b)$$

$$H_{z,m,n}(x, y) = D(m,n)cos(m\pi x/a).cos(n\pi y/b)$$

The total $z$-component of the electric field at a fixed frequencey is then

$$E_z(x, y, z) = \sum_{m,n} E_{z,m,n}(x, y)exp(-\gamma(m,n)z)$$

and the total $z$-component of the magnetic field is

$$H_z(x,y,z) = \sum_{m,n} H_{z,m,n}(x,y)exp(-\gamma(m,n)z)$$

The transverse ($x$ and $y$) components of the waveguide fields at its mouth are given by

$$E_\perp(x,y,z) = \sum_{m,n} E_{\perp,m,n}(x,y)exp(-\gamma(m,n)z),$$

$$H_\perp(x,y,z) = \sum_{m,n} H_{\perp,m,n}(x,y)exp(-\gamma(m,n)z)$$

where

$$E_{\perp,m,n}(x,y,z) = (-\gamma(m,n)/h(m,n)^2)\nabla_\perp E_{z,m,n}(x,y)$$
$$-(j\omega\mu/h(m,n)^2)\nabla_\perp H_{z,m,n}(x,y) \times \hat{z}$$
$$H_{\perp,m,n}(x,y,z) = (-\gamma(m,n)/h(m,n)^2)\nabla_\perp H_{z,m,n}(x,y)$$
$$+(j\omega\mu/h(m,n)^2)\nabla_\perp H_{z,m,n}(x,y) \times \hat{z}$$

where

$$\gamma(m,n)^2 + h(m,n)^2 = \omega^2\mu_0\epsilon_0$$

We can write

$$\mathbf{E}(x,y,z) = \sum_{m,n} C_1(m,n)\mathbf{u}_{mn}(x,y)exp(-\gamma(m,n)z)$$

$$\mathbf{E}(x,y,z) = \sum_{m,n} D_1(m,n)\mathbf{v}_{mn}(x,y)exp(-\gamma(m,n)z)$$

where $\mathbf{u}_{mn}(x,y)$ form an orthonormal sequence in $L^2([0,a]\times[0,b])\otimes\mathbb{C}^3$ and so do $\mathbf{v}_{mn}(x,y)$.

Note that the waveguide fields at its mouth are obtained by setting $z=0$. To evaluate the fields at the horn surface, we must evaluate the integral

$$\int_0^d \int_0^b \mathbf{E}(x,y,0)exp(-jd(\theta,\phi,x,y))dxdy$$

[3] Let $a_n, n \geq 1$ be a sequence of commuting annihilation operators so that

$$[a_n, a_m^*] = \delta(n,m)$$

Let $\phi_n(t,r)$ be functions of time and space and assume that they satisfy

$$\sum_n \phi_n(t,r)\phi_n(s,r')^* = min(t,s)K(r-r')$$

Construct a space-time process

$$A(t,r) = \sum_n [a_n \phi_n(t,r)],$$

so that

$$A(t,r)^* = \sum_n [a_n^* \phi_n(t,r)^*]$$

Show that

$$[A(t,r), A(s,r')^*] = min(t,s)K(r-r')$$

Hence, deduce that if $d$ denotes time differential, then

$$dA(t,r).dA(t,r')^* = K(r-r')dt$$

which is the quantum noise-field theoretic generalization of the celebrated Quantum Ito formula of Hudson and Parthasarathy. Now, assume that

$$F(t,r) = A(t,r) + A(t,r)^*$$

is the $z$-component of the magnetic vector potential in space-time. Calculate the electric and magnetic fields $\mathbf{E}(t,r), \mathbf{H}(t,r)$ corresponding to this vectro potential and evaluate in a coherent state $|\phi(u)>$, the following averages

$$<\phi(u)|F(t,r)|\phi(u)> = <\phi(u)|F(t,r)F(s,r')|\phi(u)>$$

and

$$<\phi(u)|\mathbf{E}(t,r) \otimes \mathbf{E}(s,r')|\phi(u)>$$

$$<\phi(u)|\mathbf{H}(t,r) \otimes \mathbf{H}(s,r')|\phi(u)>$$

$$<\phi(u)|\mathbf{E}(t,r) \otimes \mathbf{H}(s,r')|\phi(u)>$$

# Chapter 3

# Probability Theory

## 3.1 Syllabus

[0] Some philosophical remarks on probability theory: Why probabilistic models are required to simplify calculations involving very complex deterministic dynamical systems ? The Buffon Needle problem and its application to the Monte-Carlo calculation of $\pi$.

[1] A.N.Kolmogorov's axiomatic foundations of probability theory: The sample space, $\sigma$-algebra of events and probability measure; the notion of a classical probability space $(\Omega, \mathcal{F}, P)$. Importance of the countable additivity postulate for the probability measure.

[2] Properties of the probability measure.

[3] The notion of independence of events. The Borel-Cantelli lemmas.

[4a] The general definition of a random variable on a probability space. Joint probability distributions and their properties.

[4b] Lebesgue integration in a probability space: The notion of expectation of a random variable.

[4c] Cornerstone theorems of Lebesgue integration theory: Monotone convergence theorem, Fatou's Lemma, dominated convergence theorem.

[5] Statement of the Caratheodory extension theorem: Extension of a probability measure as a countably additive measure on an algebra of events to the $\sigma$-algebra generated by the algebra.

[6] The product of probability spaces: Notion of the product measure and its application to the construction of independent experiments, Fubini's theorem on integration w.r.t a product probability measure.

[7] Examples of probability spaces from die throwing to coin tossing.

[8] Absolute continuity of two measures and the Radon-Nikodym theorem.

[9] Application of the Radon-Nikodym theorem to the construction of the conditional expectation of a random variable given a sub $\sigma$-algebra.

[10] Another derivation of the conditional expectation using orthogonal projection operators in a Hilbert space.

[11] Properties of the conditional expectation.

[12] Application of the conditional expectation to the construction of the minimum mean square nonlinear estimate of a random variable given a family of random variables.

[13] Application of the Radon-Nikodym theorem to the construction of probability density of a finite set of random variables.

[14] Describing discrete probability distributions using the Dirac $\delta$-distribution.

[15] Estimation of parameters in linear models using linear minimum mean square methods.

[16] Joint characteristic function of a finite set of random variables.

[17] Positive definite properties of the characteristic function and Bochner's theorem.

[18] Jensen's inequality for convex functions of random variables.

[19] Chebyshev's inequality, Markov's inequality.

[20] [a] Various notions of convergence of an infinite sequence of random variables. Convergence almost surely, convergence in probability, convergence in the mean square sense, convergence in $L^p$-norm, convergence in distribution. The relationship between these modes of convergence.

[b] The weak and strong laws of large numbers for sequences of independent random variables.

[c] The Gaussian distribution and the central limit theorem: Proof based on the use of the characteristic function.

[21] Definition of a stochastic process in discrete time and continuous time.Kolmogorov's existence theorem for stochastic processes in discrete time and in continuous time.

[22] The AR,MA and ARMA time series models.

[23] Transmission of stochastic processes through linear and nonlinear filters. Derivation of differential and difference equations satisfied by the output moments and the input output cross moments in terms of the input moments.

[24] Stationary and Wide sense stationary processes.

[25] Von-Neumann's $L^2$-ergodic theorem, Birkhoff's individual ergodic theorem and ergodicity of a measure preserving transformation with applications to stochastic processes.

[26] Autocorrelation, spectrum, higher order spectra and the causal and non-Causal Wiener filters.

[27] Nonlinear filtering and the Kalman and extended Kalman filters for real time filtering.

[28] Simulation of random variables on a computer by transformation of a uniformly distributed random variable.

[29] The Brownian motion, Poisson process and some of their properties.

[30] Stochastic integration w.r.t Brownian motion and stochastic differential equations driven by the Brownian motion process.

[31] Martingales and their properties. Doob's inequality for Martingales, the Martingale downcrossing inequality and the Martingale convergence theorem.

[32] Stochastic integration w.r.t a Martingale.

[33] Examples of Martingales.

[34] An introduction to large deviation theory with applications to the diffusion exit problem and stabilization of stochastic differential equations with feedback controllers.

[35] Stochastic processes in robotics:

[a] The $d$-link robot equation.

[b] The $d$-link robot equation with 3-D links–analysis using Lie group theory.

[c] Large deviation control of 3-D link robots.

[35] Markov chains and the Chapman-Kolmogorov equations. Examples including the pure birth process, the birth-death process, the telegraph process. The stationary distribution of a Markov chain.

[36] Derivation of the Fokker-Planck equations for a continuous state space Markov process from Ito's stochastic differential equation.

[37] Approximation of the Boltzmann kinetic transport equation for a plasma by the Fokker-Planck equation.

[38] An introduction to probability in quantum mechanics.

[a] Interference of wave functions.

[b] Interpretation of quantum probabilities using Feynman's path integral formula for the probability amplitude.

[c] Transition probabilities in quantum mechanics.

[d] Transition probabilities when the system Hamiltonian is perturbed by a time varying Hamiltonian-Development of time dependent perturbation theory.

[e] The quantum mechanical harmonic oscillator and its application to the construction of the Boson Fock space.

[f] The creation, annihilation and conservation processes of Hudson and Parthasarathy in Boson Fock space.

[g] Quantum stochastic integration and quantum stochastic differential equations in hte Hudson-Parthasarathy formalism for describing the evolution of quantum systems in the presence of quantum noise.

References:

[1] A.Papoulis, "Probability Theory, Random Variables and Stochastic Processes".

[2] William Feller, "An introduction to probability theory and its applications, vol.I and II", John Wiley.

[3] K.R.Parthasarathy, "An introduction to probability and measure", Hindustan Book Agency.

[4] K.R.Parthasarathy, "An introduction to quantum stochastic calculus", Birkhauser, 1992.

[5] Harish Parthasarathy, "Developments in Mathematical and Conceptual Physics:Concepts and Applications for Engineers", Springer Nature, 2020.

[6] I.Karatzas and S.Shreve, "Brownian motion and stochastic calculus", Springer.

## 3.2 The basic axioms of Kolmogorov

A triplet $(\Omega, \mathcal{F}, P)$ is called a classical probability space for an experiment where $\Omega$ is the sample space, namely a set whose elements are called elementary outcomes of the experiment, $\mathcal{F}$ is a $\sigma$-field of subsets of $\Omega$ and $P : \mathcal{F} \rightarrow [0,1]$ is a probability measure. By a $\sigma$-field, we mean that it is closed under countable unions and complementation (and hence also under countable intersections (by De-Morgan's rule $\bigcap_n E_n = (\bigcup_n E_n^c)^c$), and therefore it also contains the sample space as well as the nullset. The elements of $\mathcal{F}$ (which are subsets of $\Omega$) are called the events of the experiment. If $E \in \mathcal{F}$, we say that the event $E$ has occurred if on performing the experiment, the elementary outcome $\omega \in E$. We say that the event $E$ has not occurred, ie $E^c$ has occurred if the elementary outcome $\omega \in E^c$, or equivalently $\omega \notin E$. If $E_n, n = 1, 2, ...$ is a finite or infinite sequence of events, then the finite/countable union $\bigcup_n E_n$ is an event by hypothesis and this event is said to have occurred if the elementary outcome $\omega$ is in at least one of the $E_n's$. Likewise, if $\omega$ is in all the $E_n's$, ie $\omega \in \bigcap_n E_n$, then we say that all the events $E_n, n = 1, 2, ...$ have occurred. Note that $\mathcal{F}$ need not be closed under arbitrary unions/intersections. The reason for this is seen when we define the probability measure $P$ as a countably additive set function on $\mathcal{F}$ (which means that $E = \bigcup_n E_n, E_n \cap E_m = \phi \forall n \neq m$ imply $P(E) = \sum_n P(E_n)$) such that $P(\Omega) = 1$ and hence $P(\phi) = 0$. Now suppose we assumed that $\mathcal{F}$ is closed under arbitrary unions, not necessarily countable and then we also make $P$ additive under uncountable unions of disjoint events. Then, we run into trouble as the following example shows: Let $P$ be the uniform distribution on the closed interval $[0,1]$. Then $P([0,1]) = 1$. However $P(\{x\}) = 0$ for any single point $x \in [0,1]$ for $P$ by definition is given by $P([a,b]) = b - a$ for $0 \leq a \leq b \leq 1$. On the other hand, uncountable additivity of $P$ would result in

$$1 = P([0,1]) = \sum_{x \in [0,1]} P(\{x\}) = \sum_{x \in [0,1]} 0 = 0$$

which is absurd. That is the reason why we have to be content with $\mathcal{F}$ being closed under countable unions and $P$ being countably additive on $\mathcal{F}$.

## 3.3 Exercises

[1] Show that if $A, B \in \mathcal{F}$ and $A \subset B$, then

$$P(A) \leq P(B)$$

[2] Show using countable additivity of $P$ on $\mathcal{F}$ that if $E_n \in \mathcal{F}, n = 1, 2, ...$ and $E_n \uparrow E$, ie, $E_n \subset E_{n+1} \forall n \geq 1$ and $\bigcup_{n \geq 1} E_n = E$, then

$$P(E_n) \uparrow P(E)$$

Conversely, show that if $P$ is finitely additive on $\mathcal{F}$ and this property holds then $P$ is countably additive on $\mathcal{F}$.

hint: Let $E_0 = \phi$ and define

$$F_n = E_n - E_{n-1} = E_n \cap E_{n-1}^c, n \geq 1$$

then the $F_n's$ are pairwise disjoint events and

$$\bigcup_n F_n = \bigcup_n E_n = E$$

Apply now the countable additivity property of $P$ and use the fact that

$$P(F_n) = P(E_n) - P(E_{n-1})$$

[3] Show that if $E_n \downarrow E$ (all being events), ie, $E_{n+1} \subset E_n \forall n$ and $E = \bigcap_n E_n$, then

$$P(E_n) \downarrow P(E)$$

hint: $E_n \downarrow E$ iff $E_n^c \uparrow E^c$. Now use the result of the previous exercise. Conversely show that if $P$ is finitely additive on $\mathcal{F}$ and this property holds good, then $P$ is countably additive.

[4] Study project on the Caratheodory extension theorem. Let $\mathcal{B}$ field, ie, a collection of $\Omega$-subsets that is closed under finite unions and complementation and if $P$ is a countably additive probability measure on $\mathcal{B}$ (ie, $E_n \in \mathcal{B}, n \geq 1$, $E_n \cap E_m = \phi \forall n \neq m$ and $E = \bigcup_n E_n \in \mathcal{B}$, then $P(E) = \sum_n P(E_n)$), then $P$ has a unique countably additive extension to a probability measure $P_0$ on the $\sigma$-field $\mathcal{F} = \sigma(\mathcal{B})$ generated by $\mathcal{B}$. By unique countable extension, we mean that (a) $P_0$ is a probability measure on $\mathcal{F}$ and (b) $P_0(E) = P(E) \forall E \in \mathcal{B}$

[5] If $X_n$ is a bounded sequence of random variables such that $X_n \leq X_{n+1} \forall n$, then $X_n$ increases to a limit $X$. Show that for $X$ to be measurable, ie, a random variable in general, we require $\mathcal{F}$ to be a $\sigma$-field just being a field will not suffice.

hint: The set $\{\omega : X(\omega) \in (a, b]\}$ is the increasing limit of the events $\{\omega : X_n(\omega) \in (a, b]\}, n = 1, 2, ...$, in particular the former is the countable union of the latter. Hence for the former to be measurable, ie, an event, the class $\mathcal{F}$ of events must be closed under countable unions. Further, if we require the continuity condition

$$P(X \in (a, b]) = lim_n P(X_n \in (a, b])$$

then $P$ must be countably additive in general, finite additivity will not suffice.

[6] Let $(\Omega_k, \mathcal{F}_k, P_k), k = 1, 2, ..., r$ be probability spaces. Let

$$\Omega = \Omega_1 \times \Omega_2 \times ... \times \Omega_r$$

and let $\mathcal{F}$ be the $\sigma$ field on $\Omega$ generated by the measurable rectangles, ie, by sets of the form $E_1 \times E_2 \times .... \times E_r$ with $E_m \in \mathcal{F}_m, m = 1, 2, ..., r$. Let $\mathcal{B}$ denote the field consisting of finite disjoint unions of such rectangles (Show that this is indeed a field). It is clear that $\mathcal{F}$ is the $\sigma$-field generated by $\mathcal{B}$. Prove using the countable additivity of $P_k$ on $\mathcal{F}_k, k = 1, 2, ..., r$ that the finitely additive set function $P$ defined on $\mathcal{B}$ by

$$P(A_1 \cup A_2 \cup ... \cup A_r) = P(A_1) + ... + P(A_r)$$

where $A_1, ..., A_r$ are disjoint measurable rectangles and if $A = E_1 \times .... \times E_r$ is a measurable rectangle, then

$$P(A) = P_1(E_1)...P_r(E_r)$$

is also countably additive on $\mathcal{B}$ and hence use Caratheodory's extension theorem to deduce that $P$ extends to a unique probability measure $P_0$ on $\mathcal{F}$ (When we say probability measure, we mean that it should be countably additive).

Note: In order to show that $P$ is countably additive on $\mathcal{B}$, it suffices to show that if $E_n \in \mathcal{B}, E_n \downarrow \phi$ then $P(E_n) \downarrow 0$.

[7] The Kolmogorov existence theorem for stochastic processes. Let $F_n(x_1, ..., x_n), n \geq 1$ be a consistent family of probability distributions on $\mathbb{R}^n, n = 1, 2, ...$ respectively. Then, if

$$\mathcal{B} = \bigcup_{n \geq 1} (\mathcal{B}(\mathbb{R}^n) \times \mathbb{R}^{\mathbb{Z}_+})$$

prove that $\mathcal{B}$ is a field in $\mathbb{R}^{\mathbb{Z}_+}$.

## 3.4 More Exercises

[1] if $X_1, ..., X_n, ...$ is a sequence of random variables on a probability space $(\Omega, \mathcal{F}, P)$, then show that if we define the joint probability distribution function of the first $n$ r.v's by

$$F_n(x_1, ..., x_n) = P(X_1 \leq x_1, ..., X_n \leq x_n) = P(\bigcap_{k=1}^{n} X_k^{-1}((-\infty, x_k])), x_1, ..., x_n \in \mathbb{R}$$

then $F_n, n \geq 1$ has the following properties:

$$lim \downarrow x_i y_i F_n(x_1, .., x_i, .., x_n) = F_n(x_1, ..., y_i, ..., x_n)$$

ie, $F_n$ is right continuous in each of its arguments. To prove this, make use of the fact that $[-\infty, y_i) = lim x_i \downarrow y_i(-\infty, x_i]$ and hence $lim x_i \downarrow y_i X_i^{-1}((-\infty, x_i]) = X_i^{-1}((-\infty, y_i])$. Hence, deduce that

$$lim x_i \downarrow y_i \bigcap_{k=1}^{n} X_k^{-1}((-\infty, x_k]) = X_1^{-1}((-\infty, x_1]) \bigcap ... \bigcap X_i^{-1}((-\infty, y_i]) \bigcap ... \bigcap X_n^{-1}((-\infty, x_n])$$

Then, make use of the continuity of the probability measure $P$ (which is a consequence of the countable its additivity) to deduce the result.

## 3.5 Exercises on stationary stochastic processes, spectra and polyspectra

[1] Let $X(t), t \in \mathbb{R}$ $(X(n), \in \mathbb{Z})$ be a stochastic process in continuous time (discrete time). The process is said to be stationary if the joint distribution of the random variables $(X(t), X(t+t_1), ..., X(t+t_k))$ does not depend on $t$ for any $k, t_1, ..., t_k$. In this case, define the $(k+1)^{th}$ order moments of the process as

$$M_X(t_1, ..., t_k) = \mathbb{E}(X(t)X(t+t_1)...X(t+t_k))$$

Show that this does not depend upon $t$, ie, the process is $(k+1)^{th}$-order stationary. Second order stationarity in particular means that the autocorrelation function $R_X(s) = \mathbb{E}(X(t)X(t+s))$ does not depend on $t$. Give an example of a process that is second order stationary but is not stationary.

Then, define its $k$-variate Fourier transform by

$$P_{X,k}(\omega_1, ..., \omega_k) = \int_{\mathbb{R}^k} M_X(t_1, ..., t_k) exp(-j(\omega_1 t_1 + ... + \omega_k t_k)) dt_1 ... dt_k$$

$P_{X,k}$ is called the $k^{th}$ order polyspectrum of the process $X$. In the discrete time case, we define it using the k-variate DTFT of the moment sequence rather than the continuous time FT, ie, CTFT.

Show that if $X(t)$ is passed through an LTI system with impulse response $h(t)$ so that its output is

$$Y(t) = \int_{\mathbb{R}} h(s)X(t-s)ds$$

or in discrete time,

$$Y(n) = \sum_{m \in \mathbb{Z}} h(m)X(n-m)$$

then

$$P_{Y,k}(\omega_1, ..., \omega_k) = H(\omega_1)...H(\omega_k)\bar{H}(\omega_1 + ... + \omega_k)P_{X,k}(\omega_1, ..., \omega_k), k = 1, 2, ...$$

In particular, show that

$$S_Y(\omega) = P_{Y,2}(\omega) = |H(\omega)|^2 S(X(\omega), S_X(\omega) = P_{X,2}(\omega)$$

A process that is both first and second order stationary is said to be wide sense stationary (WSS). Let $X(t)$ be a WSS process and define its time and ensemble average power by

$$W_X = lim_{T \to \infty} \mathbb{E} \frac{1}{T} \int_{-T/2}^{T/2} X(t)^2 dt$$

Prove using the Parseval theorem that

$$W_X = \frac{1}{2\pi}\int_{\mathbb{R}} S_X(\omega)d\omega$$

where

$$S_X(\omega) = lim_{T\to\infty}\frac{1}{T}\mathbb{E}|\hat{X}_T(\omega)|^2 = \int_{\mathbb{R}} R_X(s)exp(-j\omega s)ds$$

where

$$\hat{X}_T(\omega) = \int_{-T/2}^{T/2} X(t)exp(-j\omega t)dt$$

and

$$R_X(s) = \mathbb{E}(X(t)X(t+s))$$

For this reason $S_X(\omega)$ is called the power spectral density (PSD) of the WSS process $\{X(t) : t \in \mathbb{R}\}$. The above result, namely that the PSD of a WSS process is the Fourier transform of its autocorrelation function is called the Wiener-Khintchine theorem.

For proving this, you must assume that

$$lim_{|s|\to\infty}R_X(s) = 0$$

which in particular, is true if

$$\int_{\mathbb{R}} |R_X(s)|ds < \infty$$

[2] A research problem based on problem [1]. Explain how using measurements of the power spectral density of the input and output of an LTI system, you can estimate the magnitude $|H(\omega)|$ of the transfer function of the system and by using measurements of the polyspectrum of order $k$ where $k \geq 3$ of the input and outputs of the LTI system, we can also estimate the phase of the LTI system.

[3] Let $Z(\omega), \omega \in \mathbb{R}$ be a zero mean complex valued stochastic process on a probability space such that

$$\mathbb{E}(dZ(\omega).d\bar{Z}(\omega')) = (2\pi)^{-1}S(\omega)d\omega.\delta_{\omega,\omega'}$$

Now define a stochastic process $X(t)$ as the stochastic integral

$$X(t) = \int_{\mathbb{R}} exp(j\omega t)dZ(\omega)$$

Show that $X(t)$ is WSS with autocorrelation

$$R_X(s) = \mathbb{E}(X(t+s)\bar{X}(t)) = \int_{\mathbb{R}} exp(j\omega s)S(\omega)d\omega/2\pi$$

Hence deduce that

$$S(\omega) = 2\pi.\mathbb{E}(|dZ(\omega)|^2)/d\omega$$

is the power spectral density of the process $X(t)$. Conversely, given a WSS process $X(t)$, define a process $Z(\omega), \omega \in \mathbb{R}$ by the equation

$$Z(\omega_2)-Z(\omega_1) = (2\pi)^{-1} \int_{\mathbb{R}} \frac{(exp(-j\omega_2 t) - exp(-j\omega_1 t))}{-jt} X(t)dt, -\infty < \omega_1 < \omega_2 < \infty$$

Then show that formally we can write

$$dZ(\omega)/d\omega = (2\pi)^{-1} \int_{\mathbb{R}} exp(-j\omega t)X(t)dt, \omega \in \mathbb{R}$$

Note that when we are rigorous, this statement is true only if the complex measure on $\mathbb{R}$ defined by

$$\mu_Z((\omega_1, \omega_2]) = Z(\omega_2) - Z(\omega_1)$$

is absolutely continuous w.r.t the Lebesgue measure. Show that even if it is not so, but the limit

$$lim_{\omega_2 \to \omega_1} \mathbb{E}(|Z(\omega_2) - Z(\omega_1)|^2)/(\omega_2 - \omega_1)$$

$$= (2\pi)^{-1} S(\omega_1)$$

exists for all $\omega_1 \in \mathbb{R}$, then show also by virtue of the WSS property of $X(t)$ that we have the orthogonality relations

$$\mathbb{E}[(Z(\omega_2) - Z(\omega_1).(\bar{Z}(\omega_3) - \bar{Z}(\omega_4))] = 0$$

for

$$\omega_2 > \omega_1 \geq \omega_3 > \omega_4$$

and hence deduce the relation

$$R_X(s) = \mathbb{E}(X(t+s)\bar{X}(t)) = (2\pi)^{-1} \int_{\mathbb{R}} S(\omega)exp(j\omega s)d\omega$$

in the sense of Riemann-Stieltjes. In this case, show that we can define the integral

$$\int_{\mathbb{R}} exp(j\omega t)dZ(\omega)$$

in the $L^2$ sense as an $L^2$-limit of Riemann sums and that this $L^2$ limit equals $X(t)$.

## 3.6 Random measures

[4] This problem is a generalization of the previous problem. Let $(\Omega, \mathcal{F}, P)$ be a probability space and let $\mathcal{H} = L^2(\Omega, \mathcal{F}, P)$ denote the Hilbert space of all complex valued random variables $X$ on this probability space for which

$$\mathbb{E}|X|^2 = \int |X(\omega)|^2 dP(\omega) < \infty$$

Let $Z$ be a complex set function on a measurable space $(X, \mathcal{E})$ with the property that $Z$ is countably additive in the $L^2$-sense, ie, if $E_1, E_2, ...$ is a sequence of pairwise disjoint sets in $\mathcal{E}$, then

$$\mathbb{E}|\mu(\bigcup_n E_n) - \sum_{n=1}^{N} \mu(E_n)|^2 \to 0, N \to \infty - - - (1)$$

Let $\mu$ be a measure on $(X, \mathcal{E})$, ie $(X, \mathcal{E}, \mu)$ is a measure space. Assume that

$$\mathbb{E}(Z(A).\bar{Z}(B)) = \mu(A \cap B), A, B \in \mathcal{E}$$

Show using the countable additivity of $\mu$ on $\mathcal{E}$, that this condition automatically guarantees that $Z(.)$ will be countably additive in the $L^2$-sense, ie, the property (1) will hold.

Let $f$ be a complex valued measurable function on this measure space and assume that

$$\int_X |f(x)|^2 d\mu(x) < \infty$$

ie,

$$f \in L^2(X, \mathcal{E}, \mu)$$

We wish to define a stochastic integral

$$\int_X f(x) dZ(x) \in \mathcal{H}$$

in the $L^2$ sense and elucidate some properties of this stochastic integral. Choose a simple sequence of measurable functions $f_n$ on $(X, \mathcal{E}, \mu)$ converging in the $L^2$ sense to $f$, ie, each $f_n$ has the form

$$f_n(x) = \sum_{k=1}^{N_n} c(n,k) \chi_{E_{n,k}}(x), n \geq 1$$

By saying that this sequence converges to $f$ in the $L^2$-sense, we mean that

$$\int_X |f_n(x) - f(x)|^2 d\mu(x) \to 0, n \to \infty$$

Then define

$$I_Z(f_n) = \sum_{k=1}^{N_n} c(n,k) Z(E_{n,k})$$

Show that $\{I_Z(f_n)\}$ is a Cauchy sequence in $\mathcal{H}$, or more precisely,

$$\mathbb{E}|I_Z(f_n) - I_Z(f_m)|^2 = \int_X |f_n(x) - f_m(x)^2 d\mu(x) \to 0, n, m \to \infty$$

where the last convergence follows from the fact that every convergent sequence in a Hilbert space (or more generally, in any inner product space) is Cauchy. Deduce that there exists an element $I_Z(f) \in \mathcal{H}$ such that

$$\mathbb{E}|I_Z(f_n) - I_Z(f)|^2 \to 0$$

and that this $L^2$ limit $I_Z(f)$ does not depend upon the sequence $f_n$ of simple functions converging to $f$. We write

$$I_Z(f) = \int_X f(x)dZ(x)$$

and call it the $L^2$-stochastic integral of $f$ w.r.t $Z$.

## 3.7 Exercises on the construction of the integral w.r.t a probability measure

[1]

[a] Let $(\Omega, \mathcal{F}, P)$ be a probability space and let $X$ be a random variable on it. We say that $X$ is a simple r.v. if it assumes atmost only a finite number of distinct values, say $c_1, ..., c_n$. Define

$$E_k = X^{-`1}(\{c_k\}) = \{\omega \in \Omega : X(\omega) = c_k\}, k = 1, 2, ...n$$

Show that $E_1, ..., E_n$ are disjoint events, ie,

$$E_k \cap E_j = \phi, k \neq k, E_k \in \mathcal{F}$$

and further

$$\Omega = \bigcup_{k=1}^{n} E_k$$

Show that we can write

$$X(\omega) = \sum_{k=1}^{n} c_k \chi_{E_k}(\omega)$$

Define

$$\int X dP = \sum_{k=1}^{n} c_k P(E_k)$$

Show that if we write the same r.v $X$ in another way as

$$X(\omega) = \sum_{k=1}^{m} d_k \chi_{F_k}(\omega)$$

where the $F_k'$s need not be disjoint (but they are events), then

$$\sum_{k=1}^{m} d_k P(F_k) = \int X dP = \sum_{k=1}^{n} c_k P(E_k)$$

[b] Show that the set of simple r.v.s is a vector space over the real number, ie, it is closed under addition and scalar multiplication by real numbers. Therefore, this set is also closed under all finite real linear combinations.

[c] Show that if $X, Y$ are simple r.v's and $X(\omega) \leq Y(\omega) \forall \omega \in \Omega$, then

$$\int X dP \leq \int Y dP$$

In particular, deduce using [b] that if $X$ is a non-negative simple r.v. then

$$\int X dP \geq 0$$

[2] First we construct the integral of a non-negative r.v. Let $(\Omega, \mathcal{F}, P)$ be a probability space and $X$ a non-negative real valued random variable on this space. A simple r.v. is a r.v. that assumes only a finite number of values. For each positive integer $N$, define the simple r.v $X_N$ by

$$X_N(\omega) = \sum_{k=0}^{N.2^N} (k/2^N)\chi_{X^{-1}((k/2^N,(k+1)/2^N])}(\omega)$$

where $\chi_E(\omega)$ denotes the indicator of $E$, ie, $\chi_E(\omega) = 1$ if $\omega \in E$ and $\chi_E(\omega) = 0$ if $\omega \notin E$. Using the decomposition

$$(k/2^N, (k+1)/2^N] = (2k/2^{N+1}, (2k+1)/2^{N+1}] \cup ((2k+1)/2^{N+1}, (2k+2)/2^{N+1}]$$

of the lhs into a disjoint union, deduce that

$$0 \leq X_N(\omega) \leq X_{N+1}(\omega) \forall N \geq 1$$

ie, $X_N, N \geq 1$ is a non-decreasing sequence of simple r.v.s Show further that

$$lim_{N \to \infty} X_N(\omega) = X(\omega) \forall \omega \in \Omega$$

Deduce using the result of the previous exercise that

$$0 \leq \int X_N dP \leq \int X_{N+1} dP < \infty, \forall N \geq 1$$

and hence that

$$I = lim_{N \to \infty} \int X_N dP$$

Exists. Define

$$\int X dP = I$$

Let $Y_N, N \geq 1$

**Exercises on stationarity, dynamical systems and ergodic theory**

[1] Let $f \in L^1(\Omega, \mathcal{F}, P)$ and let $T : \Omega \to \Omega$ be a measure preserving transformation, ie,

$$T^{-1}(\mathcal{F}) \subset \mathcal{F}, PoT^{-1} = P$$

if $T$ is invertible and $X$ is a random variable on $(\Omega, \mathcal{F}, P)$, then show that the process $X(T^n \omega), n \in \mathbb{Z}$ is a stationary stochastic process on $(\Omega, \mathcal{F}, P)$.

## 3.8 Test on Probability theory

[1] Let $(\Omega, \mathcal{F}, P)$ be a probability space and let $E_n, n = 1, 2, ...$ be an infinite sequence of events on this space, ie $E_n \in \mathcal{F}, n = 1, 2, ....$ Then justify the statement that the event that an infinite number of the $E_n's$ occur is given by

$$\{E_n, i.o\} = \bigcap_{n \geq 1} \bigcup_{k \geq n} E_k$$

Show further that the probability of this event satisfies

$$P(\{E_n, i.o\}) = lim_{n \to \infty} \sum_{n \geq 1} P(\bigcup_{k \geq n} E_k)$$

$$\leq \sum_{k \geq n} P(E_k)$$

and in particular, show that if

$$\sum_{k \geq 1} P(E_k) < \infty$$

then the probability of an infinite number of $E_n's$ occurring is zero.

[2] Show that if $X_n, n \geq 1$ is an infinite sequence of random variables on the same probability space, then $X_n$ converges to zero with probability one if for each $\epsilon > 0$,

$$\sum_{n \geq 1} P(|X_n| > \epsilon) < \infty$$

hint: Show that the event that $X_n$ does not converges to zero can be expressed as

$$\{X_n \to 0\}^c = \bigcup_{k\geq 1}\{|X_n| > 1/k, i.o\}$$

and that this event has probability zero if

$$P(\{|X_n| > 1/k, i.o\}) = 0, k = 1, 2, ..$$

Now make use the result of the preceding problem.

[3] Let $(\Omega, \mathcal{F}, P)$ be a probability space and let $\phi$ be a continuous convex bounded function. By convex, we mean that

$$\phi(\lambda.x + (1-\lambda)y) \leq \lambda.\phi(x) + (1-\lambda)\phi(y) \forall x, y \in \mathbb{R}, 0 \leq \lambda \leq 1$$

Let $X \in L^1(\Omega, \mathcal{F}, P)$, ie, $\mathbb{E}|X| < \infty$. Then prove Jensen's inequality:

$$\mathbb{E}(\phi(X)) \geq \phi(\mathbb{E}X)$$

hint: First prove this result for simple random variables by using the given definition of convexity, then obtain a sequence of simple random variables that converge to the given random variable and take limits using Lebesgue's dominated convergence theorem.

## 3.9 More Assignment problems in probability theory

### 3.9.1 convergence of random walks to diffusion

[1] Let $\mathbf{X}(n), n \in \mathbb{Z}$ be random walk on the $d$ dimensional lattice, ie, $\mathbf{X}(n) \in \mathbb{Z}^d$ with transition probabilities given by

$$P(\mathbf{X}(n+1)-\mathbf{X}(n) = \mathbf{e}_k|X(n)) = p(k), P(\mathbf{X}(n+1)-\mathbf{X}(n) = -\mathbf{e}_k|\mathbf{X}(n)) = q(k), k = 1, 2, ..., d$$

where

$$\mathbf{e}_k = [0, 0, .., 0, 1, 0, ..., 0]^T \in \mathbb{Z}^d$$

with a one in the $d^{th}$ position and zeros at all the other positions and

$$p(k), q(k) \geq 0, \sum_{k=1}^{d}(p(k) + q(k)) = 1$$

Define the probability of the random walk being at the position $\mathbf{k} = \sum_{j=1}^{d} k_j \mathbf{e}_j$ at time $n$ by

$$P(n, \mathbf{k}) = Pr(\mathbf{X}(n) = \mathbf{k})$$

where

$$\mathbf{k} = [k_1, ..., k_d] = k_1\mathbf{e}_1 + .. + k_d\mathbf{e}_d, k_1, ..., k_d \in \mathbb{Z}$$

From elementary intuition, derive the recurrence relation

$$P(n+1,\mathbf{k}) = \sum_{j=1}^{d}(P(n,\mathbf{k}-\mathbf{e}_j)p(j) + P(n,\mathbf{k}+\mathbf{e}_j)q(j))$$

By elementary intuitive arguments, show that if

$$P(n,\mathbf{k}) = \sum_{r_1,...,r_d,s_1,...,s_d} \frac{n!}{r_1!...r_d!s_1!...s_d!} p(1)^{r_1}...p(d)^{r_d}q(1)^{s_1}...q(d)^{s_d}$$

where the sum is over all non-negative integers $r_1,...,r_d,s_1,...,s_d$ for which

$$r_j - s_j = k_j, j = 1,2,...,d, \sum_{j=1}^{d}(r_j+s_j) = n$$

Now, suppose we view this random walk as a space-time discretized version of a continuous time-stochastic process $\mathbf{Y}(t)$ with values in $\mathbb{R}^d$ such that if $f(t,\mathbf{x})$ is the probability density of $\mathbf{Y}(t)$ with $\mathbf{x} \in \mathbb{R}^d$ and $P(n,\mathbf{k})$ is approximated by $f(n\tau,\mathbf{k}\Delta)\Delta^d$ with $\tau$ being the time discretization step size and $\Delta$ the spatial discretization step size, then show that the above recursion can be expressed as

$$f(t+\tau,\mathbf{x}) = \sum_{j=1}^{d}(p(j)f(t,\mathbf{x}-\Delta.\mathbf{e}_j) + q(j)f(t-\tau,\mathbf{x}+\Delta\mathbf{e}_j))$$

Show that if

$$\Delta \to 0, \tau \to 0, \Delta^2/2\tau \to D,$$

$$p(j)-q(j) \to 0, (p(j)-q(j))\Delta/\tau \to v_j, p(j)+q(j) \to a(j)$$

then taking this continuum limit, $f(t,\mathbf{x})$ will satisfy the partial differential equation (a diffusion equation)

$$\frac{\partial f(t,\mathbf{x})}{\partial t} = \sum_{j=1}^{d}(-v_j\frac{\partial f(t,\mathbf{x})}{\partial x_j} + D_j\frac{\partial^2 f(t,\mathbf{x})}{\partial x_j^2})$$

where

$$D_j = Da(j), j = 1,2,...,d$$

Assuming that at time $t = 0$ the particle executing this diffusion process is located at the origin, ie,

$$f(0,\mathbf{x}) = \delta(\mathbf{x})$$

show that if we define the spatial Fourier transform of $f$ by

$$F(t,\mathbf{K}) = \int f(t,\mathbf{x})exp(i\mathbf{K}.\mathbf{x})d^d\mathbf{x}$$

where

$$\mathbf{K}.\mathbf{x} = \sum_{j=1}^{d} K_j x_j$$

then $F$ satisfies the ode

$$\frac{\partial F(t,\mathbf{K})}{\partial t} = (i\mathbf{v},\mathbf{K}) + \mathbf{K}^T\mathbf{D}\mathbf{K})F(t,\mathbf{K}), t \geq 0$$

with the initial condition,

$$F(0,\mathbf{K}) = 1$$

where

$$\mathbf{v} = [v_1, ..., v_d]^T, \mathbf{D} = diag[D_1, ..., D_d]$$

so that

$$(\mathbf{v},\mathbf{K}) = \mathbf{v}^T\mathbf{K} = \sum_{j=1}^{d} v_j K_j,$$

$$\mathbf{K}^T\mathbf{D}\mathbf{K} = \sum_{j=1}^{d} D_j K_j^2$$

Show that the solution is

$$F(t,\mathbf{K}) = exp(it\mathbf{v}^T\mathbf{K} - t\mathbf{K}^T\mathbf{D}\mathbf{K}) = exp(it\sum_{j=1}^{d} K_j v_j - t\sum_{j=1}^{d} D_j K_j^2)$$

which is the characteristic function of a $d$-dimensional Gaussian random vector having mean $\mathbf{v}t$ and covariance matrix $2t\mathbf{D}$. By Fourier inversion show that

$$f(t,\mathbf{x}) = (4\pi.det(\mathbf{D})dt)^{-1/2}.exp(-(\mathbf{x}-\mathbf{v}t)^T\mathbf{D}^{-1}(\mathbf{x}-\mathbf{v}t)/4t)$$

$$= (4\pi.D_1...D_dt)^{-1/2}.exp(-\sum_{j=1}^{d}(x_j - v_jt)^2/4D_jt)$$

## 3.10 Multiple choice questions on probability theory

Instructions:Select the most appropriate answer.

[1] Let $X, Y$ be two random variables with joint density $f(x, y)$ and marginal densities $f_X(y), f_Y(y)$ respectively. Then the probability density of $Z = X + Y$ is given by

$$[a] \int f(z + y, y)dy$$

$$[b] \int f_X(z - y)f_Y(y)dy$$

$$[c] \int f(z - y, y)dy$$

$$[d] none of the above$$

[2] Let $X$ be the outcome when a fair die is thrown. Then, the probability distribution function (CDF) of $X$ is given by

$$[a] \frac{1}{6}\sum_{k=1}^{6} \delta(x - k)$$

$$[b] \frac{1}{6}\sum_{k=1}^{6} \theta(x - k)$$

$$[c] \theta(x)/6$$

$$[d] none of the above.$$

Here, $\theta(x)$ is the unit step function.

[3] Let $X_1, X_2, ..., X_n$ be random variables defined on the same probability space that are non necessarily uncorrelated. Then the variance $Var(S_n)$ of $S_n = X_1 + ... + X_n$ is given by

$$[a] \sum_{k=1}^{n} Var(X_k)$$

$$[b] \sum_{k=1}^{n} Var(X_k) + \sum_{1\leq k<j\leq n} Cov(X_k, X_j)$$

$$[c]\sum_{k=1}^{n} Var(X_k) + 2 \sum_{1\le k<j\le n} Cov(X_k, X_j)$$

$$[d] none of the above.$$

[4] Let $X_n, n \in \mathbb{Z}$ be a stationary $L^1$ sequence in a probability space with mean $\mu$ and autocovariance $C(\tau)$. Then $|\tau| \to \infty$, and writing $S_n = X_1+...+X_n$, we have that

[a] $S_n/n$ always converges in $L^2$ to $\mu$

[b] $S_n/n$ converges with probability one to $\mathbb{E}(X_1|\mathcal{T})$ where $\mathcal{T}$ is the tail $\sigma$-field.

[c] $S_n/n$ converges in $L^2$ to $\mathbb{E}X_1$ if $C(\tau)$ is summable.

[d] Both [b] and [c].

[5] Let $(\Omega, \mathcal{F}, P)$ be a probability space and let $X$ be a random variable on this space. Then, the probability distribution of $X$ is given by

[a] $F_X(x) = P(X^{-1}((-\infty, x]))$

[b] $F_X(x) = P(X^{-1}((-\infty, x)))$

[c] $F_X(x) = P(X^{-1}(\{x\})$

[d] None of the above

[6] Let $X(t)$ be an $L^2$ stochastic process that is passed through a time varying filter with impulse response $h(t,s)$ so that the filter output is given by $Y(t) = \int h(t,\tau)X(\tau)d\tau$. Then if $R_{XX}(t,\tau) = \mathbb{E}(X(t)X(\tau))$, $R_{YY}(t,\tau) = \mathbb{E}(Y(t)Y(\tau))$ and $R_{YX}(t,\tau) = \mathbb{E}(Y(t)X(\tau))$, we have

[a] $R_{YY}(t,\tau) = \int h(t,s)R_{XX}(s,\tau)ds$

[b] $R_{YY}(t,\tau) = \int h(t,s)R_{YX}(s,\tau)ds$

[c] $R_{YY}(t,\tau) = \int h(t,s_1)h(\tau,s_2)R_{XX}(s_1,s_2)ds_1ds_2$

[d] Both [b] and [c]

[7] Let $(\Omega_k, \mathcal{F}_k), k = 1,2$ be two measurable spaces and let $T : (\Omega_1, \mathcal{F}_1) \to (\Omega_2, \mathcal{F}_2)$ be a measurable transformation. $P_1$ be a probability measure on the first space and $P_2 = P_1 o T^{-1}$. Let $X$ be an integrable random variable on the second probability space. Then,

[a] $\int X(\omega)dP_2(\omega) = \int X(T\omega)dP_1(\omega)$

[b] $\int X(\omega)dP_1(\omega) = \int X(T\omega)dP_2(\omega)$

[c] $\int X(\omega)dP_1(\omega) = \int X(\omega)dP_2(\omega)$

[d] None of the above.

[8] Let $\{X_n\}$ be a sequence of random variables defined on a probability space. Then consider the following statements: [1] The probability distribution of $X_n$ converges to the probability distribution $F$ of a r.v $X$ at all continuity points of $F$, [2] $P(|X_n - X| > \epsilon) \to 0$ for all $\epsilon > 0$, [3] $\mathbb{E}(X_n - X)^2 \to 0$, [4] $P(X_n \to X) = 1$. Then

[a] [3] $\implies$ [2]
[b] [4] $\implies$ [2] $\implies$ [1]
[c] [4] $\implies$ [3]
[d] both [a] and [b]

[9] Let $X(t)$ be a random process passed through a system having the input output relation described by the differential equation

$$d^2Y(t)/dt^2 = adY(t)/dt + bY(t) + X(t)$$

Then the cross-correlation $R_{YX}(t,s) = \mathbb{E}(Y(t)X(s))$ satisfies

[a] $\frac{\partial^2 R_{YX}(t,s)}{\partial t^2} = a\frac{\partial R_{YX}(t,s)}{\partial t} + bR_{YX}(t,s) + R_{XX}(t,s)$
[b] $\frac{\partial^2 R_{YX}(t,t)}{\partial t^2} = a\frac{\partial R_{YX}(t,t)}{\partial t} + bR_{YX}(t,t) + R_{XX}(t,t$
[c] $\frac{\partial^2 R_{YX}(t,s)}{\partial t^2} = a\frac{\partial R_{YY}(t,s)}{\partial t} + bR_{YY}(t,s) + R_{XX}(t,s)$
[d] None of the above.

[10] Let $X$ be a random variable with a strictly increasing probability distribution function $F(x)$. Then, if $U$ is a uniformly distributed random variable with values in $[0,1]$, the random variable $F^{-1}(U)$ has the probability distribution

[a] uniform
[b] $F(x)$
[c] $F^{-1}(x)$
[d] none of the above.

**Study projects in probability theory**

## 3.11 Construction of Brownian motion on $[0,1]$ using the Haar basis

Step 1: For $n \geq 1$ and $k = 0, 1, ..., 2^{n-1} - 1$, define $H_{n,k}(t)$ to be $2^{(n-1)/2}$ for $t \in [2k/2^n, (2k+1)/2^n)$ and $-2^{(n-1)/2}$ for $t \in [(2k+1)/2^n, (2k+2)/2^n)$. For all other $t \in [0,1]$ set $H_{n,k}(t) = 0$. Define $H_0(t) = 1$. Show that if $f \in L^2[0,1]$

and $< f, H_{n,k} >, < f, H_0 >= 0$ for all $n, k$, then $f = 0$. This proved by noting that $f \perp H_{n,k}$ implies

$$\int_{2k/2^n}^{(2k+1)/2^n} f(t)dt = \int_{(2k+1)/2^n}^{(2k+2)/2^n} f(t)dt, k \in I(n)$$

where

$$I(n) = \{0, 1, ..., 2^{n-1} - 1\}$$

By considering the limit of the above equations as $n \to \infty$, deduce that $f \perp H_{n,k}$ for all $n, k$ implies that $f$ is a.s a constant on $[0,1]$ and by further making use of $f \perp H_0$, deduce that this constant is zero.

Step 2: Prove the orthogonality relations

$$< H_{nk}, H_{ml} >= \delta_{nm}\delta_{kl}$$

Do this by first taking $m = n$ and noting then that if $k \neq l$, then $H_{nk}, H_{ml}$ have disjoint supports, ie non-overlapping supports and hence the two are orthogonal. Next observe that if $n > m$ then by expressing $j/2^m$ as $j.2^{n-m}/2^n$, it follows that if $H_{nk}$ and $H_{ml}$ have overlapping supports, then the support of $H_{nk}$ is contained entirely in the first half or entirely in the second half of that of $H_{ml}$ and hence by using the fact that the integral of $H_{nk}$ is zero, deduce orthogonality of these two. Finally observe the trivial result that $H_{nk}$ is orthogonal to $H_{00}$ since its integral over $[0,1]$ is zero. Conclude then that $H_{00}, \{H_{nk} : n \geq 1, k \in I(n)\}$ is an onb for $L^2[0,1]$.

Step 3: Define the Schauder functions

$$S_{nk}(t) = \int_0^t H_{nk}(s)ds$$

and using the identity

$$f(t) = \sum_{n,k} < f, H_{nk} > H_{nk}(t), f \in L^2[0,1]$$

deduce by taking $f(s) = \chi_{[0,t]}(s), t, s \in [0,1]$ that

$$\sum_{n,k} S_{nk}(t)S_{nk}(s) = min(t,s)$$

Observe that the graph of $S_{nk}(t)$ is a symmetric tent of height $2^{-(n+1)/2}$ over the interval $[(2k-1)/2^n, (2k+2)/2^n]$, the symmetry being about the mid point of this interval $(2k+1)/2^n$. Observe also that for a fixed $n$, the $S'_{nk}s$ have non-overlapping supports as $k$ varies over $I(n)$.

Step 4: Now let $\xi(n,k), n \geq 1, k \in I(n)$ be iid $N(0,1)$ r.v's. Define

$$b(n) = max(|\xi(n,k)| : k \in I(n)\}$$

and observe that

$$P(b(n) > n) \leq 2^{n-1}.P(|\xi(1,1)| > n) = 2^n.(1 - \Phi(n)) \leq 2^n.exp(-n^2/2)/n\sqrt{2\pi}$$

and hence

$$\sum_n P(b(n) > n) < \infty$$

so that by the Borel-Cantelli lemma,

$$P(b(n) > n, i.o) = 0$$

Show that this is the same as saying that for a.e.$\omega$, there exists a finite positive integer $N(\omega)$ such that

$$b(n,\omega) \leq n, \forall n > N(\omega)$$

Conclude that

$$\sum_{k\in I(n)} |\xi(n,k)(\omega)S_{nk}(t)| \leq b(n,\omega) \sum_{k\in I(n)} S_{nk}(t)$$

$$\leq b(n,\omega)2^{-(n+1)/2} \leq n.2^{-(n+1)/2}, \forall n > N(\omega)$$

Conclude that if we define the processes

$$B_N(t) = \sum_{n=1}^{N} \sum_{n\in I(n)} \xi(n,k)S_{nk}(t), N \geq 1$$

then the $B_N's$ are continuous processes that converge uniformly to a limiting process $B(t)$ over $[0,1]$ almost surely and hence the limiting process $B(t)$ has almost surely continuous sample paths. Complete the proof by showing that

$$\mathbb{E}B_N(t) = 0, Cov(B_N(t), B_N(s)) = \sum_{k\in I(n), 1\leq n\leq N} S_{nk}(t)S_{nk}(s)$$

and further that for each $t$ $B_N(t)$ is a Cauchy sequence in $L^2(\Omega,\mathcal{F},P)$ and therefore converges in $L^2$ to $B(t)$. Then using continuity properties of the inner product in $L^2(\Omega,\mathcal{F},P)$, deduce that

$$\mathbb{E}(B(t)) = lim_N\mathbb{E}(B_N(t)), \mathbb{E}(B(t)B(s)) = lim_N\mathbb{E}(B_N(t)B_N(s))$$

$$= \sum_{n\geq 1, k\in I(n)} S_{nk}(t)S_{nk}(s) = min(t,s)$$

ie, $B(.)$ is a Gaussian process with a.s. continuous sample paths having zero mean and covariance $min(t,s)$, or in other words, $B(.)$ is a Brownian motion process over $[0,1]$.

Remark: To prove that $B(.)$ is a Gaussian process, it suffices to show that if $\{t_1,...,t_k\}$ is a finite set of points in $[0,1]$, then $(B(t_1),...,B(t_k))$ is a Gaussian random vector. But this is immediately a consequence of the fact that $(B_N(t_1),...,B_N(t_k))$ is a Gaussian random vector which converges in distribution since it converges in probability since it converges a.s since the process $B_N(.)$ converges uniformly a.s.

## 3.12 The law of the iterated logarithm

This result states that if $B(.)$ is standard BM, then

$$limsup_{t\to\infty}\frac{B(t)}{\sqrt{2t.loglog(t)}}=1a.s$$

Equivalently, since $tB(1/t)$ is also a BM, we can state the law of the iterated logarithm as

$$limsup_{t\to 0}\frac{B(t)}{2t.loglog(1/t)}=1a.s$$

Equivalently since $-B(t)$ is also a BM, this law can also be stated as

$$liminf_{t\to\infty}\frac{B(t)}{2t.loglog(t)}=-1$$

Intuitively what these result state is that for very large times $t$, $B(t)$ almost surely oscillates between the two bounding curves $x=\pm\sqrt{2t.loglog(t)}$.

To prove this result, define

$$h(t)=\sqrt{2t.lnln(1/t)},0<t<1$$

so that for $0<\theta<1$ and $n=0,1,...$, we have

$$h(\theta^n)=\sqrt{2\theta^n.ln(n.ln(1/\theta)}$$

Then we use Doob's Martingale's inequality in the form

$$P(max_{0<s<t}(B(s)-\lambda s/2)>\beta)=P(max_{0<s<t}exp(\lambda.B(s)-\lambda^2 s/2)\leq exp(-\lambda\beta))$$

$$\leq exp(-\lambda\beta),\lambda>0$$

In this inequality, we choose

$$\beta=h(\theta^n)/2,\lambda=(1+\delta)\theta^{-n}h(\theta^n)$$

to get

$$P(max_{\theta^{n+1}<s<\theta^n}(B(s)-(1+\delta)\theta^{-n}h(\theta^n)s/2)>h(\theta^n))\leq exp(-(1+\delta)\theta^{-n}h(\theta^n)^2/2)$$

from which we deduce that

$$P(max_{\theta^{n+1}<s<\theta^n}B(s)>h(\theta^n))(1+\delta/2))$$

$$\leq exp(-(1+\delta)ln(nln(1/\theta))=O(n^{-(1+\delta)})$$

and the rhs is summable. So by the Borel-Cantelli lemma and using the fact that for $t\in(\theta^{n+1},\theta^n)$, we have

$$h(t)=\sqrt{2t.lnln(1/t)}\geq\sqrt{2.\theta^{n+1}.lnln(1/\theta^n)}$$

$$= \sqrt{\theta}h(\theta^n)$$

and we get

$$P(max_{\theta^{n+1}<s<\theta^n}(B(s)/h(s)) > (1+\delta/2)\sqrt{\theta}, i.o) = 0$$

Thus,

$$P(limsup_{s\to 0}(B(s)/h(s)) > (1+\delta/2)\sqrt{\theta}) = 0$$

Letting $\theta$ increase to one through a set of rationals gives us

$$P(limsup_{s\to 0}(B(s)/h(s)) > (1+\delta/2)) = 0 \forall \delta > 0$$

and letting now $\delta \downarrow 0$ gives us the result that

$$P(limsup_{s\to 0}(B(s)/h(s) \leq 1) = 1$$

This completes the proof of the first half of the law of the iterated logarithm. For the second half, we consider the events

$$E_n = \{(B(\theta^n) - B(\theta^{n+1})/\sqrt{\theta^n - \theta^{n+1}} > x_n)$$

$$= 1 - \Phi(x_n))$$

where $\Phi$ is the standard normal distribution. We choose

$$x_n = \sqrt{1-\theta}h(\theta^n)/\sqrt{\theta^n - \theta^{n+1}} = h(\theta^n)/(\theta^{n/2})$$

We have for $x > 0$, the inequality

$$1 - \Phi(x) = (2\pi)^{-1/2}\int_x^\infty exp(-y^2/2)dy$$

$$=\geq (2\pi)^{-1/2}\int_x^\infty (y/x)exp(-y^2/2)dy$$

$$= (2\pi)^{-1/2}xexp(-x^2/2), x > 0$$

In particular,

$$(1 - \Phi(x_n)) \geq (2\pi)^{-1/2}(h(\theta^n)/\theta^{n/2})exp(-h(\theta^n)^2/2\theta^n)$$

$$= K\sqrt{2ln(n.ln(1/\theta))}.exp(-ln(n.ln(1/\theta)))$$

whose sum over $n \geq 1$ is clearly divergent. Hence, by the second Borel-Cantelli lemma, the events

$$E_n = \{(B(\theta^n) - B(\theta^{n+1}))/h(\theta^n) > \sqrt{1-\theta}\}, n \geq 1$$

occur infinitely often with probability one. Further, from the previous half with the Brownian motion $B$ replaced by $-B$, we have the result that a.s. there exists an integer $N = N(\omega)$ such that

$$\{-B(\theta^{n+1})/h(\theta^{n+1}) \leq 1 + \delta\}, \forall n > N$$

This is the same as saying that

$$-B(\theta^{n+1})/h(\theta^n) \leq \sqrt{\theta}(1+\delta)\forall n > N$$

and combining this with the previous result gives us the result that the events

$$F_n = \{B(\theta^n)/h(\theta^n) > \sqrt{1-\theta} - (1+\delta)\sqrt{\theta}\}$$

occur infinitely often. This true for every $\delta > 0$ and for every $\theta \in (0,1)$. Letting first $\delta \downarrow 0$ gives us the result that

$$P(limsup_{t \to 0} B(t)/h(t) > \sqrt{1-\theta} - \sqrt{\theta}) = 1$$

Then letting $\theta \uparrow 1$ gives us the result that

$$P(limsup_{t \to 0} B(t)/h(t) \geq 1) = 1$$

and this completes the proof of the law of the iterated logarithm for Brownian motion:

$$P(limsup_{s \to 0} B(s)/h(s) = 1) = 1$$

## 3.13 Levy's modulus of continuity for Brownian motion

Let

$$g(\delta) = \sqrt{2\delta.ln(1/\delta)}, 0 < \delta < 1$$

Let

$$0 < \theta < 1$$

Define

$$E_n = \{max_{1 \leq j \leq 2^n} |B(j/2^n) - B((j-1)/2^n)|/g(1/2^n) \leq \sqrt{1-\theta}\}, n = 1, 2, ...$$

We shall prove that

$$\sum_n P(E_n) < \infty$$

and then it will follow from the Borel-Cantelli Lemma that

$$P(E_n, i.o) = 0$$

or equivalently that for a.e.$\omega$, there exists a finite positive integer $N(\omega)$ such that for all $n > N(\omega)$, we have

$$max_{1 \leq j \leq 2^n} |B(j/2^n, \omega) - B((j-1)/2^n, \omega)|/g(1/2^n) > \sqrt{1-\theta}$$

which would imply that for such $\omega$,

$$limsup_{h\downarrow 0}max_{0\leq t\leq 1-h}|B(t+h,\omega)-B(t,\omega)|/g(h)\geq\sqrt{1-\theta}$$

and hence letting $\theta\downarrow 0$ through rationals (Note that $P(F_n)=1, n=1,2,...$ implies $P(\bigcap_n F_n)=1$), we would get the first half of the Levy modulus of continuity theorem:

$$limsup_{h\downarrow 0}max_{0\leq t\leq 1-h}|B(t+h)-B(t)|/g(h)\geq 1a.s$$

To prove the summability of $P(E_n)$, we note that

$$P(E_n)=(1-\xi_n)^{2^n}\leq exp(-2^n\xi_n)$$

in view of the independence of the events $E_n$, where

$$\xi_n=P(|B(1/2^n)|>g(1/2^n)\sqrt{1-\theta})$$
$$=1-\Phi(x_n)$$

where $\Phi(x)$ is the standard normal distribution function and

$$x_n=g(1/2^n)\sqrt{1-\theta}.2^{n/2}$$

Now, for any $x>0$, we have using integration by parts,

$$1-\Phi(x)=(2\pi)^{-1/2}\int_x^\infty exp(-u^2/2)du$$

$$=(2\pi)^{-1/2}(\int_x^\infty(1/u).u.exp(-u^2/2)du$$

$$=(2\pi)^{-1/2}(exp(-x^2/2)/x-\int_x^\infty(1/u^2)exp(-u^2/2)du)$$

$$\geq(2\pi)^{-1/2}(exp(-x^2/2)/x-x^{-2}\int_x^\infty exp(-u^2/2)du)$$

and this inequality can also be expressed as

$$(1+1/x^2)(1-\Phi(x))\geq(2\pi)^{-1/2}x^{-1}.exp(-x^2/2)$$

or equivalently,

$$1-\Phi(x)\geq\frac{x}{1+x^2}.exp(-x^2/2)$$

Thus,

$$\xi_n\geq\frac{x_n}{1+x_n^2}exp(-x_n^2/2)$$

and

$$x_n^2/2=g(1/2^n)^2(1-\theta).2^{n-1}=2.(1/2^n).log(2^n).2^{n-1}(1-\theta)$$
$$=n(1-\theta)log(2)$$

and hence,

$$x_n \to \infty$$

and further,

$$exp(-x_n^2/2) = 2^{-n(1-\theta)}$$

so that

$$exp(-2^n \xi_n) \geq K_n.exp(-2^{n\theta}), K_n = (2\pi)^{-1/2} \frac{x_n}{1+x_n^2} \leq 1$$

for large $n$ and hence

$$\sum_n exp(-2^n \xi_n) \leq \sum_n exp(-2^{n\theta}) < \infty$$

from the desired summability of $P(E_n)$ follows.

To prove the second half of the Levy modulus theorem, we choose $\epsilon > 0$ and consider the events

$$E_n = \{max_{1 \leq i \leq i+k \leq 2^n, 0 \leq k \leq 2^{n\theta}} |B((i+k)/2^n) - B(i/2^n)|/g(k/2^n) > 1+\epsilon\}$$

and deduce easily that

$$P(E_n) \leq 2^n \sum_{k=1}^{2^{n\theta}} P(|B(k/2^n)| > (1+\epsilon)g(k/2^n))$$

$$= 2^n \sum_{k=1}^{2^{n\theta}} P(|Z| > (1+\epsilon)g(k/2^n)(2^n/k)^{1/2})$$

where $Z$ is a standard normal random variable. Now define

$$x(n,k) = (1+\epsilon)g(k/2^n)(2^n/k)^{1/2}$$

Then,

$$x(n,k)^2/2 = (1+\epsilon)^2(k/2^n)log(2^n/k).2^n/k = (1+\epsilon)^2 log(2^n/k)$$

$$exp(-x(n,k)^2/2) = (2^n/k)^{-(1+\epsilon)^2}$$

and hence

$$P(E_n) \leq 2^{n(1-(1+\epsilon)^2)}. \sum_{k=1}^{2^{n\theta}} k^{(1+\epsilon)^2}$$

Now,

$$\int_0^{2^{n\theta}} x^{(1+\epsilon)^2} dx = 2^{n\theta((1+\epsilon)^2+1)}/((1+\epsilon)^2+1)$$

and hence,

$$exp(-x(n,k)^2/2) \leq K.2^{n(1-(1+\epsilon)^2+n\theta(1+\epsilon)^2+n\theta}$$

$$= K.2^{n[1+\theta-(1+\epsilon)^2(1-\theta)]}$$

and hence, if we select $\epsilon$ so that

$$(1+\epsilon)^2 > (1+\theta)/(1-\theta) - - - (a)$$

then it would follow that

$$\sum_n P(E_n) < \infty$$

and hence by the Borel-Cantelli lemma,

$$P(E_n, i.o) = 0$$

We then get the result that for a.e. $\omega$, there exists a finite positive integer $N(\omega)$ such that for all $n > N(\omega)$,

$$\{max_{1\leq i\leq i+k\leq 2^n, 0\leq k\leq 2^{n\theta}}|B((i+k)/2^n) - B(i/2^n)|/g(k/2^n) \leq 1+\epsilon\}$$

and hence since $2^{n\theta}/2^n = 2^{-n(1-\theta)}$ converges to zero as $n \to \infty$, it follows from the continuity of the Brownian sample paths that

$$limsup_{h\downarrow 0} max_{0\leq t\leq 1-h}(|B(t+h) - B(t)|/g(h) \leq 1+\epsilon$$

and now letting $\epsilon$ decrease to zero yields the second half of the Levy modulus of continuity theorem.

Remark: Let $t, s \in [0,1]$, $t \geq s, t-s = h$. Let $\delta_1 > 0$ be given. Let $D_n = \{k/2^n : k = 0, 1, ..., 2^n\}$. Note that $D_n \subset D_{n+1}$. Note that $D = \bigcup_n D_n$ is the set of all dyadic rationals in $[0,1]$. Fix any $\delta > 0$ and choose $n$ large enough so that $|t-t'| < \delta, |s-s'| < \delta$ for some $t', s' \in D_n$ where $\delta$ is chosen so that $|u| < \delta$ implies $|B(t+u) - B(t)| < \delta_1$ for all $t \in [0, 1-u]$. This is possible since $B$ is uniformly continuous on $[0,1]$. Then

$$|B(t) - B(t')|, |B(s) - B(s')| < \delta_1$$

and hence,

$$|B(t) - B(s)| \leq 2\delta_1 + |B(t') - B(s')|$$

**On the number of data samples required for training a neural network for satisfactory performance**

Consider a neural network with output $y(t)$ satisfying the difference equation (RNN)

$$x(t+1) = f(x(t), \theta) + w(t+1)$$

The training to estimate $\theta$ is based on using the measurements

$$y(t) = h(x(t)) + v(t)$$

The joint probability density of $y(t), 1 \leq t \leq N$ given $\theta$ can in principle be computed from this statistical model:

$$p(y(1), ..., y(N)|\theta) = \int p(y(1), ..., y(N)|x(1), ..., x(N))p(x(1), ..., x(N)|\theta)dx(1)..dx(N)$$

$$= \int [\Pi_{t=1}^{N} p_v(y(t) - h(x(t))][\Pi_{t=0}^{N-1} p_w(x(t+1) - f(x(t), \theta))]dx(1)...dx(N)$$

and using this formula, the CRLB for $\theta$ can be computed as the inverse of the Fisher information matrix $J(\theta)$:

$$J_N(\theta) = -\mathbb{E}[\frac{\partial^2 log(p(y(1), ..., y(N)|\theta)}{\partial\theta\partial\theta^T}]$$

The question is therefore how large should $N$ be taken so that $Tr(J(\theta)^{-1})$ is smaller than a given threshold variance $\sigma(\epsilon)^2$? Although the CRLB is a lower bound on the variance of an unbiased estimator, there may not exist any such estimator. Hence, a more practical way to do this problem is to use the fact that the asymptotic variance of the MLE coincides with the CRLB and hence we may instead pose the question as how large should the data set be so that the variance of the MLE attains the CRLB ? Since however, we are using the EKF which is a very weak approximation to the exact MLE, we may also address the question in the following way:

# Chapter 4

# Models for the Refractive Index of Materials and liquids

Reference:Harish Parthasarathy and Steven A.Langford, NSUT preprint.

## 4.1 Abstract

We describe some mathematical models based on classical and quantum field theory and statistical field theory for explaining the refractive indices of materials. The first model proposes to describe the electromagnetic field interacting with the Dirac field of electrons and positrons by replacing the value of the electronic charge with a functional of the electromagnetic field. This idea is based on the fact that the nature of the singularity of the electromagnetic field completely describes the nature, ie location and value of the point charges in it. The solution to the resulting Dirac equation in a background electromanetic field will then give us the probability density function for the spatial location of the electron and by averaging the electron's electric and magnetic dipole moment operator with respect to this probability distribution, we can obtain formulas for the quantum averaged polarization and magnetization or equivalently the permittivity and permeability of the medium without having to describe the electronic field directly in terms of the electronic charge. This philosophy is in conformity with what many physicists believe today [Hans Van Leunen] that all properties of electrons should be derivable from the electromagnetic field itself. The second model describes a direct approach to computing the RI of a material based on Dirac's quantum mechanics for a system of $N$ interacting particles in an external electromagnetic field. If we solve the Dirac equation using perturbation theory for a single particle in an electromagnetic field, we could then calculate the quantum averaged electric and magnetic dipole moment of the electron which would in turn enable us to determine the permittivity and permeability of the medium in terms of the electric and magnetic fields. However, this analysis does not show how the RI depends upon the temperature of the material. In order to obtain temperature dependence, we consider Dirac's quantum mechanics for an N particle system taking interparticle interactions into account apart from interaction of the particles with an external electromagnetic field and by partial tracing the mixed state Dirac equation over the other particles and then making some approximations we derive a quantum Boltzmann equation for the quantum density operator and if this equation is solved using perturbation theory with the intial state as the Gibbs state (which has temperature dependence), then the final equilibrium state in the presence of a static electromagnetic field and interparticle interactions will also depend upon temperature. When this final density matrix is used to compute quantum averages of the electric and magnetic dipole moment, we are able to explain the dependence of the RI on both the electromagnetic field and temperature. The wavelength dependence of the RI can be explained by assuming the background electromagnetic field to be black-body radiation which has the energy density of the electromagnetic field dependent upon both frequency/wavelength and temperature. The final model described in this manuscript takes into account cosmological and background gravitational effects on the refractive index of the material. Gravity affects quantum mechanics via the spinor connection of the gravitational field which has to be introduced into Dirac's equation in order to make it invariant under local Loretnz transformations and arbitrary diffeomorphisms of space-time. Thus, this general relativistic generalization of Dirac's equation gives us the dependence of the wave function on the background metric tensor of curved space-time. If this background metric is taken to the Schwarzchild metric, the wave function would depend upon the mass of the blackhole and the gravitational constant while if it is taken to be Robertson-Walker metric for an expanding homogeneous and isotropic universe, then the wave function will also depend on the radius of the universe and hence on Hubble's constant. Calculating the average electric and magnetic dipole moments w.r.t such a wave function would then yield the dependence of the RI on the radius of the expanding universe and on its curvature. By taking fine measurements of the RI, we would then in principle be able to measure Hubble's constant and hence the radius of the universe at the present epoch.

## 4.2 Determining the charge from singularities of the potential

Let the point charge be $Q$ and let its location be $r_0$. The potential produced by it is

$$V(r) = Q/4\pi\epsilon|r - r_0|$$

Thus,

$$\nabla^2 V(r) = -Q\delta(r - r_0)/\epsilon$$

We can thus recover $Q$ and $r_0$ from $V(.)$ using the formula

$$\int f(r)\nabla^2 V(r)d^3r = (-Q/\epsilon)f(r_0)$$

for any measurable function $f$ having compact support. In particular, let $g(r)$ be another function. Then,

$$\frac{\int f(r)\nabla^2 V(r)d^3r}{\int g(r)\nabla^2 V(r)d^3r} = f(r_0)/g(r_0)$$

Therefore, if $f, g$ are functions that are zero outside a compact subset $K$ of $\mathbb{R}^3$ and are such that $f(x, y, z)/g(x, y, z) = x$, then

$$x_0 = \frac{\int f(r)\nabla^2 V(r)d^3r}{\int g(r)\nabla^2 V(r)d^3r}$$

Likewise $y_0, z_0$ can be recovered from $V(.)$. In this way, we can recover $r_0 = (x_0, y_{0,0})$ from $V(.)$ Then, $Q$ is also determined using

$$Q = (-\epsilon/f(r_0))\int f(r)\nabla^2 V(r)d^3r$$

Now consider the problem of determining the point charges and their locations given their number from the electrostatic potential generated by them. Let these charges be $Q_1, ..., Q_n$ and let their locations be $r_1, ..., r_n$. If $V(r)$ is the electrostatic potential generated by them, then Poisson's equation gives

$$\epsilon\nabla^2 V(r) = \sum_{k=1}^{n} Q_k\delta(r - r_k)$$

and hence, if $f(r)$ is a bounded measurable function, we have that

$$\sum_{k=1}^{n} Q_k f(r_k) = \epsilon\int f(r)\nabla^2 V(r)d^3r$$

Now choose the functions $f_1(r) = x^m, f_2(r) = y^m, f_3(r) = z^m, m = 0, 1, 2, ...$ and derive from the above, the following system of equations

$$\sum_{k=1}^{n} Q_k x_k^m = \int x_k^m \nabla^2 V(r)d^3r,$$

$$\sum_{k=1}^{n} Q_k y_k^m = \int y_k^m \nabla^2 V(r)d^3r,$$

$$\sum_{k=1}^{n} Q_k z_k^m = \int z_k^m \nabla^2 V(r)d^3r$$

for $m = 0, 1, 2, ...N - 1$. Thus we get a system of $3N$ equations which can be solved for $Q_k, x_k, y_k, z_k, k = 1, 2, ..., n$ or a least squares solution can be obtained provided that $N \geq 4n$. In this way a finite discrete charge distribution in space can be completely determined from the potential field. We could also do this using measurements of the electric field only using Gauss' law:

$$\epsilon divE(r) = \sum_{k=1}^{n} Q_k \delta(r - r_k)$$

Thus,

$$\epsilon \int f(r) divE(r) d^3r = \sum_{k=1}^{n} Q_k f(r_k)$$

Another way to express the point charge distribution as a functional of the potential is to assume that the distance between the locations of any two charges in the set is greater than $2\delta$. Let $Q_1, ..., Q_n$ denote the point charges with locations $r_1, ..., r_n$ so that the charge density is

$$\rho(r) = \sum_{k=1}^{n} Q_k \delta(r - r_k)$$

with

$$|r_k - r_j| > \delta \forall k \neq j$$

Then let $B(\delta)$ denote the open ball in $\mathbb{R}^3$ with the origin as centre and radius $\delta$:

$$B(\delta) = \{r \in \mathbb{R}^3 : |r| < \delta\}$$

Then given an arbitary point $r \in \mathbb{R}^3$, we have that $B(r, \delta) = r + B(\delta)$ can contain at most only one of the $r_k's$. It follows that for any $r$,

$$-\epsilon \int_{B(r,\delta)} \nabla^2 V(r') d^3r'$$

equals either zero or $Q_k$ for some $k = 1, 2, ..., n$. It equals $Q_k$ iff $|r - r_k| < \delta$. In other words, by moving the center ball $B(\delta)$ to different points, we get a result either equal to zero or $Q_k$ and from the location of the centre of the ball, we can determine $r_k$ upto an accuracy of $\delta$. This result can also be stated as

$$lim_{r \to r_k, \delta \to 0} \int_{B(r,\delta)} (-\epsilon \nabla^2 V(r')) d^3r'$$

$$= Q_k$$

## 4.3 Dirac's equation with charge replaced by a functional of the electromagnetic field

Now consider an electron of charge $-e$ interacting with the atomic nucleus of charge $Ze$. Let $A_q(t,r)$ denote the free quantum electromagnetic field in space-time and let $\psi(t,r)$ denote the second quantized Dirac wave function of the electron. It satisfies the equation

$$[(\gamma^\mu(i\partial_\mu + eA_{q\mu}(t,r) + eA_{N\mu}(r)) - m]\psi(t,r) = 0$$

where

$$A_{N0}(r) = -Ze^2/|r|, A_{Nj}(r) = 0, j = 1,2,3$$

is the classical nuclear potential. According to our theory, $A_q$, the free quantum electromagnetic field does not have any singularity and hence if $\delta$ is sufficiently small, we have

$$\epsilon \int_{B(\delta)} \nabla^2(A_{q\mu}(t,r) + A_{N\mu}(r))d^3r = -Ze\delta_{\mu,0}$$

This equation then determines the electron charge $-e$ from the **total** electromagnetic potential. One way to write down Dirac's equation without introducing explicitly the electronic charge $-e$ is then to write it as

$$[\gamma^\mu(i\partial_\mu - Z^{-1}[(\int_{B(\delta)} \nabla^2 A_0(t,r')d^3r')]A_\mu(t,r)]) - m]\psi(t,r) = 0 - -- (1)$$

where

$$A_\mu = A_{qmu} + A_{N\mu}$$

is the total electromagnetic four potential comprising of the free field part described in term of photon creation and annihilation operators and the classical nuclear part.

Remark: It should be noted that in our formalism, the electron exists only because it is a part of the atom having an atomic nucleus. The existence of the electron without a corresponding nucleus is not meaningful.

Now the total electromagnetic field $A_\mu$ in the region $|r| > 0$, ie, in $\mathbb{R}^3 - \{0\}$ satisfies the wave equation

$$\nabla^2 A_\mu - \frac{1}{c^2}\partial_t^2 A_\mu = 0, \partial^\mu A_\mu = 0 - -- (2)$$

The question is , "Is it possible to derive all the consequences of conventional quantum field theory from equns (1) and (2) only ? To answer this question, let us assume first that the total electromagnetic field $A_\mu$ is given and we have to solve (1). Perturbatively solving it gives us to a first order approximation in the interaction term

$$\psi = \psi_0 + \psi_1,$$

$$[i\gamma^\mu\partial_\mu - m]\psi_0 = 0,$$

$$[i\gamma^{\mu}\partial_{\mu} - m]\psi_1 =$$

$$Z^{-1}[\int_{B(\delta)} \nabla^2 A_0(t,r')d^3r')]A_{\mu}(t,r)\gamma^{\mu}\psi_0$$

$\psi_0$ therefore represents the free Dirac field expressible as a superposition of electron and positron creation and annihilation operators. The first order perturbation $\psi_1$ to the free Dirac field is then

$$\psi_1 = Z^{-1}\int S(x-x')[\int_{B(\delta)} \nabla^2 A_0(t,r')d^3r')]A_{\mu}(x')\gamma^{\mu}\psi_0(x')d^4x' \quad ---(3)$$

We now use (3) to compute radiative corrections to the electron propagator in terms of the photon propagator:

$$< T(\psi(x)\psi(x')^*) > \approx < T(\psi_0(x)\psi_0(x')^*) > + < T(\psi_0(x)\psi_1(x')^*) >$$

$$+ < T(\psi_1(x)\psi_0(x')^*) >$$

where now

$$< T(\psi_0(x)\psi_0(x')^*) >= S_0(x-x')$$

is the free Dirac field electron propagator known to be given by

$$S_0(x-x') = K.\int (\gamma^{\mu}p_{\mu} - m)^{-1}exp(ip.x)d^4p$$

## 4.4 Determining charges and their velocities from singularities in the electromagnetic field

Now come to the time varying case for calculating charges and their velocities from the singularities in the electromagnetic field. A point charge $Q$ moving along the trajectory $R(t), t \geq 0$ with non-relativistic velocity generates an electromagnetic field given approximately by

$$E(t,r) = \frac{Q(r-R(t))}{4\pi\epsilon|r-R(t)|^3},$$

$$B(t,r) = \frac{\mu QV(t) \times (r-R(t))}{|r-R(t)|^3}$$

Equivalently in terms of the Maxwell equations,

$$divE(t,r) = Q\delta(r-R(t))/\epsilon,$$

$$curlB(t,r) = \mu QV(t)\delta(r-R(t)) + \epsilon\partial_t E(t,r)$$

and hence we deduce that for a smooth function $f(r)$ of space coordinates,

$$\int f(r) divE(t,r)d^3r = (Q/\epsilon)f(R(t)),$$

$$\int f(r) curlB(t,r)d^3r - \epsilon \int f(r)\partial_t E(t,r)d^3r = \mu QV(t)f(R(t))$$

and by selecting $f$ appropriately, it is clear from these two equations how to determine the charge and its trajectory including velocity at each time from the total electromagnetic field in space.

We apply this idea to quantize the electromagnetic field and Dirac field when the nucleus having charge $Q = Ze$ that binds the electron moves along a trajectory $R(t)$. The magnetic vector potential generated by such a nucleus is

$$A_N(t,r) = \frac{\mu QV(t)}{4\pi|r - R(t)|},$$

and the electric scalar potential is given by

$$A_{N0}(t,r) = \frac{Q}{4\pi\epsilon|r - R(t)|}$$

in the non-relativistic approximation. Equivalently, in the non-relativistic approximation, we have

$$\nabla^2 A_N(t,r) = -\mu QV(t)\delta(r - R(t)),$$

$$\nabla^2 A_{N0}(t,r) = -Q\delta(r - R(t))/\epsilon$$

so that for any test function $f(r)$, we have

$$\int f(r)\nabla^2 A_N(t,r)d^3r = -\mu QV(t)f(R(t)),$$

$$\int f(r)\nabla^2 A_{N0}(t,r)d^3r = -Qf(R(t))/\epsilon$$

By taking the ratio of these two equations, we obtain the charge velocity vector

$$V(t) = (\mu\epsilon)^{-1}\frac{\int f(r)\nabla^2 A_N(t,r)d^3r}{\int f(r)\nabla^2 A_{N0}(t,r)d^3r}$$

The charge $Q$ can be calculated in terms of the field $A_{N0}$ by integrating the Laplacian applied to it over a small neighbourhood of its position $R(t)$. However, to do so, we require first to estimate $R(t)$ from the field. That can be done by taking $f(r) = |r|$ giving thereby

$$\int |r|.\nabla^2 A_{N0}(t,r)d^3r = -Q|R(t)|/\epsilon,$$

and then taking $f(r) = |r|^2$, we get

$$\int |r|^2 \nabla^2 A_{N0}(t,r) d^3r = -Q|R(t)|^2/\epsilon$$

Eliminating $|R(t)|$ between these two equations gives us the charge as

$$Q = -\epsilon \frac{(\int |r|.\nabla^2 A_{N0}(t,r)d^3r)^2}{\int |r|^2 \nabla^2 A_{N0}(t,r)d^3r}$$

$R(t)$ may now be calculated using

$$\int r\nabla^2 A_{N0}(t,r)d^3r = -QR(t)/\epsilon$$

and $V(t)$ using

$$\int \nabla^2 A_N(t,r)d^3r = -\mu QV(t)$$

If we assume that the quantum electromagnetic field $A_{q\mu}$ fluctuates rapidly in space, then its spatial average over any small open ball of finite radius will be negligible and hence we can to a good degree of approximation write

$$\int f(r)(\frac{\int_{B(r,\delta)} \nabla^2 A_0(t,r')d^3r'}{V(B(\delta))})d^3r = -Qf(R(t))/\epsilon,$$

$$\int f(r)(\frac{\int_{B(r,\delta)} \nabla^2 A(t,r')d^3r'}{V(B(\delta))})d^3r = -\mu QV(t)f(R(t))$$

where

$$A = A_N + A_q, A_0 = A_{N0} + A_{q0}$$

or equivalently,

$$A_\mu = A_{N\mu} + A_{q\mu}$$

These are respectively the total magnetic vector potential due to the nucleus and the quantum field and the total electrostatic field due to the same. Note that works because the nuclear potential has a singularity at the origin and at other spatial points, it varies slowly in space, while the quantum field is smooth thereby ensuring that

$$\frac{\int_{B(r,\delta)} A_{q\mu}(t,r')d^3r'}{V(B(\delta))} \approx 0$$

and since

$$\frac{\int_{B(r,\delta)} A_{N\mu}(t,r')d^3r'}{V(B(\delta))} \approx A_{N\mu}(t,r)$$

for small positive $\delta$. Therefore,

$$\frac{\int_{B(r,\delta)} A_\mu(t,r')d^3r'}{V(B(\delta))} \approx A_{N\mu}(t,r)$$

Dirac's equation for the electron wave function is now expressible entirely in terms of the total electromagnetic field without even bringing in the electronic charge parameter. Formally, this equation is therefore expressible as

## 4.5 Relativistic considerations involved in the determination of charge and velocities from the electromagnetic field

Suppose that the nucleus is moving with relativistic velocities. Then, we replace the Laplacian operator by the wave operator in the above equations thereby obtaining

$$\Box A_N(t,r) = -\mu Q V(t)\delta(r - R(t)),$$
$$\Box A_{N0}(t,r) = -Q\delta(r - R(t))/\epsilon$$

where

$$\Box = \nabla^2 - \mu\epsilon\partial_t^2$$

It is then clear how all the parameters of the moving nucleus, namely it charge, position trajectory and velocity can be computed as functions of weighted integrals of the total electromagnetic field. Specifically, we find that

$$\int f(r)(\frac{\int_{B(r,\delta)} \Box A(t,r')d^3r'}{V(B(\delta))}d^3r = -\mu Q V(t) f(R(t))$$

$$\int f(r)(\frac{\int_{B(r,\delta)} \Box A_0(t,r')d^3r'}{V(B(\delta))}d^3r = -Qf(R(t))/\epsilon$$

The Dirac equation is now of the form

$$[\gamma^\mu(i\partial_\mu + F(A_{nu}(t,r), r \in \mathbb{R}^3)A_\mu(t,r)) - m]\psi(t,r) = 0$$

where $e = F(A_\nu(t,r), r \in \mathbb{R}^3)$ is the electronic charge value determined as above as spatial functional of the electromagnetic field. The electromagnetic field on the other hand satisfies Maxwell's equations in the form

$$\Box A_\mu(t,r) = 0, r \neq R(t)$$

The whole point of this exercise is that by measuring data about the Dirac wave function, or equivalently the Dirac four current density, we can in principle calculate the electromagnetic field $A_\mu$ and hence from the singularity theory mentioned above, calculate the nuclear charge as well as its trajectory.

## 4.6 Calculating the masses of $N$ gravitating particles and their positions and their trajectories from measurement of the gravitational potential distribution in space-time using the Newtonian theory

Let $m_1, ..., m_N$ denote the masses of $N$ point particles moving under their mutual gravitation along trajectories $r_1(t), ..., r_N(t)$. The Newtonian equations of motion are

$$r_j''(t) = \sum_{k=1,k\neq j}^{N} Gm_k(r_j - r_k)/|r_k - r_k|^3, j = 1,2,...,N$$

The gravitational potential generated by these masses is then

$$\Phi(r) = \sum_{j=1}^{N} Gm_j/|r - r_j|$$

and this potential satisfies Poisson's equation

$$\nabla^2\Phi(t,r) = 4\pi G\sum_{j=1}^{N} m_j\delta(r - r_j(t))$$

Thus, we get for a test function $f(r)$,

$$\int f(r)\nabla^2\Phi(t,r)d^3r = 4\pi G\sum_{j=1}^{N} m_j f(r_j(t))$$

By choosing test functions $f_1, ..., f_N$ appropriately, we get the following linear system of equations for the masses given their positions:

$$\int_{j=1}^{N} f_k(r_j(t))m_j = \int f_k(r)\nabla^2\Phi(t,r)d^3r, k = 1,2,...,N$$

Define the $N \times N$ matrix valued function of time

$$A(t) = ((f_k(r_j(t))))_{1\leq k,j\leq N}$$

and let

$$B(t) = A(t)^{-1} = ((b_{ij}(t)))$$

Then,

$$m_k = \sum_{j=1}^{N} b_{kj}(t)\int f_j(r)\nabla^2\Phi(t,r)d^3r$$

This formula will work even if the masses are functions of time. Now the $b_{jk}(t)'s$ are functions of the $r_j(t)'s$. So the $r_j(t)'s$ also have to be estimated from the potential distribution. Choose $N$ vectors $\xi_1, ..., \xi_N$ in $\mathbb{R}^3$. Then, we have

$$\int <\xi_k, r> \nabla^2\Phi(t,r)d^3r = 4\pi G\sum_{j=1}^{N} m_j <\xi_k, r_j>, k = 1,2,...,N$$

This is a system of $N$ linear equations for the $N$ masses and defining the matrix

$$((c_{kj}(r_j)))_{1\leq k,j\leq N} = C(r_1, ..., r_N) = 4\pi G((<\xi_k, r_j>))_{1\leq k,j\leq N}$$

gives us

$$m_k = \sum_{j=1}^{N} e_{kj}(r_1, ..., r_N) \int <\xi_j, r> \nabla^2\Phi(t,r)d^3r, k = 1, 2, ..., N$$

where

$$((e_{kj})) = C^{-1}$$

Thus we obtain the following $N$ equations for $r_1, ..., r_N$:

$$\sum_{j=1}^{N} b_{kj}(t) \int f_j(r)\nabla^2\Phi(t,r)d^3r$$

$$= \sum_{j=1}^{N} e_{kj}(r_1, ..., r_N) \int <\xi_j, r> \nabla^2\Phi(t,r)d^3r, k = 1, 2, ...., N$$

and by varying the vectors $\xi_k$ in these equations, we can derive at least $3N$ equations for the $N$ vectors $r_1, ..., r_N$ which can in principle be solved.

## 4.7 Masses and trajectories of particles from the metric field in general relativity

Now we address the same problem in Einsteinian gravity. The energy-momentum tensor for $N$ point particles of masses $m_1, ..., m_N$ is given by

$$T^{\mu\nu}(x) = \sum_k m_k(-g(x))^{-1/2}\delta^3(x - x_k(t))(dx_k^\mu(t)/dt)(dx_k^\nu/d\tau_k)$$

where $\tau_k$ is the proper time for the $k^{th}$ particle. It is given by

$$d\tau^2 = g_{\mu\nu}(x_k(t))dx_k^\mu(t)dx_k^\nu(t)$$

where

$$x_k^0(t) = t$$

is the universal coordinate time. The Einstein field equations corresponding to this energy-momentum tensor are

$$G_{\mu\nu} = R_{\mu\nu} - (1/2)Rg_{\mu\nu} = -KT_{\mu\nu}, K = 8\pi G$$

We find that

$$\int T^{\mu\nu}(t,r)\sqrt{-g(t,r)}f(t,r)dtd^3r = \sum T^{\mu\nu}(x)\sqrt{-g(x)}f(x)d^4x$$

$$= \sum_k m_k \int f(x_k(t))v_k^\mu(t)v_k^\nu(t)d\tau_k(t)$$

where

$$v_k^\mu(t) = dx_k^\mu/d\tau_k$$

is the four velocity of the $k^{th}$ particle. From this equation, we can infer by choosing different functions $f : \mathbb{R}^4 \to \mathbb{R}$, the particle trajectories as functions of coordinate time as well as their masses.

## 4.8 Application of the same ideas to Superstring Theory

A superstring comprising of a Bosonic and a Fermionic part is given by

$$X^\mu(\tau,\sigma) = x^\mu + p^\mu\tau - i\sum_{n\neq 0}(\alpha^\mu(n)/n)exp(in(\tau-\sigma)) - i\sum_{n\neq 0}(\tilde{\alpha}^\mu(n)/n)exp(in(\tau+\sigma))$$

$$\psi^\mu(\tau,\sigma) = \psi_+(\tau,\sigma) + \psi_-(\tau,\sigma)$$

$$= \sum_n S_n^\mu exp(in(\tau-\sigma)) + \sum_n \tilde{S}_n^\mu exp(in(\tau+\sigma))$$

since these satisfy the string field equations

$$\partial_+\partial_- X^\mu = 0, \partial_-\psi_+^\mu = 0, \partial_+\psi_-^\mu = 0$$

where

$$\partial_+ = \partial_\tau + \partial_\sigma, \partial_- = \partial_\tau - \partial_\sigma$$

so that

$$\partial_+\partial_- = \partial_\tau^2 - \partial_\sigma^2$$

Note that the Lagrangian for the Bosonic part of the string is

$$L_B = (1/2)\partial_+ X^\mu.\partial_- X_\mu$$

while that of the Fermionic part is

$$L_F = -i\psi_+^T\partial_-\psi_- - i\psi_-^T\partial_+\psi_-$$

Note that $\psi_+^T\partial_-\psi_-$ is an abbreviation for $\psi_+^\mu\partial_-\psi_+\mu$ and likewise for the other term. If $\psi_+$ and $\psi_-$ denote the canonical position fields for the Fermionic component of the superstring, then the corresponding canonical momenta are

$$\pi_+ = \partial L_F/\partial\partial_\tau\psi_+ = -i\psi_+,$$

$$\pi_- = \partial L_F.\partial_\tau\psi_- = -i\psi_-$$

so that the canonical anticommutation relations are

$$[\psi_+(\tau,\sigma), \psi_+(\tau,\sigma')]_+ = -\delta(\sigma-\sigma')$$

$$[\psi_-(\tau,\sigma),\psi_-(\tau,\sigma')]_+ = -delta(\sigma-\sigma')$$

These equations give

$$[S_n^\mu, S_m^\nu]_+ = \eta^{\mu\nu}\delta(n+m),$$
$$[\tilde{S}_n^\mu, \tilde{S}_m^\nu]_+ = \eta^{\mu\nu}\delta(n+m),$$

To obtain the Noether conserved currents for the Fermionic sector, we first observe that $L_F$ is invariant under the infinitesimal transformations

$$\delta\psi_+ = \epsilon.\psi_-, \delta\psi_- = -\epsilon\psi_+$$

where $\epsilon$ is an infinitesimal parameter. The first conserved Noether current corresponding to this symmetry is then given by

$$J^- = (\partial L_F/\partial\partial_-\psi_+)\delta\psi_+ + (\partial L_F/\partial\partial_-\psi_-)\delta\psi_- = \psi_+^T\psi_+$$

which is obeys the conservation law

$$\partial_- J^- = 0$$

when the field equations are satisfied. Likewise, the second conserved current corresponding to this symmetry is

$$J^+ = (\partial L_F/\partial\partial_+\psi_-)\delta\psi_- = \psi_-^T\psi_-$$

which satisfies the conservation law

$$\partial_+ J^+ = 0$$

when the field equations are satisfied. Likewise, $J^+ = \psi_-^T\psi_-$ satisfies the conservation law

$$\partial_+ J^+ = 0$$

when the field equations are satisfied. The problem is can we calculate $p^\mu$, the translational D-momentum of the string from measurements on the string observables ? More generally, if we introduce perturbation terms to the string field Lagrangian depending on a finite set of parameters, can we express these parameters as functionals of the string field ? For example, if we introduce a generalized gauge field interacting with the string field with the strength of this interaction being defined by a coupling constant analogous to the charge being the coupling constant for the interaction between the electromagnetic field and the four velocity of a point particle, then can be estimate this coupling constant as a functional of the string field and the generalized gauge field ? If so, then can we eliminate the dependence of the string field equations upon this coupling constant ?

**Acknowledgements:**I am grateful to Prof.Hans Van Leunen, Prof.Andre Michaud and Prof.Steven Arthur Langford for encouraging me to work on this problem and apply the method of determining all the charges and their locations from electromagnetic field measurements to Dirac's relativistic wave equation, by replacing the electronic charge which appears in this equation with functionals of the quantum electromagnetic field.

## 4.9 The quantum Boltzmann equation for a plasma

In this section, we derive an approximate nonlinear evolution equation for the density operator of a single particle when the quantum plasma consists of $N$ identical particles interacting with each other and also with an external electromagnetic field. The joint density operator of the $N$ particles satisfies the quantum Liouville or Schrodinger-Von-Neumann equation with the Hamiltonian consisting of a sum of identical Hamiltonians each acting in a single particle Hilbert space plus the sum of identical pairwise interacting potentials of two particles with each one acting in the tensor product of two identical Hilbert spaces. By taking the partial trace of this quantum Liouville equation and making approximations (which in the classical Boltzmann kinetic transport theory are called the molecular chaos approximation), we derive an approximate quadratic nonlinear evolution equation for the single particle density operator in an external electromagnetic field. The single particle Hamiltonians can either be the single particle Schrodinger equation in an external electromagnetic field or a single particle Dirac Hamiltonian or even a single particle Dirac Hamiltonian in curved space-time interacting with an external electromagnetic field. The quadratic nonlinear terms which arise due to the pairwise interaction of particles represent quantum generalization of the so called "collsion term" that appears in the classical Boltzmann equation in kinetic transport theory and which are usually evaluated using classical scattering theory or more specifically using binary elastic collision theory of two particles. It should be noted that our method of deriving the quantum Boltzmann equation by partial tracing is the quantum analogue of the classical BBGKY theory in which one writes down the classical Liouville equation for the distribution function of $N$ particles in phase space (ie, in the joint position-velocity space of all the $N$ particles) and then integrates this equation over the phase space variables of all but the first particles and then makes the molecular chaos approximation in which the joint distribution of two particles is approximated by a product of the individual distributions.

Suppose that the joint density matrix of $N$ particles is $\rho(123...N)$. It satisfies the Schrodinger equation

$$i\partial_t\rho_t(12...N) = [\sum_{a=1}^{N} H_a + \sum_{1\leq a<b\leq N} V_{ab}, \rho_t(12..N)]$$

In this equation, if we take a partial traced over $2, 3, ..., N$, we get

$$'i\partial_t\rho_{1t} = [H_1, \rho_{1t}] + (N-1)Tr_2[V_{12}, \rho_{12}]$$

and if we take the trace of the same over $3, 4, ..., N$, we get

$$i\partial_t\rho_{12t} = [H_1 + H_2 + V_{12}, \rho_{12t}] + (N-2)Tr_3[V_{13} + V_{23}, \rho_{123t}]$$

We write

$$\rho_{123} = (1/3)(\rho_{12} \otimes \rho_3 + \rho_{13} \otimes \rho_2 + \rho_1 \otimes \rho_{23}) + g_{123}$$

where $g_{123}$ is small. Then, neglecting second order of smallness terms like $V$ multiplied with $g_{123}$ gives us the approximate equation

$$i\partial_t\rho_{12t} = [H_1+H_2+V_{12}, \rho_{12}]+((N-2)/3)Tr_3[V_{13}+V_{23}, \rho_{12}\otimes\rho_3+\rho_{13}\otimes\rho_2+\rho_1\otimes\rho_{23}]$$

This is a bit hard to handle. So we content ourselves with the approximation

$$\rho_{12} = \rho_1 \otimes \rho_1 + g_{12}$$

where $g_{12}$ is small. We then get approximately,

$$i\partial_t\rho_{1t} = [H_1, \rho_{1t}] + (N-1)Tr_2[V_{12}, \rho_1 \otimes \rho_1]$$

Writing

$$V_{12} = \sum_a W_{1a} \otimes W_{2a}$$

gives us

$$Tr_2[V_{12}, \rho_1 \otimes \rho_2] = \sum_a Tr(\rho_1 W_{2a})[W_{1a}, \rho_1]$$

and the our Boltzmann equation becomes

$$i\partial_t\rho_1 = [H_1, \rho_1] + (N-1)\sum_a Tr(\rho_1 W_{2a})[W_{1a}, \rho_1]$$

Suppose we make the approximation

$$\rho_{123} = \rho_1 \otimes \rho_1 \otimes \rho_1 + g_{123}$$

where $g_{123}$ is small. Then we get

$$i\partial_t\rho_{12t} = [H_1 + H_2 + V_{12}, \rho_{12}] + (N-2)Tr_3[V_{13} + V_{23}, \rho_1 \otimes \rho_1 \otimes \rho_1]$$

Even this equation is hard to manipulate further without assuming some specific form of the interaction potential $V_{12}$. We consider

$$\rho_{12} = \rho_1 \otimes \rho_1 + g_{12},$$

$$\rho_{123} = (1/3)(\rho_{12} \otimes \rho_3 + \rho_{13} \otimes \rho_2 + \rho_1 \otimes \rho_{23}) + g_{123}$$
$$= \rho_1 \otimes \rho_1 \otimes \rho_1 + (1/3(g_{12} \otimes \rho_3 + g_{13} \otimes \rho_2 + \rho_1 \otimes g_{23}) + g_{123}$$

We first derive a differential equation for $g_{12}$ after neglecting second order of smallness terms:

$$i\partial_t\rho_{12} = i\partial_t\rho_1 \otimes \rho_1 + i\rho_1 \otimes \partial_t\rho_1$$
$$+i\partial_t g_{12}$$
$$= [H_1, \rho_1] \otimes \rho_1 + (N-1)Tr_2[V_{12}, \rho_1 \otimes \rho_1] \otimes \rho_1 + \rho_1 \otimes [H_1, \rho_1]$$
$$+(N-1)\rho_1 \otimes Tr_2[V_{12}, \rho_1 \otimes \rho_1] + i\partial_t g_{12}$$
$$= [H_1 + H_2 + V_{12}, \rho_{12}] + (N-2)Tr_3[V_{12} + V_{13}, \rho_{123}]$$

$$= [H_1 + H_2, \rho_1 \otimes \rho_1] + [V_{12}, \rho_1 \otimes \rho_1]$$
$$+(N-2)Tr_3[V_{13} + V_{23}, \rho_1 \otimes \rho_1 \otimes \rho_1]$$

After making the appropriate cancellations, we get

$$i\partial_t g_{12} =$$
$$= [V_{12}, \rho_1 \otimes \rho_1] + (N-2)Tr_3[V_{13} + V_{23}, \rho_1 \otimes \rho_1 \otimes \rho_1]$$
$$-(N-1)Tr_2[V_{12}, \rho_1 \otimes \rho_1] \otimes \rho_1$$
$$-(N-1)\rho_1 \times Tr_2[V_{12}, \rho_1 \otimes \rho_1]$$

Note that on writing

$$V_{12} = \sum_a W_{1a} \otimes W_{2a}$$

and using the fact that the $V'_{jk}s$ are identical copies of each other acting on different copies of the tensor product of two identical copies a Hilbert space just as the $H'_k s$ are identical copies of each other acting on different copies of the same Hilbert space, we get

$$Tr_3[V_{13} + V_{23}, \rho_1 \otimes \rho_1 \otimes \rho_1]$$
$$= \sum_a [Tr(\rho_1 W_{2a})([W_{1a}, \rho_1] \otimes \rho_1 + \rho_1 \otimes [W_{1a}, \rho_1])]$$

A better approximation to the quantum Boltzmann equation can then be obtained by solving this equation for $g_{12}(t)$ and substituting it into the equation

$$i\partial_t \rho_1 = [H_1, \rho_1] + (N-1)Tr_2[V_{12}, \rho_{12}]$$
$$= [H_1, \rho_1] + (N-1)Tr_2[V_{12}, \rho_1 \otimes \rho_1 + g_{12}]$$

Formally this equation has the form

$$i\partial_t \rho_1(t) = [H_1, \rho_1(t)] + \delta.F_1(\rho_1(s), s \leq t)$$

where $F$ is an operator valued nonlinear functional of $\rho_1(s), s \leq t$. This equation can be solved upto $O(\delta)$ using first order perturbation theory:

$$\rho_1(t) = U(t)\rho_1(0)U(t)^* + \delta. \int_0^t U(t-\tau)F_\tau(\rho_1(s), s \leq \tau)U(t-\tau)^* d\tau$$

where

$$U(t) = exp(-itH_1)$$

If we consider the Hamiltonian to comprise of an interaction between the particles and an electromagnetic field, then we can write

$$H_1 = (p_1 + eA(t,r))^2/2m - e\Phi(t,r) \approx p_1^2/2m - e\Phi(t,r) + (e/2m)((p_1, A) + (A, p_1))$$

and the particle interaction potential as

$$V_{12} = V(|r_1 - r_2|)$$

In that case, in the position representation, we have

$$[p_1^2, \rho_1] = [p_1, \rho_1].p_1 + p_1.[p_1, \rho_1]$$

and noting that $p_1$ is represented by the kernel

$$p_1(r, r') = -i\nabla_r \delta^3(r - r') = i\nabla'_r \delta^3(r - r')$$

we get

$$[p_1, \rho_1](r, r') = -i\nabla_r \rho_1(r, r') + i\nabla'_r \rho_1(r, r')$$

$$[p_1, \rho_1].p_1(r, r') = \nabla_r.\nabla'_r \rho_1(r, r') - \nabla_r'^2 \rho_1(r, r')$$

and likewise,

$$p_1.[p_1, \rho_1](r, r') = \nabla_r.\nabla'_r \rho_1(r, r') - \nabla_r^2 \rho_1(r, r')$$

$$[(A, p_1), \rho_1](r, r') = [A_k p_{1k}, \rho_1](r, r') =$$

$$A_k(t, r) p_{1k} \rho_1(r, r') - \int \rho_1(r, r'') A_k(t, r'') p_{1k}(r'', r') dr''$$

$$= -i(A(t, r), \nabla_r)\rho_1(r, r') - i(\nabla'_r, A(t, r'))\rho_1(r, r'))$$

Therefore with neglect of nonlinear terms in the electromagnetic field, our position space representation of the quantum Boltzmann dynamics of the single particle density operator is given by

$$i\partial_t \rho_1(t, r, r') = (2m)^{-1}(2\nabla_r.\nabla'_r \rho_1(t, r, r') - (\nabla_r^2 + \nabla_r'^2)\rho_1(t, r, r'))$$

$$-(i/m)(A(t, r), \nabla_r)\rho_1(t, r, r') - i(\nabla'_r, A(t, r'))\rho_1(t, r, r')$$

$$-e\Phi(t, r)\rho_1(t, r, r') + e\Phi(t, r')\rho_1(t, r, r')$$

$$+ nonlinearterms.$$

Remark on the nonlinear terms:

$$Tr_2[V_{12}\rho_1 \otimes \rho_1](r_1, r'_1) =$$

$$V(r_1, r_2)\delta(r_1 - r''_1)\delta(r_2 - r'_2)\rho_1(t, r''_1, r'_1)\rho_1(t, r'_2, r_2) d^3r'_2 d^3r_2 d$$

$$= \int V(r_1, r_2)\rho_1(t, r_1, r'_1)\rho_1(t, r_2, r_2) d^3r_2$$

and likewise,

$$Tr_2[(\rho_1 \otimes \rho_1)V_{12}](r_1, r'_1) =$$

$$\int \rho_1(t,r_1,r_1')\rho_1(t,r_2,r_2)V_{12}(r_1',r_2)d^3r_2$$

In principle, using perturbation theory, this quantum Boltzmann equation can be solved for to obtain the single particle density operator kernel $\rho_1(t,r,r')$ as a function of the electromagnetic field and then the average dipole moment of the electron in this electromagnetic field will be given by

$$\mathbf{p}(t) = \int (-e\mathbf{r})\rho_1(t,\mathbf{r},\mathbf{r})d^3r$$

and its average magnetic moment by

$$\mathbf{m}(t) = \int (-e\mathbf{L}(r,r')/2m)\rho_1(t,\mathbf{r}',\mathbf{r})d^3rd^3r'$$

where $\mathbf{L}(r,r')$ is the kernel of the orbital angular momentum operator in the position representation. This would then solve the problem of explaining the origin of permittivity and permeability of the plasma from the quantum statistical mechanical point of view.

## 4.10 Perturbative solution of the Boltzmann equation

$$i\partial_t\rho_1(t) = [H_1,\rho_1(t)] + \delta(N-1)Tr_2[V_{12},\rho_1(t)\otimes\rho_1(t)]$$

where $\delta$ is a perturbation parameter. We've also observed that an additional term can be added to this equation to improve its accuracy, namely,

$$\delta^2(N-1)Tr_2[V_{12},g_{12}]$$

where

$$g_{12}(t) =$$

$$= -i\int_0^t [[V_{12},\rho_1\otimes\rho_1] + (N-2)Tr_3[V_{13}+V_{23},\rho_1\otimes\rho_1\otimes\rho_1]$$

$$-(N-1)Tr_2[V_{12},\rho_1\otimes\rho_1]\otimes\rho_1$$

$$-(N-1)\rho_1\times Tr_2[V_{12},\rho_1\otimes\rho_1]](s)ds$$

$$= \int_0^t F_2(\rho_1(s)\otimes\rho_1(s),\rho_1(s)\otimes\rho_1(s)\otimes\rho_1(s))ds$$

so our perturbative solution would have the form

$$\rho_1(t) = U(t)\rho_1(0)U(t)^* + \delta.\int_0^t U(t-s)F_1(\rho_1(s)\otimes\rho_1(s))U(t-s)^*ds$$

$$+\delta^2(N-1)\int_0^t U(t-s)[\int_0^s F_2(\rho_1(\tau)\otimes\rho_1(\tau),\rho_1(\tau)\otimes\rho_1(\tau)\otimes\rho_1(\tau))d\tau]U(t-s)^*ds$$

$$= \rho_1(t) = U(t)\rho_1(0)U(t)^* + \delta.\int_0^t U(t-s)F_1(\rho_1(s)\otimes\rho_1(s))U(t-s)^*ds$$

$$+\delta^2(N-1)\int_{0<\tau<s<t}[U(t-s)F_2(\rho_1(\tau)\otimes\rho_1(\tau),\rho_1(\tau)\otimes\rho_1(\tau)\otimes\rho_1(\tau))d\tau]U(t-s)^*]dsd\tau$$

where

$$F_1(\rho_1\otimes\rho_1) = (N-1)Tr_2[V_{12},\rho_1\otimes\rho_1]$$

## 4.11 The quantum Boltzmann equation derived from the Dirac relativistic wave equation

$$H(t) = (\alpha, -i\nabla + eA) + \beta m - e\Phi = H_0 + H_I(t)$$

where

$$H_0 = (\alpha, -i\nabla) + \beta m, H_I(t) = e(\alpha, A) - e\Phi$$

Let $V_{12} = V(r_1, r_2)$ be the interaction potential between two particles. Our aim is to formulate the Quantum Boltzmann Equation

$$i\partial_t\rho(t) = [H(t),\rho(t)] + (N-1)Tr_2[V_{12},\rho(t)\otimes\rho(t)]$$

in the position space representation. We first note that $\rho(t,r,r')$ for each $t,r,r'$ is a $4\times 4$ matrix and

$$[H_0,\rho] = [\alpha_k p_k + \beta m, \rho](r,r')$$

$$= \alpha_k(-i\partial_k\rho(t,r,r')) - i\partial'_k\rho(t,r,r')\alpha_k$$

and

$$[\beta,\rho](r,r') = \beta.\rho(t,r,r') - \rho(t,r,r')\beta$$

Further,

$$((\alpha,A)\rho)(r,r') = A_k(t,r)\alpha_k\rho(t,r,r') = (\alpha, A(t,r))\rho(t,r,r')$$

$$(\rho(\alpha,A))(r,r') = \rho(t,r,r')(\alpha, A(t,r'))$$

Thus,

$$[(\alpha,A),\rho](r,r') = (\alpha, A(t,r))\rho(t,r,r') - \rho(t,r,r')(\alpha, A(t,r'))$$

and likewise,

$$[\Phi,\rho](r,r') = (\Phi(t,r) - \Phi(t,r'))\rho(t,r,r')$$

## 4.12 Quantum electrodynamics in a background medium described by a permittivity and permeability function

We have to quantize the Maxwell equations

$$[\epsilon(\mu\nu\alpha\beta, x) * F^{\alpha\beta}(x)]_{,\nu} = 0$$

where $*$ denotes space-time convolution and

$$F_{\mu\nu}(x) = A_{\nu,\mu}(x) - A_{\mu,\nu}(x)$$

The Lagrangian for density for the electromagnetic field is

$$L = (-1/4)F_{\mu\nu}(x).(\epsilon(\mu\nu\rho\sigma, x) * F^{\rho\sigma}(x))$$

The above Maxwell equations can be expressed in the space-time frequency domain as

$$\hat{\epsilon}(\mu\nu\alpha\beta, k)k_\nu \hat{F}^{\alpha\beta}(k) = 0$$

where

$$\hat{F}^{\mu\nu}(k) = \int F^{\mu\nu}(x)exp(-ik.x)d^4x$$

$$\hat{\epsilon}(\mu\nu\alpha\beta, k) = \int \epsilon(\mu\nu\alpha\beta, x)exp(-ik.x)d^4x$$

where

$$k.x = k^0x^0 - \sum_{r=1}^{3} k^r x^r, x^0 = t, (x^r)_{r=1}^3 = \mathbf{r}$$

Further,

$$\hat{F}_{\mu\nu}(k) = k_\mu \hat{A}_\nu(k) - k_\nu \hat{A}_\mu(k)$$

and so our Maxwell equations assume the form

$$\hat{\epsilon}(\mu\nu\alpha\beta, k)k_\nu(k^\alpha \hat{A}^\beta(k) - k^\beta \hat{A}^\alpha(k)) = 0$$

We can adopt the gauge condition

$$\hat{\epsilon}(\mu\nu\alpha\beta, k)k_\nu \hat{A}^\beta(k) = 0$$

and then we get the generalized wave equation in the four wave vector domain

$$\hat{\epsilon}(\mu\nu\alpha\beta, k)k_\nu k^\beta \hat{A}^\alpha(k) = 0 - - - (1)$$

from which the dispersion relation can easily be obtained

$$det((\hat{\epsilon}(\mu\nu\alpha\beta, k)k_\nu k^\beta))_{0\leq\mu,\alpha\leq 3} = 0 - - - (2)$$

Note that this equation is an eighth degree polynomial equation in the $k^\mu$ and hence it will generally have eight solutions for $k^0$ in terms of $(k^r)_{r=1}^3$. In the special case when the medium is the vacuum, we have

$$\hat{\epsilon}(\mu\nu\alpha\beta, k) = \delta_{\mu\alpha}.\delta_{\nu\beta}$$

and the dispersion relation reduces to the standard one

$$k_\nu k^\nu = 0$$

Assume that the solution to the above dispersion relation (1) can be expressed as

$$k^0 = \omega_m(K), K = (k^r)_{r=1}^3, m = 1, 2, ..., 8$$

Also assume that a set of linearly independent eigenvectors $\hat{A}^\mu(k)$ that satisfy both the gauge condition and the dispersion relation are given by

$$e^\mu(K, s), s = 1, 2, ..., 8$$

Then it is clear that the electromagnetic four potential can be expanded as

$$A^\mu(x) = \int [a(K, s)e^\mu(K, s)exp(-ik^{(s)}.x)]d^3K$$

where the sum is over $s = 1, 2, ..., 8$ and

$$k^{(s)} = (\omega_s(K), K), K = (k^r)_{r=1}^3$$

More precisely, we can write

$$A^\mu(x) = \int [a(K, s)e^\mu(K, s)exp(-i(\omega_s(K)t - K.r))d^3K$$

In order that this be a real field we must assume that for each $s$,

$$a(K, s)^* e^\mu(K, s)^* = a(-K, s')e^\mu(-K, s')$$

for some $s'$ and that for this $s'$,

$$-\omega_s(K) = \omega_{s'}(-K)$$

The energy of this electromagnetic field can be obtained in the frequency domain as follows. First, the Lagrangian density in the four wave vector domain is

$$\hat{L} = (-1/4)\epsilon(\mu\nu\alpha\beta, k)\hat{F}_{\mu\nu}(k)\hat{F}^{\alpha\beta}(k)$$

Then the position fields in the four wave vector domain being $\hat{A}_\mu(k)$, it follows that the corresponding momentum fields in the four wave vector domain are

$$\hat{\pi}^\nu(k) = \partial\hat{L}/\partial(k_0\hat{A}_\nu(k))$$

$$= \epsilon(0\nu\alpha\beta, k)\hat{F}^{\alpha\beta}(k)$$

and hence the Hamiltonian density in the four vector domain is given by applying the wave vector domain Legendre transformation to the Lagrangian density:

$$\mathcal{H}(k) = \hat{\pi}^{\nu}(k)k_0\hat{A}_\nu(k) - \hat{L}(k)$$

$$= \epsilon(0\nu\alpha\beta, k)\hat{F}^{\alpha\beta}(k)k_0\hat{A}_\nu(k) + (1/4)\hat{\epsilon}(\mu\nu\alpha\beta, k)\hat{F}_{\mu\nu}(k)\hat{F}^{\alpha\beta}(k)$$

From these calculations, it is clear that the total field Hamiltonian, ie field energy can be expressed as

$$U = \int C_1(\mu\nu\alpha\beta, k)\hat{F}_{\mu\nu}(k)^*\hat{F}_{\alpha\beta}(k)d^3K$$

here it is understood thqat we substitute for $k^0$, its dispersion relation values in terms of $K = (k^r)_{r=1}^3$ and then sum up over all the roots. Here, $C_1(\mu\nu\alpha\beta, k)$ is a function of the permittivity-permeability tensor $\epsilon(\mu\nu\alpha\beta, k)$ (not a functional, simply an ordinary function). Alternately, substituting for $\hat{F}_{\mu\nu}(k)$ its value in term of $\hat{A}_\mu(k)$, we can express the field energy in the form

$$U = \int C(K, s)a(K, s)^*a(K, s)d^3K$$

where the function $C(K, s)$ is derived from $C_1(\mu\nu\alpha\beta, k)$ and the polarization vectors $e^\mu(K, s)$. Note that

$$\hat{F}_{\mu\nu}(k) = k_\mu\hat{A}_\nu(k) - k_\nu\hat{A}_\mu(k)$$

which in turn equals

$$a(K, s)(k_\mu e_\nu(K, s) - k_\nu e_\mu(K, s))\delta(k^0 - \omega_s(K))$$

the sum over $s$ being understood.

## 4.13 Models for the refractive index of a material based on classical and quantum physics

### 4.13.1 Classical physics

Consider an electron of effective mass $m(r)$ and charge $e(r)$ moving around its equilibrium position w.r.t the atomic nucleus. Let $\gamma(r)$ denote the damping coefficient during the electron's motion and let $E(t, r)$ denote the applied external electric field. If $\xi(t)$ is the displacement of the electron relative to the nucleus, then we have from classical Newtonian mechanics,

$$m(r + \xi(t))\xi''(t) + \gamma(r + \xi(t))\xi'(t) + K(r + \xi(t))\xi(t) = -eE(t, r + \xi(t))/m$$

and since $\xi(t)$ is very small, we can assume that

$$K(r+\xi) \approx K(r), \gamma(r+\xi) \approx \gamma(r), m(r+\xi) \approx m(r)$$

Then if the electric field has frequency $\omega$, we can write

$$E(t,r) = Re(E_0(r)exp(i\omega t))$$

and writing

$$\xi(t) = Re(\xi_0 exp(i\omega t))$$

we find on substituting into the equation of motion

$$[m(r)(\omega_0(r)^2 - \omega^2) + i\gamma(r)\omega]\xi_0 = -eE_0(r), (K(r)/m(r))^{1/2} = \omega_0(r)$$

so the dipole moment of the electron can be expressed in the form

$$p(t,r) = -e\xi(t) = Re(p_0(r)exp(i\omega t))$$

where

$$p_0(r) = e^2 E_0(r)/[m(r)(\omega_0(r)^2 - \omega^2) + i\gamma(r)\omega]$$

If there are $N(r)$ atoms per unit volume at $r$, then the polarization phasor (ie dipole moment per unit volume) is given by

$$P(\omega,r) = N(r)e^2 E_0(r)/[m(r)(\omega_0(r)^2 - \omega^2) + i\gamma(r)\omega]$$

from which, we deduce that the complex refractive index of the material as a function of the location is given by

$$n(\omega,r) = (1 + Ne^2\epsilon_0^{-1}/[m(r)(\omega_0(r)^2 - \omega^2) + i\gamma(r)\omega])^{1/2}$$

$$\approx 1 + N(r)e^2\epsilon_0^{-1}/2[m(r)(\omega_0(r)^2 - \omega^2) + i\gamma(r)\omega]$$

for low electron densities $N(r)$

Remark: The complex permittivity is given by

$$\epsilon(\omega,r) = \epsilon_0(1 + n(r))$$

so that if we separate out the real and imaginary parts as

$$\epsilon(\omega,r) = \epsilon_R(\omega,r) + i\epsilon_I(\omega,r)$$

then the true permittivity of the medium is given by $\epsilon_R(\omega,r)$ while the true conductivity is given by

$$\sigma(\omega,r) = -\omega\epsilon_I(\omega,r)$$

In case, the medium in which the electron moves is anisotropic, $m(r), \gamma(r), K(r)$ become $3 \times 3$ matrices and we get the result that the permittivity becomes a $3 \times 3$ matrix given by

$$\epsilon(\omega,r) = \epsilon_0(I_3 + N(r)e^2((K(r) - m(r)\omega^2) + i\gamma(r)\omega)^{-1})$$

This is independent of the temperature and also of the electric. However, it depends on the frequency and hence the wavelength of electromagnetic radiation. Further, this formula can be generalized to the situation in which there are $N(\omega_0, r)d\omega_0$ atoms per unit volume whose electrons are bound to them by spring constants having natural frequencies in the range $[\omega_0, \omega_0 + d\omega_0]$. Then, the total dipole moment per unit volume at frequency $\omega$ is given by

$$P(\omega, r) = [\int N(\omega_0, r)e^2[m(\omega_0, r)(\omega_0^2 - \omega^2) + i\gamma(\omega_0, r)]^{-1}d\omega_0]E_0(\omega, r)$$

which results in a permittivity tensor given by

$$\epsilon(\omega, r) = \epsilon_0 I_3 + \int N(\omega_0, r)e^2[m(\omega_0, r)(\omega_0^2 - \omega^2) + i\gamma(\omega_0, r)]^{-1}d\omega_0$$

In this formula, $m(\omega_0, r)$ and $\gamma(\omega_0, r)$ denote respectively the mass tensor and the damping coefficient tensor for an electron whose nucleus is located at $r$ and which is bound to the nucleus with a spring constant of value $K(\omega_0, r) = m(\omega_0, r)\omega_0^2$.

## 4.13.2 Quantum statistical physics

To get temperature dependence and also field dependence of the refractive index, we have to solve Schrodinger's equation for the electron bound to the nucleus with a Gibbs distribution over the different energy eigenstates.

Specifically suppose that we solve Schrodinger's equation for the unperturbed atom having a single electron. Let its stationary energy eigenstates be denoted by $|u_n(r)>, n = 1, 2, ....$ Let $E_n$ denote the energy of the eigenstate $u_n(r)$. Then, the initial mixed state of the system at temperature $T$ is given by

$$\rho_0(T) = \sum_{n\geq 1} p_n(T)|u_n >< u_n|$$

or more precisely, in the position representation, it is given by

$$\rho_0(T)(r, mr') = \sum_{n\geq 1} p_n(T)u_n(r)u_n(r')^*$$

where

$$p_n(T) = exp(-\beta E_n)/Z(\beta), \beta = 1/kT, Z(\beta) = \sum_n exp(-\beta E_n)$$

Next, we solve the quantum Boltzmann's kinetic transport equation with this initial condition in the presence of an external electromagnetic field $(E, B)$ to obtain the one particle mixed state at time $\tau$ as a function of this external field :

$$\rho_\tau(T) = \mathcal{N}(\tau, E, B\rho_0(T))$$

where $N(.)$ is a nonlinear operator obtained by solving the quantum Boltzmann equation and this nonlinear operator is applied to the initial state $\rho_0(T)$. So after time $\tau$, we can evaluate the average electric and magnetic dipole moment of the electron in this state as a function of the temperature and the electromagnetic field and on noting that the strength of the electomagnetic field is a function of the wavelength/frequency, we obtain the quantum averaged polarization and magnetization as a function of temperature and wavelength. The cumulative distribution function of the refractive index can also be derived using our mixed state at time $\tau$. In fact, if $X$ is any observable, its cumulative distribution function in the state $\rho$ is given by

$$F_\rho(x) = Tr[\rho.\chi_{(-\infty,x]}(X)]$$

where $\chi_E(x)$ is the indicator function of the set $E \subset \mathbb{R}$.

## 4.14 Quantum statistical field theory

Let $\psi(t,r)$ denote the wave operator field of second quantized matter. The second quantized Hamiltonian in the Dirac picture is given by

$$H = \int \psi(t,r)^*((\alpha, -i\nabla + eA(r)) + \beta m)\psi(t,r)d^3r$$
$$+ \int V_{\mu\nu}(r,r')\psi(t,r)^*\alpha^\mu\psi(t,r)\psi(t,r')^*\alpha^\nu\psi(t,r')$$

just as in the Hartree-Fock theory. The wave operator fields satisfy the canonical equal time anticommutation relations

$$\{\psi_l(t,r), \psi_m(t,r')^*\} = \delta_{lm}\delta^3(r-r')$$

and the second term in the second quantized Hamiltonian represents the interaction between Dirac charges and currents at two different spatial points. The wave operator fields $\psi(t,r)$ evolve according to the Heisenberg dynamics

$$\partial_t\psi(t,r) = i[H, \psi(t,r)]$$

The electronic polarization operator field is given by

$$P(t,r) = -e\mathbf{r}\psi(t,r)^*\psi(t,r)$$

This is the dipole moment operator per unit volume. Let $\rho_0$ denote the initial state of the quantum system, say the Gibbs state:

$$\rho_0 = exp(-\beta H)/Z(\beta), Z(\beta) = Tr(exp(-\beta H)), \beta = 1/kT$$

Since we are adopting the Heisenberg picture, this state does not evolve with time, only the observables evolve with time. The average Polarization of the medium at time $t$ is therefore given by

$$< P > (t,r) = Tr(\rho_0.P(t,r))$$

The magnetic dipole moment operator field per unit volume is given by

$$M(t,r) = \psi(t,r)^*(-e(\mathbf{L}+g\sigma)/2m)\psi(t,r)$$

and its average value is given by

$$< M > (t,r) = Tr(\rho_0 M(t,r))$$

To calculate these averages, we must first determine the dynamics of the temperature Green's function

$$G(t,r|t',r') = Tr(\rho_0 T\{\psi(t,r)\psi(t',r')^*\})$$

where $T$ is the time ordering operator. Note that since the magnetic vector potential is not assumed to vary with time, the total Hamiltonian operator is a constant of the motion and hence so is the density operator $\rho_0$. Note that from the canonical anticommutation rules,

$$[\psi(t,r')^*\alpha^\mu\psi(t,r'),\psi_k(t,r)] =$$

$$= [\alpha^\mu(l,m)\psi_l(t,r')^*\psi_m(t,r'),\psi_k(t,r)] =$$

$$-\alpha^\mu(l,m)\delta_{lk}\delta^3(r'-r)\psi_m(t,r')$$

$$= -\delta^3(r-r')\alpha^\mu(k,m)\psi_m(t,r) = -\delta^3(r-r')[\alpha^\mu\psi(t,r)]_k$$

Equivalently, in vector notation,

$$[\psi(t,r')^*\alpha^\mu\psi(t,r'),\psi(t,r)] = -\delta^3(r-r')\alpha^\mu\psi(t,r)$$

It follows that if we define

$$J^\mu(t,r) = \psi(t,r)^*\alpha^\mu\psi(t,r)$$

then

$$[J^\mu(t,r'),\psi(t,r)] = -\delta^3(r-r')\alpha^\mu\psi(t,r)$$

and therefore,

$$[J^\mu(t,r')J^\nu(t,r''),\psi(t,r)] =$$

$$-[\delta^3(r-r'')J^\mu(t,r')\alpha^\nu\psi(t,r) + \delta^3(r'-r)\alpha^\mu\psi(t,r)J^\nu(t,r'')]$$

Then,

$$[\int V_{\mu\nu}(r',r'')J^\mu(r')J^\nu(r'')d^3r'd^3r'',\psi(t,r)] =$$

$$-[\int V_{\mu\nu}(r',r)J^\mu(t,r')d^3r']\alpha^\nu\psi(t,r)$$

$$-\int \alpha^\mu\psi(t,r)[\int V_{\mu\nu}(r,r')J^\nu(t,r')d^3r']$$

We may assume without loss of generality that

$$V_{\mu\nu}(r, r') = V_{\nu\mu}(r', r)$$

and then deduce that

$$[\int V_{\mu\nu}(r', r'')J^\mu(r')J^\nu(r'')d^3r'd^3r'', \psi(t, r)] =$$

$$= -\{\alpha^\mu\psi(t, r), \int V_{\mu\nu}(r, r')J^\nu(t, r')d^3r'\}$$

We therefore obtain from the Heisenberg dynamics the following dynamical equation for the wave field operator $\psi(t, r)$.

$$i\partial_t\psi(t, r) = H_{D0}\psi(t, r) + \{\alpha^\mu\psi(t, r), \int V_{\mu\nu}(r, r')J^\nu(t, r')d^3r'\}$$

where

$$H_{D0} = (\alpha, -i\nabla) + \beta m$$

is the first quantized free particle Dirac Hamiltonian. We can further approximate this equation by replacing $J^\mu(t, r')$ on the rhs by its quantum average

$$< J^\mu(t, r) >= Tr(\rho_0(T)J^\mu(t, r))$$

It should be noted that this average current density can be expressed in terms of the temperature Green's function as

$$< J^\mu(t, r) >= -Tr[\alpha^\mu G(t, r|t, r')]|_{r' \to r}$$

where the trace here is an ordinary matrix trace for $4 \times 4$ matrices.

Reference:Fetter and Walecka, "Quantum Theory of Many Particle Systems", Dover, 1971.

## 4.15 Relating the refractive index of a material to the metric tensor of space-time

The curvature of space-time affects quantum phenomena. For example, in order to take into account the space-time curvature, we have to write down Dirac's equation in curved space-time and then formulate the quantum Boltzmann equation by starting from such a generalized Dirac equation. This is accomplished as follows. Let $V_a^\mu$ be a tetrad basis for our curved space-time and let $\Gamma_\mu = \Gamma^\mu_{ab}[\gamma^a, \gamma^b]$ denote the spinor connection of the gravitational field. Then, the four component wave function satisfies

$$[V_a^\mu\gamma^a(i\partial_\mu + eA_\mu + i\Gamma_\mu) - m]\psi(x) = 0$$

From this equation, we can infer what the generalized Dirac Hamiltonian must be. This is achieved by separating the time derivative component from the spatial derivative components:

$$iV_a^0\gamma^a\partial_0\psi + [iV_a^r\gamma^a\partial_r + eV_a^\mu\gamma^a A_\mu + eV_a^\mu\gamma^a A_\mu + iV_a^\mu\gamma^a\Gamma_\mu - V_a^\mu\gamma^a m]\psi = 0$$

Now multiplying both sides of this equation by $V_b^0\gamma^b$ and using

$$V_a^0 V_b^a\gamma^a\gamma^b = (1/2)\eta^{ab}V_a^0 V_b^0 = (1/2)g^{00}$$

where $\eta^{ab}$ is the Minkowski metric of flat space-time and $g^{\mu\nu}$ is the exact contravariant metric of our curved space-time, we get

$$ig^{00}\partial_0\psi + [iV_b^0 V_a^r\gamma^b\gamma^a\partial_r + eV_b^0 V_a^\mu\gamma^b\gamma^a A_\mu + iV_b^0 V_a^\mu\gamma^b\gamma^a\Gamma_\mu - V_b^0 V_a^\mu\gamma^b\gamma^a m]\psi = 0$$

This equation can be expressed in the standard Hamiltonian form by defining the curved space time Dirac Hamiltonian in an electromagnetic field as

$$H = (g^{00})^{-1}V_b^0 V_a^r\gamma^b\gamma^a(-i\partial_r) - (g^{00})^{-1}(eV_b^0 V_a^\mu\gamma^b\gamma^a A_\mu) + (g^{00})^{-1}V_b^0 V_a^\mu\gamma^b\gamma^a(-i\Gamma_\mu)$$
$$+(g^{00})^{-1}V_b^0 V_a^\mu\gamma^b\gamma^a m]$$

As an example of this calculation, consider the Schwarzchild metric in which

$$g_{00} = \alpha(r) = 1 - 2m/r, g_1 = -\alpha(r)^{-1}, g_{22} = -r^2, g_{33} = -r^2 sin^2(\theta)$$

We have

$$d\tau^2 = g_{\mu\nu}dx^\mu dx^\nu = (\omega_0)^2 - \omega_1^2 - \omega_2^2 - \omega_3^2$$

where

$$\omega_0 = \sqrt{\alpha(r)}dt, \omega_1 = \sqrt{\alpha(r)^{-1}}dr,$$
$$\omega_2 = rd\theta, \omega_3 = rsin(\theta)d\phi$$

Thus, since

$$g^{\mu\nu} = \eta^{ab}V_a^\mu V_b^\nu, g_{\mu\nu} = \eta_{ab}V_\mu^a V_\nu^b$$

we get

$$d\tau^2 = \eta_{ab}V_\mu^a V_\nu^b dx^\mu dx^\nu$$
$$= (V_\mu^0 dx^\mu)^2 - (V_\mu^1 dx^\mu)^2 - (V_\mu^2 dx^\mu)^2 - (V_\mu^3 dx^\mu)^2$$

Thus,

$$\omega_0 = \sqrt{\alpha(r)}dt = V_\mu^0 dx^\mu = V_0^0 dt + V_1^0 dr + V_2^0 d\theta + V_3^0 d\phi,$$

so that

$$\sqrt{\alpha(r)} = V_0^0, V_1^0 = V_2^0 = V_3^0 = 0,$$

Note that in these conventions,

$$(V_\mu^0) = (V_0^0, V_1^0, V_2^0, V_3^0)$$

Likewise,

$$\omega_1 = V_\mu^1 dx^\mu = \alpha(r)^{-1/2}dr$$

and therefore,

$$V_0^1 = 0, V_1^1 = \alpha(r)^{-1/2}, V_2^1 = 0, V_3^1 = 0$$

Note that

$$(V_\mu^1) = (V_0^1, V_1^1, V_2^1, V_3^1)$$

$$\omega_2 = V_\mu^2 dx^\mu = rd\theta, \omega_3 = V_\mu^3 dx^\mu = r.sin(\theta)d\phi$$

so that

$$V_0^2 = 0, V_1^2 = 0, V_2^2 = r, V_3^2 = 0,$$

$$V_0^3 = 0, V_1^3 = 0, V_2^3 = 0, V_3^3 = r.sin(\theta)$$

The spinor connection of the gravitational field is given by

$$\Gamma_\mu = V_{a\nu} V_{b:\mu}^\nu [\gamma^a, \gamma^b] = \omega_\mu^{ab} [\gamma^a, \gamma^b]$$

This can be derived in various ways, one by using the fact that the covariant derivative of the tetrad is zero, ie

$$V_{\mu,\nu}^a - \Gamma_{\mu\nu}^\rho V_\rho^a + \omega_\nu^{ab} V_{b\mu} = 0$$

or equivalently,

$$V_{\mu:\nu}^a + \omega_\nu^{ab} V_{b\mu} = 0$$

The second derivation is based on starting with the Einstein-Hilbert Lagrangian for the gravitational field, namely the curvature tensor in spinor notation:

$$R = R_{\mu\nu}^{ab} V_a^\mu V_b^\nu,$$

$$R_{\mu\nu}^{ab} = \omega_{\nu,\mu}^{ab} - \omega_{\mu,\nu}^{ab} - [\omega_\mu, \omega_\nu]^{ab}$$

and setting the variational derivative of $\int R d^4x$ w.r.t $\omega_\mu^{ab}$ to zero to arrive at an algebraic equation for $\omega_\mu^{ab}$ which when solved yields the desired form of the gravitational spinor connection. The third and most interesting way to derive the form of the spinor connection of the gravitational field is based on the use of non-Abelian gauge group theory with the gauge group now being the spinor representation of the Lorentz group. In other words, the transformation law of the spinor connection of the gravitational field under a local Lorentz transformation should be such that the Dirac equation remains invariant under it. Let $\Gamma_\mu(x)$ denote the spinor connection in one frame. The corresponding covariant derivative is $\nabla_\mu = \partial_\mu + \Gamma_\mu$. Now suppose we apply a local Lorentz transformation $\Lambda(x)$. Then the Dirac wave function transforms from $\psi(x)$ to $D(\Lambda(x))\psi(x)$ and for the Dirac equation to remain invariant under this transformation, we require that $\Gamma_\mu(x)$ should transform to $\Gamma'_\mu(x)$ where

$$D(\Lambda(x))(\partial_\mu + \Gamma_\mu(x))D(\Lambda(x))^{-1} =$$

$$\partial_\mu + \Gamma'_\mu(x)$$

This is equivalent to requiring that

$$\Gamma'_\mu(x) = D(\Lambda(x))\Gamma_\mu(x)D(\Lambda(x))^{-1} + D(\Lambda(x))(\partial_\mu D(\Lambda(x))^{-1})$$

or equivalently that

$$\Gamma'_\mu(x) = D(\Lambda(x))\Gamma_\mu(x)D(\Lambda(x))^{-1} - (\partial_\mu D(\Lambda(x)))D(\Lambda(x))^{-1}$$

Equivalently, if

$$\Lambda(x) = I + \omega(x)$$

is an infinitesimal local Lorentz transformation, then we require that

$$\delta\Gamma_\mu(x) = \Gamma'_\mu(x) - \Gamma_\mu(x) =$$

$$[D(\omega(x)), \Gamma_\mu(x)] + \partial_\mu D(\omega(x))$$

where

$$D(\omega(x)) = (1/4)[\gamma^a, \gamma^b]\omega_{ab}(x)$$

is the differential of the spinor representation (interpreted in terms of Lie algebra representations) of the Lorentz group evaluated at $\omega(x)$. On the other hand, if we take $\Gamma_\mu(x) = V_{a\nu:\mu}V_b^\nu[\gamma^a, \gamma^b]/2$, then under this infinitesimal local Lorentz transformation, it changes by

$$\delta\Gamma_\mu(x) = ([\gamma^a, \gamma^b]/4)[(\delta V_{a\nu:\mu})V_b^\nu + V_{a\nu:\mu}\delta V_b^\nu]$$

where

$$\delta V_{a\nu} = \omega_a^b(x)V_{b\nu}$$

so that

$$\delta V_{a\nu:\mu} = \omega_a^b V_{b\nu:\mu} + \omega^b_{a,\mu}V_{b\nu}$$

since the $\omega_b^a$'s and $\omega_{ab}$'s transform as scalar fields under diffeomorphisms of space-time. By comparing these two transformation laws, we can show that they are identical using the canonical commutation relations for the Loretnz algebra generators.

## 4.16 Cosmological effects on the refractive index

[a] Classical analysis, main idea: The metric of space time is assumed to be the Robertson-Walker metric corresponding to a homogeneous, isotropic expanding universe:

$$d\tau^2 = dt^2 - S^2(t)f(r)dr^2 - r^2(d\theta^2 + sin^2(\theta)d\phi^2)$$

so that

$$g_{00} = 1, g_{11} = -S^2(t)f(r), g_{22} = -r^2, g_{33} = -r^2 sin^2(\theta)$$

This metric describes a comoving frame, ie, a frame in which a particle at rest, ie having fixed spatial coordinates $r, \theta, \phi$ satisfies the geodesic equations. In order to study the effect of the expanding universe on the refractive index of a body, we have to formulate the Vlasov equations for the particle distribution function and the electromagnetic field in this background metric and then compute statistical averages of the electric and magnetic dipole moments of a charge. The Boltzmann equation in this metric is expressed as

$$\partial_t f(t,\mathbf{r},\mathbf{v})+(\mathbf{v},\nabla_e)d(t,\mathbf{r},\mathbf{v})+(\mathbf{F}(t,\mathbf{r},\mathbf{v}),\nabla_v)f(t,\mathbf{r},\mathbf{v}) = (f_0(\mathbf{v})-f(t,\mathbf{r},\mathbf{v}))/\tau(\mathbf{v}$$

where

$$\mathbf{v} = d\mathbf{r}/dt, v^r = dx^r/dt, r = 1,2,3$$

and $\mathbf{F}(t,\mathbf{r},\mathbf{v})$ is determined from the equation of motion of a charged particle in the curved space-time metric of Robertson and Walker and also under the influence of an electromagentic field:

$$du^\mu/d\tau + \Gamma^\mu_{\alpha\beta}(x)u^\alpha u^\beta = eF^{\mu\nu}u_\nu$$

where

$$u^\mu = dx^\mu/d\tau, d\tau = dt(g_{00} + 2g_{0r}v^r + g_{rs}v^r v^s)^{1/2}$$

The Maxwell equations are expressed as

$$F_{\mu\nu} = A_{\nu,\mu} - A_{\mu,\nu},$$

$$F^{\mu\nu}_{:\nu} = \mu_0 J^\mu$$

where

$$J^0 = q\int f(t,\mathbf{r},\mathbf{v})d^3v, J^r = q\int v^r f(t,\mathbf{r},\mathbf{v})d^3v$$

It should be noted that if $m_0$ is the mass of each charge, then the energy-momentum tensor of this charged matter fluid is given by

$$T^{\mu\nu}(x) = m_0\int f(t,\mathbf{r},\mathbf{v})u^\mu u^\nu d^3v$$

where

$$u^0 = dt/d\tau = (g_{00}(t,r) + 2g_{0r}(t,r)v^r + g_{rs}(t,r)v^r v^s)^{-1/2},$$

$$u^r = dx^r/d\tau = v^r dt/d\tau = v^r u^0$$

Remarks on the quantization of the energy-momentum tensor of a system of $N$ particles and the Einstein field equations.

[a] Let $r_1,...,r_N$ denote the position operators of $N$ particles and let $p^\mu_k, k = 1,2,...,N$ denote their four momentum operators. Thus, $p^0_k = H_{0k}$ is the energy operator of the $k^{th}$ particle and $\mathbf{p}^r = (p^r_k, r = 1,2,3)$ are the Cartesian components of the three momentum vector of the $k^{th}$ particle. According to Dirac's relativistic theory of the electron, if the particles do not interact, then

$$H_{0k} = (\alpha,\mathbf{p}_k) + \beta m_k = H_{0k}(\mathbf{p}_k)$$

assuming that the $k^{th}$ particle has mass $m_k$. More precisely, we should use the curved space-time Dirac Hamiltonian for $H_{0k}$ in terms of the background metric and an associated tetrad basis. We denote this Hamiltonian by $H_{0k}(\mathbf{r}_k\mathbf{p}_k, g_{\mu\nu})$.

Then the energy-momentum tensor operator field of the matter field comprising the $N$ particles is given by

$$T^{\mu\nu} = \sum_{k=1}^{N} m_k \delta^3(r - r_k)(-g(r_k))^{-1/2} p_k^\mu p_k^\nu / H_{0k}$$

We are assuming that although the particle's motion is quantized, the metric of space-time is classical which means that the Einstein field equations are

$$G^{\mu\nu}(x) = R^{\mu\nu}(x) - (1/2)Rg^{\mu\nu}(x) = -8\pi G < T^{\mu\nu}(x) >$$

where the average $< . >$ is a quantum average that is taken with respect to the evolving wave function or mixed state of the system of $N$ particles. In this formalism, we are working in the Schrodinger picture in which states evolve with time but observables remain constant in time. The mixed state of the system of $N$ particles $\rho(t)$ satisfies the quantum Liouville equation:

$$i\rho'(t) = [\sum_{k=1}^{N} H_{0k}, \rho(t)]$$

where in each component $H_{0k}, k = 1, 2, ..., N$, identical copies of the same Dirac matrices are taken but acting on different tensor product Hilbert space components. Thus, the Hilbert space of the system of $N$ particles is given by

$$\mathcal{H} = \bigotimes_{k=1}^{N} \mathcal{H}_k, \mathcal{H}_k = L^2(\mathbb{R}^3) \otimes \mathbb{C}^4$$

If all the particles are identical then we can use the Boltzmann equation approximation by considering the marginal state $\rho_1(t)$ of just one particle which may be interacting with the other particles also. $\rho_1(t)$ will approximately satisfy an equation of the form

$$i\rho_1'(t) = [H_{01}(r_1, p_1, g_{\mu\nu}), \rho_1(t)] + (N-1)Tr_2[V(r_1, p_1, r_2, p_2), \rho_1(t) \otimes \rho_1(t)]$$

With this quantum Boltzmann equation approximation, we have the following approximation for the quantum averaged energy-momentum tensor:

$$< T^{\mu\nu}(t, r) >= NTr(\rho_1(t)\delta^3(r - r_1)(-g(r)^{-1/2}[p_1^\mu p_1^\nu / H_{01}(r_1, p_1, g_{\mu\nu})]$$

Note that to keep this average real, we may interpret $p_1^\mu p_1^\nu / H_{01}$ as

$$(1/2)(p_1^\mu H_{01}^{-1} p_1^\nu + p_1^\nu H_{01}^{-1} p_1^\mu]$$

## 4.17 Glossary of symbols

:

$A_\mu(x)$ Covariant components of the electromagnetic four potential.

$A^\mu(x)$ Contravariant components of the electromagnetic four potential.

$F_{\mu\nu}(x)$ Covariant components of the antisymmetric electromagnetic field tensor. $F_{0r} = -F_{r0}, r = 1, 2, 3$ are the electric field components while $F_{12}, F_{23}, F_{31}$ are the magnetic field components.

$\rho(t)$ density matrix representing a mixed state of a quantum system at time $t$.

$\rho(t, r, r') =< r|\rho(t)|r' >$ position space representation of the density matrix of a mixed state of a quantum system.

$\rho_{12..N}(t)$ Joint mixed state of $N$ particles of a quantum system.

$Tr_{23...N}\rho_{12...N}(t) = \rho_1(t)$ marginal mixed state of the first particle of an $N$ particle quantum system. $Tr_{23...N}$ denotes the partial trace operation. In the position space representation,

$$[Tr_{23...N}\rho_{12..N}](t, r_1, r_1') = \int \rho_{12...N}(t, r_1, r_2..., r_N, r_1', r_2..., r_N)d^3r_2...d^3r_N$$

or equivalently in terms of countable orthonormal bases,

$$< e_{i_1}|[Tr_{23...N}\rho_{12...N}](t)|e_{j_1} >=$$

$$\sum_{i_2,...,i_N} < e_{i_1} \otimes e_{i_2} \otimes ... \otimes e_{i_N}|\rho_{12...N}|e_{j_1} \otimes e_{2,i_2} \otimes ... \otimes e_{N,i_N} >$$

$\psi(x)$: four component Dirac wave function, also called a bispinor.

$\Gamma_\mu(x)$: Spinor connection of the gravitational field. Used to determine the effect of gravity on Dirac's wave function for relativistic quantum mechanics.

$S(t)$ Radius of the expanding universe at the epoch $t$.

$V_a^\mu(x)$ Tetrad basis for the metric of space-time. This enables us to express the metric locally in Minkowski form. In other words, it can be used to describe a locally inertial frame.

## 4.18 A Model for the refractive index of materials and liquids based on cosmological and quantum mechanical considerations

**Abstract**

Abstract: We describe a mathematical model based on classical and quantum field theory and statistical field theory for correlating the refractive indices of materials with experiment. The model described in this paper gives a direct approach to computing the RI of a material based on Dirac's quantum mechanics for a system of $N$ interacting particles in an external electromagnetic field taking corrections due to gravitational effects into account. If we solve the Dirac equation using perturbation theory for a single particle in an electromagnetic field we obtain the wave function in the position space representation. We could calculate the quantum averaged electric and magnetic dipole moment of the electron taken with respect to the probability density of the position of the electron described according to Max Born's interpretation of the wave function. This would in turn enable us to determine the permittivity and permeability of the medium in terms of the electric and magnetic fields. However, this analysis does not show how the RI depends upon the temperature of the material. In order to obtain temperature dependence, we consider Dirac's quantum mechanics for an N particle system taking interparticle interactions into account apart from interaction of the particles with an external electromagnetic field. By partial tracing the mixed state Dirac equation over the other particles and then making some approximations we derive a quantum Boltzmann equation for the quantum density operator. This equation is solved using perturbation theory with the initial state as the Gibbs state (which has temperature dependence). The final equilibrium state in the presence of a static electromagnetic field and interparticle interactions will then also depend upon temperature. When this final density matrix is used to compute quantum averages of the electric and magnetic dipole moment, we are able to explain the dependence of the RI on both the electromagnetic field and temperature. The wavelength dependence of the RI can be explained by assuming the background electromagnetic field to be black-body radiation which has the energy density of the electromagnetic field dependent upon both frequency/wavelength and temperature. To this model of the RI, we add cosmological and background gravitational correction effects based on the follwing idea: Gravity affects quantum mechanics via the spinor connection of the gravitational field. This to be introduced into Dirac's equation in order to make it invariant under local Loretnz transformations and arbitrary diffeomorphisms of space-time. Thus, this general relativistic generalization of Dirac's equation gives us the dependence of the wave function on the background metric tensor of curved space-time. We give three independent derivations for the spinor connection of the gravitational field based on standard arguments in gauge field theory and spinor forms of the Riemann curvature tensor. If this background metric is taken to be the Schwarzchild metric, the wave function would depend upon the mass of the blackhole and the gravitational constant while if it is taken to be Robertson-Walker metric for an expanding homogeneous and isotropic universe, then the wave function will also depend on the radius of the universe and hence on Hubble's constant. Calculating the average electric and magnetic dipole moments w.r.t such a wave function would then yield the dependence of the RI on the radius of the expanding universe and on its curvature. By taking fine measurements of the RI, we would then in principle be able to measure Hubble's constant and hence the radius of the universe at the present epoch. It should be noted that using the quantum Boltzmann equation or its gravitationally modified version is based on the quantum theory for a finite number of indistinguishable particles, ie, it is a first quantization approach. If the number of particles is infinite in number, we then have to adopt a second quantization approach based on Fermionic field operators for the Dirac wave function. We explain how to set up such a second quantized Hamiltonian and thereby calculate temperature Green's functions for these by assuming that the state of the second quantized field is given by the Gibbs density with the unperturbed second quantized Hamiltonian. Using this temperature Green's function, we evaluate the average polarization and magnetization of the field as a function of the electromagnetic field, the temperature and the background gravitational field. The final parts of the manuscript focus on deriving the basic cosmological equations for the expanding universe using Newtonian and Eulerian fluid mechanics in terms of the scale factor/radius of the universe and the propagation of inhomogeneities in the matter, temperature and electromagnetic field in such a uniformly expanding universe. The idea is that the perturbations to an initially applied electromagnetic field will depend on the scale factor and temperature and hence if we use classical statistical mechanics to calculate the average electric and magnetic dipole moments in such an electromagnetic field, then the permittivity, permeability and hence the refractive index will also depend upon the scale factor and the temperature. Further, we observe that if the dynamical equations of the expanding universe are quantized (just as Hawking determined the wave function of the radius of the expanding universe) using the Lindlbad open quantum system formalism, then we can in principle calculate the evolving state of the matter velocity, density, temperature and electromagnetic fields using which we can determine the quantum fluctuations in the electric and magnetic dipole moments of a system of charges which would in turn give us the mean square quantum fluctuations in the refractive index. An appendix has been included containing a brief and elementary derivation of the Lindblad master equation for an open quantum system, ie, a system interacting with a bath. The manuscript also contains a short look at what the quantum Boltzmann equation will look like for an open quantum system described by Lindlbad operators.

**1. The quantum Boltzmann equation for a plasma**

In this section, we derive an approximate nonlinear evolution equation for the density operator of a single particle when the quantum plasma consists of $N$ identical particles interacting with each other and also with an external electromagnetic field. The joint density operator of the $N$ particles satisfies the quantum Liouville or Schrodinger-Von-Neumann equation with the Hamiltonian consisting of a sum of identical Hamiltonians each acting in a single particle Hilbert space plus the sum of identical pairwise interacting potentials of two particles with each one acting in the tensor product of two identical Hilbert spaces. By taking the partial trace of this quantum Liouville equation and making approximations (which in the classical Boltzmann kinetic transport theory are called the molecular chaos approximation), we derive an approximate quadratic nonlinear evolution equation for the single particle density operator in an external electromagnetic field. The single particle Hamiltonians can either be the single particle Schrodinger equation in an external electromagnetic field or a single particle Dirac Hamiltonian or even a single particle Dirac Hamiltonian in curved space-time interacting with an external electromagnetic field. The quadratic nonlinear terms which arise due to the pairwise interaction of particles represent quantum generalization of the so called "collsion term" that appears in the classical Boltzmann equation in kinetic transport theory and which are usually evaluated using classical scattering theory or more specifically using binary elastic collision theory of two particles. It should be noted that our method of deriving the quantum Boltzmann equation by partial tracing is the quantum analogue of the classical BBGKY theory in which one writes down the classical Liouville equation for the distribution function of $N$ particles in phase space (ie, in the joint position-velocity space of all the $N$ particles) and then integrates this equation over the phase space variables of all but the first particles and then makes the molecular chaos approximation in which the joint distribution of two particles is approximated by a product of the individual distributions.

Suppose that the joint density matrix of $N$ particles is $\rho(123...N)$. It satisfies the Schrodinger equation

$$i\partial_t \rho_t(12...N) = [\sum_{a=1}^{N} H_a + \sum_{1\leq a<b\leq N} V_{ab}, \rho_t(12..N)]$$

In this equation, if we take a partial traced over $2, 3, ..., N$, we get

$$'i\partial_t \rho_{1t} = [H_1, \rho_{1t}] + (N-1)Tr_2[V_{12}, \rho_{12}]$$

and if we take the trace of the same over $3, 4, ..., N$, we get

$$i\partial_t \rho_{12t} = [H_1 + H_2 + V_{12}, \rho_{12t}] + (N-2)Tr_3[V_{13} + V_{23}, \rho_{123t}]$$

We write

$$\rho_{123} = (1/3)(\rho_{12} \otimes \rho_3 + \rho_{13} \otimes \rho_2 + \rho_1 \otimes \rho_{23}) + g_{123}$$

where $g_{123}$ is small. Then, neglecting second order of smallness terms like $V$ multiplied with $g_{123}$ gives us the approximate equation

$$i\partial_t \rho_{12t} = [H_1 + H_2 + V_{12}, \rho_{12}] + ((N-2)/3)Tr_3[V_{13} + V_{23}, \rho_{12} \otimes \rho_3 + \rho_{13} \otimes \rho_2 + \rho_1 \otimes \rho_{23}]$$

This is a bit hard to handle. So we content ourselves with the approximation

$$\rho_{12} = \rho_1 \otimes \rho_1 + g_{12}$$

where $g_{12}$ is small. We then get approximately,

$$i\partial_t \rho_{1t} = [H_1, \rho_{1t}] + (N-1)Tr_2[V_{12}, \rho_1 \otimes \rho_1]$$

Writing

$$V_{12} = \sum_a W_{1a} \otimes W_{2a}$$

gives us

$$Tr_2[V_{12}, \rho_1 \otimes \rho_2] = \sum_a Tr(\rho_1 W_{2a})[W_{1a}, \rho_1]$$

and the our Boltzmann equation becomes

$$i\partial_t \rho_1 = [H_1, \rho_1] + (N-1)\sum_a Tr(\rho_1 W_{2a})[W_{1a}, \rho_1]$$

Suppose we make the approximation

$$\rho_{123} = \rho_1 \otimes \rho_1 \otimes \rho_1 + g_{123}$$

where $g_{123}$ is small. Then we get

$$i\partial_t \rho_{12t} = [H_1 + H_2 + V_{12}, \rho_{12}] + (N-2)Tr_3[V_{13} + V_{23}, \rho_1 \otimes \rho_1 \otimes \rho_1]$$

Even this equation is hard to manipulate further without assuming some specific form of the interaction potential $V_{12}$. We consider

$$\rho_{12} = \rho_1 \otimes \rho_1 + g_{12},$$

$$\rho_{123} = (1/3)(\rho_{12} \otimes \rho_3 + \rho_{13} \otimes \rho_2 + \rho_1 \otimes \rho_{23}) + g_{123}$$
$$= \rho_1 \otimes \rho_1 \otimes \rho_1 + (1/3(g_{12} \otimes \rho_3 + g_{13} \otimes \rho_2 + \rho_1 \otimes g_{23}) + g_{123}$$

We first derive a differential equation for $g_{12}$ after neglecting second order of smallness terms:

$$i\partial_t \rho_{12} = i\partial_t \rho_1 \otimes \rho_1 + i\rho_1 \otimes \partial_t \rho_1$$

$$+i\partial_t g_{12}$$

$$= [H_1, \rho_1] \otimes \rho_1 + (N-1)Tr_2[V_{12}, \rho_1 \otimes \rho_1] \otimes \rho_1 + \rho_1 \otimes [H_1, \rho_1]$$

$$+(N-1)\rho_1 \otimes Tr_2[V_{12}, \rho_1 \otimes \rho_1] + i\partial_t g_{12}$$

$$= [H_1 + H_2 + V_{12}, \rho_{12}] + (N-2)Tr_3[V_{12} + V_{13}, \rho_{123}]$$

$$= [H_1 + H_2, \rho_1 \otimes \rho_1] + [V_{12}, \rho_1 \otimes \rho_1]$$

$$+(N-2)Tr_3[V_{13} + V_{23}, \rho_1 \otimes \rho_1 \otimes \rho_1]$$

After making the appropriate cancellations, we get

$$i\partial_t g_{12} =$$

$$= [V_{12}, \rho_1 \otimes \rho_1] + [H_1 + H_2, g_{12}] + (N-2)Tr_3[V_{13} + V_{23}, \rho_1 \otimes \rho_1 \otimes \rho_1]$$

$$-(N-1)Tr_2[V_{12}, \rho_1 \otimes \rho_1] \otimes \rho_1$$

$$-(N-1)\rho_1 \times Tr_2[V_{12}, \rho_1 \otimes \rho_1]$$

Note that on writing

$$V_{12} = \sum_a W_{1a} \otimes W_{2a}$$

and using the fact that the $V'_{jk}s$ are identical copies of each other acting on different copies of the tensor product of two identical copies a Hilbert space just as the $H'_k s$ are identical copies of each other acting on different copies of the same Hilbert space, we get

$$Tr_3[V_{13} + V_{23}, \rho_1 \otimes \rho_1 \otimes \rho_1]$$

$$= \sum_a [Tr(\rho_1 W_{2a})([W_{1a}, \rho_1] \otimes \rho_1 + \rho_1 \otimes [W_{1a}, \rho_1])]$$

A better approximation to the quantum Boltzmann equation can then be obtained by solving this equation for $g_{12}(t)$ and substituting it into the equation

$$i\partial_t \rho_1 = [H_1, \rho_1] + (N-1)Tr_2[V_{12}, \rho_{12}]$$

$$= [H_1, \rho_1] + (N-1)Tr_2[V_{12}, \rho_1 \otimes \rho_1 + g_{12}]$$

Formally this equation has the form

$$i\partial_t \rho_1(t) = [H_1, \rho_1(t)] + \delta.F_1(\rho_1(s), s \le t)$$

where $F$ is an operator valued nonlinear functional of $\rho_1(s), s \le t$. This equation can be solved upto $O(\delta)$ using first order perturbation theory:

$$\rho_1(t) = U(t)\rho_1(0)U(t)^* + \delta. \int_0^t U(t-\tau)F_\tau(\rho_1(s), s \le \tau)U(t-\tau)^* d\tau$$

where

$$U(t) = exp(-itH_1)$$

If we consider the Hamiltonian to comprise of an interaction between the particles and an electromagnetic field, then we can write

$$H_1 = (p_1+eA(t,r))^2/2m-e\Phi(t,r) \approx p_1^2/2m-e\Phi(t,r)+(e/2m)((p_1,A)+(A,p_1))$$

and the particle interaction potential as

$$V_{12} = V(|r_1 - r_2|)$$

In that case, in the position representation, we have

$$[p_1^2, \rho_1] = [p_1, \rho_1].p_1 + p_1.[p_1, \rho_1]$$

and noting that $p_1$ is represented by the kernel

$$p_1(r, r') = -i\nabla_r \delta^3(r - r') = i\nabla'_r \delta^3(r - r')$$

we get

$$[p_1, \rho_1](r, r') = -i\nabla_r \rho_1(r, r') + i\nabla'_r \rho_1(r, r')$$

$$[p_1, \rho_1].p_1(r, r') = \nabla_r.\nabla'_r \rho_1(r, r') - \nabla_r'^2 \rho_1(r, r')$$

and likewise,

$$p_1.[p_1, \rho_1](r, r') = \nabla_r.\nabla'_r \rho_1(r, r') - \nabla_r^2 \rho_1(r, r')$$

$$[(A, p_1), \rho_1](r, r') = [A_k p_{1k}, \rho_1](r, r') =$$

$$A_k(t, r)p_{1k}\rho_1(r, r') - \int \rho_1(r, r'')A_k(t, r'')p_{1k}(r'', r')dr''$$

$$= -i(A(t, r), \nabla_r)\rho_1(r, r') - i(\nabla'_r, A(t, r'))\rho_1(r, r'))$$

Therefore with neglect of nonlinear terms in the electromagnetic field, our position space representation of the quantum Boltzmann dynamics of the single particle density operator is given by

$$i\partial_t \rho_1(t, r, r') = (2m)^{-1}(2\nabla_r.\nabla'_r \rho_1(t, r, r') - (\nabla_r^2 + \nabla_r'^2)\rho_1(t, r, r'))$$

$$-(i/m)(A(t, r), \nabla_r)\rho_1(t, r, r') - i(\nabla'_r, A(t, r'))\rho_1(t, r, r')$$

$$-e\Phi(t, r)\rho_1(t, r, r') + e\Phi(t, r')\rho_1(t, r, r')$$

$$+nonlinearterms.$$

A remark on the nonlinear terms:

$$Tr_2[V_{12}\rho_1 \otimes \rho_1](r_1, r'_1) =$$

$$V(r_1,r_2)\delta(r_1-r_1'')\delta(r_2-r_2')\rho_1(t,r_1'',r_1')\rho_1(t,r_2',r_2)d^3r_2'd^3r_2d$$

$$=\int V(r_1,r_2)\rho_1(t,r_1,r_1')\rho_1(t,r_2,r_2)d^3r_2$$

and likewise,

$$Tr_2[(\rho_1\otimes\rho_1)V_{12}](r_1,r_1')=$$

$$\int\rho_1(t,r_1,r_1')\rho_1(t,r_2,r_2)V_{12}(r_1',r_2)d^3r_2$$

In principle, using perturbation theory, this quantum Boltzmann equation can be solved for to obtain the single particle density operator kernel $\rho_1(t,r,r')$ as a function of the electromagnetic field and then the average dipole moment of the electron in this electromagnetic field will be given by

$$\mathbf{p}(t)=\int(-e\mathbf{r})\rho_1(t,\mathbf{r},\mathbf{r})d^3r$$

and its average magnetic moment by

$$\mathbf{m}(t)=\int(-e\mathbf{L}(r,r')/2m)\rho_1(t,\mathbf{r}',\mathbf{r})d^3rd^3r'$$

where $\mathbf{L}(r,r')$ is the kernel of the orbital angular momentum operator in the position representation. This would then solve the problem of explaining the origin of permittivity and permeability of the plasma from the quantum statistical mechanical point of view.

**2.Perturbative solution of the Boltzmann equation**

$$i\partial_t\rho_1(t)=[H_1,\rho_1(t)]+\delta(N-1)Tr_2[V_{12},\rho_1(t)\otimes\rho_1(t)]$$

where $\delta$ is a perturbation parameter. We've also observed that an additional term can be added to this equation to improve its accuracy, namely,

$$\delta^2(N-1)Tr_2[V_{12},g_{12}]$$

where

$$g_{12}(t)=$$

$$=-i\int_0^t exp(-i(t-s)ad(H_1+H_2))([V_{12},\rho_1\otimes\rho_1]+(N-2)Tr_3[V_{13}+V_{23},\rho_1\otimes\rho_1\otimes\rho_1])$$

$$-(N-1)Tr_2[V_{12},\rho_1\otimes\rho_1]\otimes\rho_1$$

$$-(N-1)\rho_1\times Tr_2[V_{12},\rho_1\otimes\rho_1]])(s)ds$$

$$=\int_0^t exp(-i(t-s)ad(H_1+H_2))(F_2(\rho_1(s)\otimes\rho_1(s),\rho_1(s)\otimes\rho_1(s)\otimes\rho_1(s)))ds$$

so our perturbative solution would have the form

$$\rho_1(t)=U(t)\rho_1(0)U(t)^*+\delta.\int_0^t U(t-s)F_1(\rho_1(s)\otimes\rho_1(s))U(t-s)^*ds$$

$$+\delta^2(N-1)\int_0^t U(t-s)Tr_2[V_{12},[\int_0^s exp(-i(s-\tau)ad(H_1$$
$$+H_2))(F_2(\rho_1(\tau)\otimes\rho_1(\tau),\rho_1(\tau)\otimes\rho_1(\tau)\otimes\rho_1(\tau)))d\tau]]U(t-s)^*ds$$
$$=\rho_1(t)=U(t)\rho_1(0)U(t)^*+\delta.\int_0^t U(t-s)F_1(\rho_1(s)\otimes\rho_1(s))U(t-s)^*ds$$
$$+\delta^2(N-1)\int_{0<\tau<s<t} U(t-s)Tr_2[V_{12},[exp(-i(s-\tau)ad(H_1+H_2))(F_2(\rho_1(\tau)\otimes\rho_1(\tau),$$
$$\rho_1(\tau)\otimes\rho_1(\tau)\otimes\rho_1(\tau)))]]U(t-s)^*]dsd\tau$$

where

$$F_1(\rho_1\otimes\rho_1)=(N-1)Tr_2[V_{12},\rho_1\otimes\rho_1]$$

## 3.The quantum Boltzmann equation derived from the Dirac relativistic wave equation

$$H(t)=(\alpha,-i\nabla+eA)+\beta m-e\Phi=H_0+H_I(t)$$

where

$$H_0=(\alpha,-i\nabla)+\beta m, H_I(t)=e(\alpha,A)-e\Phi$$

Let $V_{12}=V(r_1,r_2)$ be the interaction potential between two particles. Our aim is to formulate the Quantum Boltzmann Equation

$$i\partial_t\rho(t)=[H(t),\rho(t)]+(N-1)Tr_2[V_{12},\rho(t)\otimes\rho(t)]$$

in the position space representation. We first note that $\rho(t,r,r')$ for each $t,r,r'$ is a $4\times 4$ matrix and

$$[H_0,\rho]=[\alpha_k p_k+\beta m,\rho](r,r')$$
$$=\alpha_k(-i\partial_k\rho(t,r,r'))-i\partial'_k\rho(t,r,r')\alpha_k$$

and

$$[\beta,\rho](r,r')=\beta.\rho(t,r,r')-\rho(t,r,r')\beta$$

Further,

$$((\alpha,A)\rho)(r,r')=A_k(t,r)\alpha_k\rho(t,r,r')=(\alpha,A(t,r))\rho(t,r,r')$$
$$(\rho(\alpha,A))(r,r')=\rho(t,r,r')(\alpha,A(t,r'))$$

Thus,

$$[(\alpha,A),\rho](r,r')=(\alpha,A(t,r))\rho(t,r,r')-\rho(t,r,r')(\alpha,A(t,r'))$$

and likewise,

$$[\Phi,\rho](r,r')=(\Phi(t,r)-\Phi(t,r'))\rho(t,r,r')$$

Remark: The cumulative distribution function of an observable $X$ in the state $\rho$ is given by

$$F_\rho(x) = Tr[\rho.\chi_{(-\infty,x]}(X)]$$

where $\chi_E(x)$ is the indicator function of the set $E \subset \mathbb{R}$ and this formula can be applied to calculate the average electric and magnetic dipole moments of the electron in the evolving Boltzmann state.

**4. Quantum statistical field theory**

Rather than using the quantum Boltzmann equation to calculate the average polarization and magnetization of a medium in an external electromagnetic field, we can use quantum statistical field theory based on second quantization principles of the wave function operator field. The quantum Boltzmann equation is based on first quantized quantum mechanics for a system comprising of a finite number of particles. However, quantum statistical field theory does not put any restriction on the number of particles. The number of particles here is assumed to be infinite and therefore its wave function is an operator valued Fermionic field. Such a Fermionic wave function field operator can equivalently be looked upon as being generated by a countably infinite number of Fermionic creation and annihilation operators modulated by a by a sequence of orthonormal stationary state ordinary wave functions.

Let $\psi(t,r)$ denote the wave operator field of second quantized matter. The second quantized Hamiltonian in the Dirac picture is given by

$$H = \int \psi(t,r)^*((\alpha, -i\nabla + eA(r)) + \beta m)\psi(t,r)d^3r$$
$$+ \int V_{\mu\nu}(r,r')\psi(t,r)^*\alpha^\mu\psi(t,r)\psi(t,r')^*\alpha^\nu\psi(t,r')$$

just as in the Hartree-Fock theory. The wave operator fields satisfy the canonical equal time anticommutation relations

$$\{\psi_l(t,r), \psi_m(t,r')^*\} = \delta_{lm}\delta^3(r-r')$$

and the second term in the second quantized Hamiltonian represents the interaction between Dirac charges and currents at two different spatial points. The wave operator fields $\psi(t,r)$ evolve according to the Heisenberg dynamics

$$\partial_t\psi(t,r) = i[H, \psi(t,r)]$$

The electronic polarization operator field is given by

$$P(t,r) = -e\mathbf{r}\psi(t,r)^*\psi(t,r)$$

This is the dipole moment operator per unit volume. Let $\rho_0$ denote the initial state of the quantum system, say the Gibbs state:

$$\rho_0 = exp(-\beta H)/Z(\beta), Z(\beta) = Tr(exp(-\beta H)), \beta = 1/kT$$

Since we are adopting the Heisenberg picture, this state does not evolve with time, only the observables evolve with time. The average Polarization of the medium at time $t$ is therefore given by

$$< P > (t,r) = Tr(\rho_0.P(t,r))$$

The magnetic dipole moment operator field per unit volume is given by

$$M(t,r) = \psi(t,r)^*(-e(\mathbf{L} + g\sigma)/2m)\psi(t,r)$$

and its average value is given by

$$< M > (t,r) = Tr(\rho_0 M(t,r))$$

To calculate these averages, we must first determine the dynamics of the temperature Green's function

$$G(t,r|t',r') = Tr(\rho_0 T\{\psi(t,r)\psi(t',r')^*\})$$

where $T$ is the time ordering operator. Note that since the magnetic vector potential is not assumed to vary with time, the total Hamiltonian operator is a constant of the motion and hence so is the density operator $\rho_0$. Note that from the canonical anticommutation rules,

$$[\psi(t,r')^*\alpha^\mu\psi(t,r'), \psi_k(t,r)] =$$

$$= [\alpha^\mu(l,m)\psi_l(t,r')^*\psi_m(t,r'), \psi_k(t,r)] =$$

$$-\alpha^\mu(l,m)\delta_{lk}\delta^3(r'-r)\psi_m(t,r')$$

$$= -\delta^3(r-r')\alpha^\mu(k,m)\psi_m(t,r) = -\delta^3(r-r')[\alpha^\mu\psi(t,r)]_k$$

Equivalently, in vector notation,

$$[\psi(t,r')^*\alpha^\mu\psi(t,r'), \psi(t,r)] = -\delta^3(r-r')\alpha^\mu\psi(t,r)$$

It follows that if we define

$$J^\mu(t,r) = \psi(t,r)^*\alpha^\mu\psi(t,r)$$

then

$$[J^\mu(t,r'), \psi(t,r)] = -\delta^3(r-r')\alpha^\mu\psi(t,r)$$

and therefore,

$$[J^\mu(t,r')J^\nu(t,r''), \psi(t,r)] =$$

$$-[\delta^3(r-r'')J^\mu(t,r')\alpha^\nu\psi(t,r) + \delta^3(r'-r)\alpha^\mu\psi(t,r)J^\nu(t,r'')]$$

Then,

$$[\int V_{\mu\nu}(r',r'')J^\mu(r')J^\nu(r'')d^3r'd^3r'', \psi(t,r)] =$$

$$-[\int V_{\mu\nu}(r',r)J^{\mu}(t,r')d^3r']\alpha^{\nu}\psi(t,r)$$

$$-\int \alpha^{\mu}\psi(t,r)[\int V_{\mu\nu}(r,r')J^{\nu}(t,r')d^3r']$$

We may assume without loss of generality that

$$V_{\mu\nu}(r,r') = V_{\nu\mu}(r',r)$$

and then deduce that

$$[\int V_{\mu\nu}(r',r'')J^{\mu}(r')J^{\nu}(r'')d^3r'd^3r'',\psi(t,r)] =$$

$$= -\{\alpha^{\mu}\psi(t,r),\int V_{\mu\nu}(r,r')J^{\nu}(t,r')d^3r'\}$$

We therefore obtain from the Heisenberg dynamics the following dynamical equation for the wave field operator $\psi(t,r)$.

$$i\partial_t\psi(t,r) = H_{D0}\psi(t,r) + \{\alpha^{\mu}\psi(t,r),\int V_{\mu\nu}(r,r')J^{\nu}(t,r')d^3r'\}$$

where

$$H_{D0} = (\alpha,-i\nabla) + \beta m$$

is the first quantized free particle Dirac Hamiltonian. We can further approximate this equation by replacing $J^{\mu}(t,r')$ on the rhs by its quantum average

$$< J^{\mu}(t,r) >= Tr(\rho_0(T)J^{\mu}(t,r))$$

It should be noted that this average current density can be expressed in terms of the temperature Green's function as

$$< J^{\mu}(t,r) >= -Tr[\alpha^{\mu}G(t,r|t,r')]|_{r'\to r}$$

where the trace here is an ordinary matrix trace for $4\times 4$ matrices.

Reference:Fetter and Walecka, "Quantum Theory of Many Particle Systems", Dover, 1971.

**5. Relating the refractive index of a material to the metric tensor of space-time**

[a]**Brief outline of the approach**

The curvature of space-time affects quantum phenomena. For example, in order to take into account the space-time curvature, we have to write down Dirac's equation in curved space-time and then formulate the quantum Boltzmann equation by starting from such a generalized Dirac equation. This is accomplished by assuming that the spinor connection of the gravitational field is chosen so that the Dirac equation based on such covariant derivative remains invariant under local Lorentz transformations. If such gravitational considerations are to be incorporated into the quantum Boltzmann equation, then one must write down Dirac's equation in curved space-time in Hamiltonian form and identify the extra terms in the Hamiltonian of each particle that come due to its interaction with gravity ie those extra terms will involve the tetrad basis and the spinor connection of the gravitational field. The quantum Boltzmann equation will have the same structure but with these additional terms in the single particle Hamiltonian involving the gravitational field. Likewise, if one has to incorporate these gravitational terms in the temperature Green's function based on statistical quantum field theory, then again one must replace the second quantized Hamiltonian term with $\int \psi(t,r)^* H\psi(t,r)d^3r$ where

$$H = (\alpha, -\nabla + eA) + \beta m + \delta H_g$$

where $\delta H_g$ is the correction term in the Dirac Hamiltonian coming from the tetrad and spinor connection terms of the gravitational field.

**The main computations**

Let $V_a^\mu$ be a tetrad basis for our curved space-time and let $\Gamma_\mu = \Gamma_{ab}^\mu[\gamma^a, \gamma^b]$ denote the spinor connection of the gravitational field. Then, the four component wave function satisfies

$$[V_a^\mu \gamma^a (i\partial_\mu + eA_\mu + i\Gamma_\mu) - m]\psi(x) = 0$$

From this equation, we can infer what the generalized Dirac Hamiltonian must be. This is achieved by separating the time derivative component from the spatial derivative components:

$$iV_a^0\gamma^a\partial_0\psi + [iV_a^r\gamma^a\partial_r + eV_a^\mu\gamma^a A_\mu + eV_a^\mu\gamma^a A_\mu + iV_a^\mu\gamma^a\Gamma_\mu - V_a^\mu\gamma^a m]\psi = 0$$

Now multiplying both sides of this equation by $V_b^0\gamma^b$ and using

$$V_a^0 V_b^a \gamma^a \gamma^b = (1/2)\eta^{ab} V_a^0 V_b^0 = (1/2)g^{00}$$

where $\eta^{ab}$ is the Minkowski metric of flat space-time and $g^{\mu\nu}$ is the exact contravariant metric of our curved space-time, we get

$$ig^{00}\partial_0\psi + [iV_b^0 V_a^r\gamma^b\gamma^a\partial_r + eV_b^0 V_a^\mu\gamma^b\gamma^a A_\mu + iV_b^0 V_a^\mu\gamma^b\gamma^a\Gamma_\mu - V_b^0 V_a^\mu\gamma^b\gamma^a m]\psi = 0$$

This equation can be expressed in the standard Hamiltonian form by defining the curved space time Dirac Hamiltonian in an electromagnetic field as

$$H = (g^{00})^{-1}V_b^0 V_a^r\gamma^b\gamma^a(-i\partial_r) - (g^{00})^{-1}(eV_b^0 V_a^\mu\gamma^b\gamma^a A_\mu) + (g^{00})^{-1}V_b^0 V_a^\mu\gamma^b\gamma^a(-i\Gamma_\mu)$$
$$+ (g^{00})^{-1}V_b^0 V_a^\mu\gamma^b\gamma^a m]$$

As an example of this calculation, consider the Schwarzchild metric in which

$$g_{00} = \alpha(r) = 1 - 2m/r, g_1 = -\alpha(r)^{-1}, g_{22} = -r^2, g_{33} = -r^2 sin^2(\theta)$$

We have

$$d\tau^2 = g_{\mu\nu}dx^\mu dx^\nu = (\omega_0)^2 - \omega_1^2 - \omega_2^2 - \omega_3^2$$

where

$$\omega_0 = \sqrt{\alpha(r)}dt, \omega_1 = \sqrt{\alpha(r)^{-1}}dr,$$
$$\omega_2 = rd\theta, \omega_3 = rsin(\theta)d\phi$$

Thus, since

$$g^{\mu\nu} = \eta^{ab}V_a^\mu V_b^\nu, g_{\mu\nu} = \eta_{ab}V_\mu^a V_\nu^b$$

we get

$$d\tau^2 = \eta_{ab}V_\mu^a V_\nu^b dx^\mu dx^\nu$$
$$= (V_\mu^0 dx^\mu)^2 - (V_\mu^1 dx^\mu)^2 - (V_\mu^2 dx^\mu)^2 - (V_\mu^3 dx^\mu)^2$$

Thus,

$$\omega_0 = \sqrt{\alpha(r)}dt = V_\mu^0 dx^\mu = V_0^0 dt + V_1^0 dr + V_2^0 d\theta + V_3^0 d\phi,$$

so that

$$\sqrt{\alpha(r)} = V_0^0, V_1^0 = V_2^0 = V_3^0 = 0,$$

Note that in these conventions,

$$(V_\mu^0) = (V_0^0, V_1^0, V_2^0, V_3^0)$$

Likewise,

$$\omega_1 = V_\mu^1 dx^\mu = \alpha(r)^{-1/2}dr$$

and therefore,

$$V_0^1 = 0, V_1^1 = \alpha(r)^{-1/2}, V_2^1 = 0, V_3^1 = 0$$

Note that

$$(V_\mu^1) = (V_0^1, V_1^1, V_2^1, V_3^1)$$
$$\omega_2 = V_\mu^2 dx^\mu = rd\theta, \omega_3 = V_\mu^3 dx^\mu = r.sin(\theta)d\phi$$

so that

$$V_0^2 = 0, V_1^2 = 0, V_2^2 = r, V_3^2 = 0,$$
$$V_0^3 = 0, V_1^3 = 0, V_2^3 = 0, V_3^3 = r.sin(\theta)$$

The spinor connection of the gravitational field is given by

$$\Gamma_\mu = V_{a\nu}V_{b:\mu}^\nu[\gamma^a, \gamma^b] = \omega_\mu^{ab}[\gamma^a, \gamma^b]$$

This can be derived in various ways, one by using the fact that the covariant derivative of the tetrad is zero, ie

$$V_{\mu,\nu}^a - \Gamma_{\mu\nu}^\rho V_\rho^a + \omega_\nu^{ab}V_{b\mu} = 0$$

or equivalently,

$$V_{\mu:\nu}^a + \omega_\nu^{ab}V_{b\mu} = 0$$

The second derivation is based on starting with the Einstein-Hilbert Lagrangian for the gravitational field, namely the curvature tensor in spinor notation:

$$R = R^{ab}_{\mu\nu} V^{\mu}_{a} V^{\nu}_{b},$$

$$R^{ab}_{\mu\nu} = \omega^{ab}_{\nu,\mu} - \omega^{ab}_{\mu,\nu} - [\omega_{\mu}, \omega_{\nu}]^{ab}$$

and setting the variational derivative of $\int R d^4 x$ w.r.t $\omega^{ab}_{\mu}$ to zero to arrive at an algebraic equation for $\omega^{ab}_{\mu}$ which when solved yields the desired form of the gravitational spinor connection. The third and most interesting way to derive the form of the spinor connection of the gravitational field is based on the use of non-Abelian gauge group theory with the gauge group now being the spinor representation of the Lorentz group. In other words, the transformation law of the spinor connection of the gravitational field under a local Lorentz transformation should be such that the Dirac equation remains invariant under it. Let $\Gamma_{\mu}(x)$ denote the spinor connection in one frame. The corresponding covariant derivative is $\nabla_{\mu} = \partial_{\mu} + \Gamma_{\mu}$. Now suppose we apply a local Lorentz transformation $\Lambda(x)$. Then the Dirac wave function transforms from $\psi(x)$ to $D(\Lambda(x))\psi(x)$ and for the Dirac equation to remain invariant under this transformation, we require that $\Gamma_{\mu}(x)$ should transform to $\Gamma'_{\mu}(x)$ where

$$D(\Lambda(x))(\partial_{\mu} + \Gamma_{\mu}(x))D(\Lambda(x))^{-1} =$$

$$\partial_{\mu} + \Gamma'_{\mu}(x)$$

This is equivalent to requiring that

$$\Gamma'_{\mu}(x) = D(\Lambda(x))\Gamma_{\mu}(x)D(\Lambda(x))^{-1} + D(\Lambda(x))(\partial_{\mu} D(\Lambda(x))^{-1})$$

or equivalently that

$$\Gamma'_{\mu}(x) = D(\Lambda(x))\Gamma_{\mu}(x)D(\Lambda(x))^{-1} - (\partial_{\mu} D(\Lambda(x)))D(\Lambda(x))^{-1}$$

Equivalently, if

$$\Lambda(x) = I + \omega(x)$$

is an infinitesimal local Lorentz transformation, then we require that

$$\delta\Gamma_{\mu}(x) = \Gamma'_{\mu}(x) - \Gamma_{\mu}(x) =$$

$$[D(\omega(x)), \Gamma_{\mu}(x)] + \partial_{\mu} D(\omega(x))$$

where

$$D(\omega(x)) = (1/4)[\gamma^a, \gamma^b]\omega_{ab}(x)$$

is the differential of the spinor representation (interpreted in terms of Lie algebra representations) of the Lorentz group evaluated at $\omega(x)$. On the other hand, if we take $\Gamma_{\mu}(x) = V_{a\nu:\mu} V^{\nu}_{b} [\gamma^a, \gamma^b]/2$, then under this infinitesimal local Lorentz transformation, it changes by

$$\delta\Gamma_{\mu}(x) = ([\gamma^a, \gamma^b]/4)[(\delta V_{a\nu:\mu})V^{\nu}_{b} + V_{a\nu:\mu}\delta V^{\nu}_{b}]$$

where

$$\delta V_{a\nu} = \omega_a^b(x)V_{b\nu}$$

so that

$$\delta V_{a\nu:\mu} = \omega_a^b V_{b\nu:\mu} + \omega_{a,\mu}^b V_{b\nu}$$

since the $\omega_b^a$'s and $\omega_{ab}$'s transform as scalar fields under diffeomorphisms of space-time. By comparing these two transformation laws, we can show that they are identical using the canonical commutation relations for the Loretnz algebra generators.

**6.Relating the refractive index to Lindblad noise operators when the universe as a system interacts with a bath**

Let $H$ be the Hamiltonian of an open quantum system consisting of $N$ identical particles having a density operator $\rho(t) = \rho_{12...N}(t)$. Its dynamical equations are

$$i\rho'(t) = [H, \rho(t)] + \theta(\rho(t))$$

where

$$\theta(\rho) = (-1/2)\sum_{k=1}^{p}[L_k^* L_k \rho(t) + \rho(t)L_k^* L_k - 2L_k \rho(t)L_k^*]$$

$L_k, k = 1, 2, ..., p$ are the Lindblad operators. We wish to derive a quantum Boltzmann equation from this by partial tracing out over all but the first particle Hilbert space. As usual, let

$$H = \sum_{a=1}^{N} H_a + \sum_{1 \le a < b \le N} V_{ab}$$

Then we get on partial tracing,

$$i\rho_1'(t) = [H_1, \rho_1(t)] + (N-1)Tr[V_{12}, \rho_1(t) \otimes \rho_1(t)]$$

$$+Tr_{23...N}(\theta(\rho(t)))$$

Since the particles are assumed to have identical interactions with the bath, The Lindblad operators can be expressed as

$$L_k = \sum_{i_1,...,i_N} c_k(i_1, ..., i_N) L_{k,i_1} \otimes ... \otimes L_{k,i_N}$$

where $c_k$ is permutation invariant. Thus, we can even write

$$L_k = \sum_{i=1}^{q} L_{k,i}^{\otimes N}$$

and then observe that

$$L_k^* L_k = \sum_{i,j=1}^{q} (L_{k,i}^* L_{k,j})^{\otimes N}$$

and

$$Tr_{23...N}(L_k^* L_k \rho(t)) \approx \sum_{i,j} (L_{k,i}^* L_{k,j} \rho_1(t))(Tr(L_{k,i}^* L_{k,j} \rho_1(t)))^{N-1}$$

$$Tr_{23..N}(L_k \rho(t) L_k^*) \approx L_{k,i} \rho_1(t) L_{k,j}^* (Tr(L_{k,i} \rho_1(t) L_{k,j}^*))^{N-1}$$

and

$$Tr_{23...N}(\rho(t) L_k^* L_k) \approx \sum_{i,j} (\rho_1(t) L_{k,i}^* L_{k,j})(Tr(\rho_1(t) L_{k,i}^* L_{k,j}))^{N-1}$$

Thus our approximate quantum Boltzmann equation for an open quantum system for $N$ particles contains polynomial terms in the one particle densities.

**7. Heat and mass transfer equations in fluid dynamics with cosmological applications** $(1/2)v^2$ is the fluid kinetic energy per unit mass. Let $\epsilon$ denote the internal energy of the fluid per unit mass. Then if $s$ denotes the entropy per unit mass of the fluid, we have the basic first law of thermodynamics:

$$Tds = d\epsilon + pd(1/\rho)$$

and if $w$ denotes the fluid enthalpy per unit mass, then

$$w = \epsilon + p/\rho$$

Thus,

$$dw = d\epsilon + pd(1/\rho) + dp/\rho = Tds + dp/\rho$$

In particular, if no heat is being pumped into the fluid, then $ds = 0$ and the above equation reduces to

$$dw = dp/\rho$$

The total energy per unit mass of the fluid is

$$u = \epsilon + v^2/2$$

and $\rho.u$ is the fluid energy per unit volume. We use the above first law of thermodynamics along with the fluid momentum equation

$$\rho dv/dt = \rho((v, \nabla)v + v_{,t}) = -\nabla p - \rho \nabla \Phi + div(\sigma)$$

and the equation of continuity

$$\rho_{,t} + div(\rho v) = 0$$

to derive the energy equation, ie, the differential equation satisfied by $u$.

Remark: Let $\sigma_{ij}$ denote the stress tensor of fluid. The rate at which is stress does work on the fluid or equivalently, the rate at which heat is dissipated in the fluid due to frictional/viscous and thermal effects is given by

$$\int_S n_i\sigma_{ij}v_j dS = \int_V (\sigma_{ij}v_j)_{,i}d^3r$$

where $V$ is the volume of the fluid and $S = \partial V$ is the surface of the fluid that bounds the volume $V$. Thus, the rate of dissipation of heat in the fluid per unit volume is given by

$$(\sigma_{ij}v_j)_{,i} = div(\sigma.v)$$

This dissipation of heat contributes to a decrease in the fluid energy. Thus, the energy equation of the fluid can be expressed as

$$\partial(\rho u)/\partial t + div(\rho w\mathbf{v}) = div(\sigma.\mathbf{v}) + \nabla.(D\nabla T)$$

where

$$w = u + p/\rho$$

is the fluid enthalpy per unit mass. Note that $w$ describes the fluid internal energy plus its kinetic energy plus its pressure energy and therefore the energy flux in the fluid should be taken as $\rho w\mathbf{v}$ rather than $\rho uv$. This is in accordance with Bernoulli's equation which states that in the absence of external forces, the enthalpy per unit mass $u + p/\rho$ is a constant along a streamline.

It should be noted that the viscous effects on the dynamics of the fluid are completely contained in the stress tensor which means that the momentum equation of the fluid should be taken as

$$\rho(v_{,t} + (v, \nabla)v) = -\nabla p - \rho\nabla\Phi + div(\sigma)$$

where

$$(div\sigma)_i = \sigma_{ij,j}$$

is the stress force per unit volume of the fluid. Note that the stress force on a volume $V$ of the fluid bounded by a surface $S$ is given by

$$\int_S \sigma_{ij}n_j dS = \int_V \sigma_{ij,j}d^3r$$

and hence $\sigma_{ij,j}$ is to be interpreted as the $i^{th}$ component of the stress force on the fluid per unit volume.

Remark: $div(\sigma.\mathbf{v})$ equals the rate at which energy is pumped into a unit volume of the the fluid by the surrounding fluid layers due to viscous friction while $\mathbf{q} = -D\nabla T$ is the heat flux caused by temperature gradients. Therefore, $-div\mathbf{q} = div(D\nabla T)$ equals the rate at which heat flows into unit volume of the fluid from the surrounding fluid layers due to temperature gradients. $\int_S(-\hat{n}).\sigma.(-\mathbf{v})dS = \int_V div(\sigma.\mathbf{v})d^3r$ is the rate at which thermal energy is pumped into the volume $V$ bounded by the surface $S$ due to friction between the fluid layers just outside $S$ with the layers just within $S$. Note that the stress tensor $\sigma$ which is the frictional force per unit area acts just outside $S$ where the fluid velocity is $\mathbf{v}$ and hence the velocity of the fluid just inside $S$ is $-\mathbf{v}$ relative to that just outside, the rate at which the $i^{th}$ component of the viscous forces from outside do work per unit area on the fluid layers just within $S$ having unit normal along the $j^{th}$ coordinate direction is $(-v_i\sigma_{ij})$ and since this force per unit area acts from the outside of $S$ towards the inside of $S$, the rate at which these forces do work on a unit area of the surface having outward normal $\mathbf{n}$ and hence unit inward normal $-\mathbf{n}$ is given by $(-v_i\sigma_{ij}(-n_j)) = (-\mathbf{n}).\sigma.(-\mathbf{v})$. A better way to say the same thing is as follows: $\sigma_{ij}$ is the $i^{th}$ component of the viscous stress force acting per unit area of the fluid having normal along the $j^{th}$ direction within the fluid just inside $S$ on the layer just outside. So $-\sigma_{ij}$ is the stress force per unit area exerted by the layers just outside on the layers just inside. Thus the rate at which these outside layers to work on the layers just within is $(v_i(-\sigma_{ij})(-n_j) = v.\sigma.n = n.\sigma.v$. Thus the total rate at which the viscous forces do work on the volume of the fluid within $V$ bounded by the surface $S$ equals $\int n.\sigma.vd^3r = \int_V div(\sigma.v)d^3r$.

In studying heat transfer within a fluid, our aim is to obtain an equations for the rate of change of the entropy with time per unit mass of the fluid. We compute

$$T.ds/dt = d\epsilon/dt - p\rho^{-2}d\rho/dt$$

$$d/dt(\rho u) = d/dt(\rho(\epsilon + v^2/2))$$

$$(\epsilon + v^2/2)d\rho/dt + \rho(d\epsilon/dt + (v, dv/dt))$$

since

$$d\rho/dt = \partial\rho/\partial t + (v, \nabla)\rho = -\rho.divv$$

by the equation of continuity,

$$d(\rho u)/dt = -\rho u divv + \rho(Tds/dt - p.divv + \rho(v, dv/dt)$$

$$= -\rho w divv + \rho Tds/dt + (v, \rho.dv/dt)$$

where

$$w = u + p/\rho$$

is the enthalpy per unit mass of the fluid. Substituting into this the momentum equation for $\rho.dv/dt$ gives us

$$d(\rho u)/dt = -\rho w divv + \rho.Tds/dt + (v, -\nabla p + F + div\sigma)$$

where $F$ is the external force per unit volume. Using the energy equation

$$\partial(\rho u)/\partial t + div(\rho w\mathbf{v}) = div(\sigma.\mathbf{v}) + \nabla.(D\nabla T)$$

and observing that

$$div(\rho w\mathbf{v}) = div(\rho u\mathbf{v}) + div(p\mathbf{v})$$

so that the energy equation can be rewritten as

$$d(\rho u)/dt + \rho u.div(\mathbf{v}) + div(p\mathbf{v}) = div(\sigma.\mathbf{v}) + \nabla.(D\nabla T) + (F, v)$$

in this gives us

$$div(\sigma.\mathbf{v}) + \nabla.(D\nabla T) - div(p\mathbf{v}) - \rho u div\mathbf{v} = -\rho w div v + \rho.T.ds/dt + (v, -\nabla p + div\sigma)$$

Now

$$div(\sigma.v) = (\sigma_{ij}v_j)_{,i} = \sigma_{ij}v_{j,i} + \sigma_{ij,i}v_j$$

and

$$(v, div\sigma) = v_i\sigma_{ij,j} = v_j\sigma_{ij,i}$$

by the symmetry of the stress tensor. Thus, we get after making a cancellation,

$$\sigma_{ij}v_{j,i} + \nabla.(D\nabla T) - div(p\mathbf{v}) - \rho u div v + \rho w div v - \rho T ds/dt + (v, \nabla p) = 0$$

Now,

$$\begin{aligned} \rho w.div v + (v, \nabla p) &= (\rho u + p) div v + div(pv) - p.div v \\ &= \rho u.div v + div(pv) \end{aligned}$$

and hence we get

$$\rho.T ds/dt = \sigma_{ij}v_{j,i} + \nabla.(D\nabla T)$$

Writing

$$T.ds/dt = C_p dT/dt$$

where $C_p$ is the specific heat at constant pressure

$$C_p = T.\partial s/\partial T|_p$$

we can express the above heat transfer equation in the form

$$\rho C_p dT/dt = \sigma_{ij}v_{j,i} + \nabla.(D\nabla T)$$

which is an equation that describes both heat convection and heat diffusion.

**8. Cosmological considerations**

Consider the Newtonian model for cosmic expansion: Energy/mass conservation for matter within the entire sphere gives us the equation

$$d(4\pi\rho S^3c^2/3)/dt = 0$$

or equivalently,

$$d(\rho S^3)/dt = 0$$

provided that we neglect the pressure contribution to the energy. The energy conservation equation of a point mass on the surface of sphere gives

$$S'^2/2 - GM/S = k, M = 4\pi\rho S^3/3$$

This second equation can be expressed as

$$S'^2/2 - 4\pi\rho S^2/3 = k$$

Let us now take pressure into consideration and formulate our equations. A sphere having comoving radius $r$ contains matter of mass $4\pi\rho(Sr)^3/3$ and the rate at which the energy of this sphere increases with time equals the total rate at which pressure forces from outside it do work on it. This rate is clearly $-p.4\pi(Sr)^2(S'r)$. Thus we obtain the equation of motion

$$d/dt(4\pi\rho(Sr)^3c^2/3) = -p.4\pi(Sr)^2S'r$$

or equivalently,

$$d(\rho S^3)/dt + 3pS^2S'/c^2 = 0$$

using Einstein's energy-mass relationship of special relativity. In the limit when $p/c^2 << \rho$, this equation reduces to the previously obtained equation of matter conservation. Suppose we wish to quantize these dynamical equations. Let us first neglect the pressure so that we are left with just a single equation

$$S'^2/2 - GM/S = k$$

with $M$ a constant. The Hamiltonian for this system with $S$ as the canonical position coordinate and $P$ as the canonical momentum coordinate is given by

$$H(S,P) = P^2/2 - GM/S$$

and if we write down the equation of the evolution operator $U(t)$ taking bath noise outside our universe into account, then we get the HP-qsde

$$dU(t) = [-(iH + P)dt + LdA(t) - L^*dA(t)^*]U(t), P = LL^*/2$$

More precisely, we assume that $S(t)$ is a classical solution to the expansion dynamics and $\delta S(t) = q(t)$ is a small quantum fluctuation. Then the slowly time varying Hamiltonian is given by

$$H(t,q,\pi) = \pi^2/2 - GM/(S(t) + q)$$

where $\pi$ the canonical momentum is given by

$$\pi = S'(t) + q'$$

In the absence of noise, the wave function of $q$ $\psi(t,q)$ therefore satisfies the Schrodinger equation

$$i\partial_t\psi(t,q) = -\partial^2\psi(t,q)/\partial q^2 - (GM/(S(t) + q))\psi(t,q)$$

and this is one way of determining the wave function of the expanding universe.

**How to calculate the permittivity and permeability in an expanding universe:** Let

$$S_{\mu\nu} = (1/4)F_{\alpha\beta}F^{\alpha\beta}g_{\mu\nu} - F_{\mu\alpha}F^{\alpha}_{\nu}$$

denote the energy-momentum tensor of the electromagnetic field. This is a random quantity. The average value of the energy density $S_{00}$ divided by 3 is known to be the pressure $p(t)$ if it is isotropic. We note that in special relativity,

$$S_{00} = (1/4)(-2F_{0r}^2 + F_{rs}^2) + F_{0r}^2 = (1/2)(F_{0r}^2 + F_{rs}^2/2) = (1/2)(E^2 + B^2)$$

which is the energy density and further,

$$F_{\alpha\beta}F^{\alpha\beta} = -2F_{0r}^2 + F_{rs}^2 = -2(E^2 - B^2)$$

so that $(-1/4)F_{\alpha\beta}F^{\alpha\beta} = (1/2)(E^2 - B^2)$ is the Lagrangian density of the field.

$$S_{rs} = (1/2)(E^2 - B^2)\delta_{rs} - (F_{r0}F_{s0} - F_{rk}F_{sk})$$

Thus,

$$S_{11} = (1/2)(E^2 - B^2) - (E_1^2 - B_2^2 - B_3^2)$$

and likewise for $S_{22}$ and $S_{33}$. It is clear that in the case of isotropic radiation fields,

$$< S_{00} >= U =< E^2 + B^2)/2 >$$

is the average energy density of the field and

$$< S_{11} >=< S_{22} >=< S_{33} >= U/3$$

since for isotropic radiation fields (as in the case of black-body radiation),

$$< E_k^2 >=< B_k >^2= U/3, k = 1, 2, 3$$

It is for this reason, that when we consider the expanding universe along with the cosmic microwave background radiation that it encloses, in the energy momentum tensor of the matter field

$$\rho v^\mu v^\nu - pg^{\mu\nu}$$

we replace the second term, ie, the pressure term by the energy-momentum tensor of the radiation field because radiation accounts for the pressure The total energy-momentum tensor of the matter plus radiation field is therefore taken as

$$T^{\mu\nu} = \rho v^\mu v^\nu + S^{\mu\nu}$$

where

$$S_{\mu\nu} = (1/4)F_{\alpha\beta}F^{\alpha\beta}g_{\mu\nu} - F_{\mu\alpha}F^{\alpha}_{\nu}$$

is the energy-momentum tensor of the radiation field. Now suppose we have a single electric dipole **p** that interacts with a slowly time varying electric field.

Suppose that this this dipole also has a magnetic moment $\mathbf{m}$ that interacts with a slowly time varying magnetic field. For example, an atom or an ion or a molecule or an ionized molecule can be regarded as having both an electric and a magnetic dipole moment. The interaction energy is given by

$$E = -\mathbf{p}.\mathbf{E}(t) - \mathbf{m}.\mathbf{B}(t)$$

and this is a function of the angles $\theta_1, \theta_2$ between $\mathbf{p}$ and $\mathbf{E}(t)$ as well as between $\mathbf{m}$ and $\mathbf{B}(t)$. If there is an apriori relationship between $\theta_1$ and $\theta_2$ defined by a measure $d\mu(\theta_1, \theta_2)$, then the average electric and magnetic dipole moments in this slowly time varying electric and magnetic fields are given by

$$<\mathbf{p}> (\mathbf{E}(t), \mathbf{B}(t)) = \frac{exp(\beta(\mathbf{p}.\mathbf{E}(t) + \mathbf{m}.\mathbf{B}(t))\mathbf{p}d\mu(, \mathbf{m})}{Z(\beta)},$$

$$<\mathbf{m}> (\mathbf{E}(t), \mathbf{B}(t)) = \frac{exp(\beta(\mathbf{p}.\mathbf{E}(t) + \mathbf{m}.\mathbf{B}(t))\mathbf{m}d\mu(, \mathbf{m})}{Z(\beta)},$$

where

$$Z(\beta) = \int exp(\beta(\mathbf{p}.\mathbf{E}(t) + \mathbf{m}.\mathbf{B}(t)))d\mu(\mathbf{p}, \mathbf{m})$$

is the classical partition function. It should be noted that while solving the Maxwell equations in a background gravitational field corresponding to the expanding universe, we take as our unperturbed fields $\mathbf{E}_0(t, \mathbf{r}), \mathbf{B}_0(t, \mathbf{r})$ and using perturbation theory with the metric perturbations from flat space-time being regarded as first order of smallness quantities, or equivalently in the cosmic expansion scenario, the scale factor of the expanding universe minus one $\delta S(t)$ as being of the first order of smallness, we solve the Maxwell equations to express the first order change in the electric and magnetic fields in terms of the unperturbed fields and the scale factor. These gravitationally perturbed fields we denote by

$$\mathbf{E}(t, \mathbf{r}) = \mathbf{E}_0(t, \mathbf{r}) + \delta\mathbf{E}(t, \mathbf{r}, \delta S)$$

$$\mathbf{B}(t, \mathbf{r}) = \mathbf{B}_0(t, \mathbf{r}) + \delta\mathbf{B}(t, \mathbf{r}, \delta S)$$

and it is precisely these perturbed fields which we use to calculate the interaction energy between the electric and magnetic dipoles and the fields. It follows immediately, that the average electric and magnetic dipole moment per unit volume at time $t$ will also be functions of the scale factor perturbations $\delta S(s), s \leq t$. Strictly speaking, while solving the Maxwell equations, we should take into account the current density produced by the matter fluid, ie, we should be solving the MHD equations in general relativity with the unperturbed electromagnetic fields being the ones that we have applied and the perturbations to these being obtained by approximately solving the MHD equations. More accurately, if the expansion of the universe is to be taken into consideration whilst doing this computation, we should assume that the background metric is the Robertson-Walker

$$d\tau^2 = dt^2 - S^2(t)dr^2/(1 - kr^2) - S^2(t)r^2(d\theta^2 + sin^2(\theta)d\phi^2)$$

and then consider small metric perturbations $\delta g_{\mu\nu}(x)$ around this background metric. We should also consider small perturbations to the zero comoving velocity field that satisfies the geodesic equations for the Robertson-Walker space-time. We should then set up the linearized Einstein-Maxwell equations for these metric perturbations with the electromagnetic field being small perturbations of the homogeneous and isotropic field corresponding to the cosmic microwave background radiation field plus the initial electromagnetic field applied locally in our laboratory and the velocity and density fields being small perturbations of the zero comoving field and the spatially constant density solution that satisfies the RW-metric based Einstein field equations. The resulting equations on solving would yield the spatio-temporally metric, velocity density and electromagnetic field perturbations. In particular we would obtain the electromagnetic field as a function of the scale factor $S(t)$ and the initial electromagnetic field from which we could in principle calculate the average electric and magnetic dipole moments using the method of classical statistical mechanics outlined above and hence derive the permittivity and permeability of the medium.

**9. Applications of heat and mass transfer equations to cosmology**
In the expanding universe, we write

$$v(t,r) = H(t)\mathbf{r} + \delta\mathbf{v}(t,\mathbf{r})$$

where $H(t) = S'(t)/S(t)$ is Hubble's constant and $\delta\mathbf{v}(t,\mathbf{r})$ is the inhomogeneous perturbation to the velocities that represent the velocity field of clumps of matter like galaxies. Likewise, we write

$$T(t,\mathbf{r}) = T_0(t) + \delta T(t,\mathbf{r})$$

and

$$\rho(t,\mathbf{r}) = \rho_0(t) + \delta\rho(t,\mathbf{r})$$

and substitute into the Navier-Stokes equations for heat and mass transfer:

$$\rho(\partial_t\mathbf{v}(t,\mathbf{r}) + (\mathbf{v},\nabla)\mathbf{v}(t,\mathbf{r}))$$

$$= -\nabla p(t,\mathbf{r}) - \rho(t,\mathbf{r})\nabla\delta\Phi(t,\mathbf{r}) + div(\sigma(t,\mathbf{r})) - - - (1)$$

$$\rho.[C_p\partial_t T(t,\mathbf{r}) + \mathbf{v},\nabla)T(t,\mathbf{r})]$$

$$= (\sigma,\nabla)\mathbf{v}(t,\mathbf{r}) + \nabla.(D\nabla T(t,\mathbf{r}))$$

where

$$\sigma_{ij} = \eta(v_{i,j} + v_{j,i})$$

and the equations of mass conservation and gravity:

$$div(\rho\mathbf{v}) + \partial_t\rho = 0,$$

$$\nabla^2\delta\Phi(t,\mathbf{r}) = 4\pi G\delta\rho(t,\mathbf{r})$$

After linearizing these equations, we obtain differential equations which are first order in time for the variables $\delta\rho(t,\mathbf{r}), \delta\mathbf{v}(t,\mathbf{r})$ and $\delta T(t,\mathbf{r})$. If we wish in addition to take electromagnetic fields into consideration, then in the above Navier-Stokes equation (1), we must add an MHD term on the right

$$\mathbf{J} \times \mathbf{B}$$

where

$$\mathbf{J} = \sigma_0(\mathbf{E} + \mathbf{v} \times \mathbf{B})$$

with $\sigma_0$ representing the electrical conductivity of the matter fluid. These equations are to be solved along with the Maxwell equations. For the present, if we are not bothered about electromagnetic fields and we wish to quantize the above perturbation equations, then we must look at quantizing a differential equation of the form

$$d\xi(t)/dt = F(t, \xi(t)) - - - (2)$$

where $\xi(t)$ is a vector that at time $t$ represents the velocity perturbations, the density perturbations and the temperature perturbations at the different spatial pixels that fill up the volume of the universe. An equation of the form (2) is generally not Hamiltonian as we could easily guess by looking at the temperature equation which contains a diffusion term. Thus, we cannot generally obtain (2) from Hamiltonian mechanics. We must instead supplement a Hamiltonian with Lindblad terms which we postulate arise owing to a connection between our universe and a surrounding bath.

**10. Appendix: A derivation of the Lindlbad master equation**

Let $H(t)$ denote the system Hamiltonian and $\rho(t)$ the density operator. It satisfies

$$i\rho'(t) = [H(t), \rho(t)]$$

which gives

$$i\rho(t+\tau) - i\rho(t) = \int_t^{t+\tau} [H(s_1), \rho(s_1)]ds_1$$

$$= \int_t^{t+\tau} [H(s_1), \rho(t) - i\int_t^{s_1} [H(s_2), \rho(s_2)]ds_2]ds_1$$

$$= [\int_t^{t+\tau} H(s_1)ds_1, \rho(t)] - i\int_{t<s_2<s_1<t+\tau} [H(s_1), [H(s_2), \rho(s_2)]ds_2ds_1$$

$$\approx \tau[H(t), \rho(t)] - i\int_{t<s_2<s_1<t+\tau} [H(s_1), [H(s_2), \rho(t)]]ds_2ds_1$$

From this equation, by assuming $H(s)$ to be a random Hermitian operator valued variable, it is easy to take averages and arrive at the GKSL equation. For example, if $(H_(s_1), H(s_2))$ takes the value $(L_k, L_j)$ with probability $p(k,j)(s_1, s_2)$

so that $H(t)$ takes the value $L_k$ with probability $p_k(s) = \sum_j p_{k,j}(s, s,')$, then assuming $\rho(t)$ to be independent of $H(s), s \geq t$, we get

$$< [H(t), \rho(t)] >= [\sum_k p_k(t) L_k, < \rho(t) >],$$

$$< [H(s_1), [H(s_2), \rho(s_2)]] >= \sum_{k,j} [p_{k,j}(s_1, s_2)(L_k L_j < \rho(s_2) >$$

$$+ < \rho(s_2) > L_k L_j - L_k < \rho(s_2) > L_j - L_j < \rho(s_2) > L_k]$$

which when substituted into the above equation and allowing $\tau \to 0$ so that $\tau.\sqrt{L_k} \to B_k$ results in the GKSL equation. It should be noted that although our derivation is based on a classical probabilistic averaging, we could easily extend this derivation to a quantum probabilistic averaging method by use of tensor products. Thus we approximate $\rho(t)$ by $\rho_S(t) \otimes \rho_B(t)$ where $\rho_S(t)$ is the system state and $\rho_B(t)$ the bath state and likewise,

$$H(t) = H_S(t) \otimes I_B + I_B \otimes H_B(t) + H_{SB}(t)$$

where $H_s(t)$ is the system Hamiltonian acting in the system Hilbert space, $H_B(t)$ is the bath Hamiltonian acting in the bath Hilbert space and $H_{SB}(t)$ is the interaction Hamiltonian acting in the tensor product of the system and bath Hilbert spaces. We leave it as an exercise to substitute these expressions into the above equation, take partial trace over the bath space and arrive at the GKSL equation for the system Hamiltonian.

## 11. Brief Highlights of this Research

[1] Derivation of the quantum Boltzmann equation by partial tracing the quantum Schrodinger-Liouville-Von-Neumann equation for $N$ particles interacting with themselves and with external electromagnetic fields over other particles.

[2] Calculating the quantum averaged electric and magnetic dipole moments of a single particle using the one particle density operator obtained as an approximate solution to the quantum Boltzmann equation using perturbation theory.

[3] Specializing the computation of average dipole moments to Dirac Hamiltonians.

[4] From the quantum averaged electric and magnetic dipole moment per unit volume, derivation of the electric permittivity and magnetic permeability and hence the refractive index.

[5] Highlighting the dependence on temperature and the frequency of the electromagnetic field of the refractive index. Noting that temperature dependence arises due to solving the quantum Boltzmann equation with the initial density operator being the Gibbs state.

[6] Attempting to derive an equilibrium solution to the quantum Boltzmann equation in the presence of a static electromagnetic field by an iterative algorithm with the initialization of the algorithm being given by the Gibbs state.

[7] Applying statistical quantum field theory of infinite particle Fermionic systems to derive the temperature Green's function and from there the average dipole moments and thence the refractive index as a function of temperature and the electromagnetic field. The method is the standard Hartree-Fock method and works when we model our quantum liquid as an infinite particle Fermi system based on canonical second quantization of Fermi fields. This is superior to the quantum Boltzmann method since the latter is based on finite particle approximations using first quantization.

[8] An independent classical probabilisitic and quantum probabilistic derivation of the Lindblad master equation and its application to the the derivation of the quantum Boltzmann equation to obtain bath corrections to the refractive index.

[9] Derivation of the cosmological equations for the expanding universe and for the propagation of inhomogeneities in the form of matter field density, velocity and temperature fluctuations and electromagnetic field fluctuations from Newtonian particle mechanics and Eulerian fluid mechanics.

[10] Evaluation of the effects of cosmological expansion and the propagation of inhomogeneities on the measurement of refractive index.

[11] Hints about how to quantize the cosmological equations using the Lindblad formalism of open quantum systems.

[12] Hints about how to calculate the mean square fluctuations in the quantum electromagnetic field in an expanding universe with applications to its effect on the refractive index of materials.

**12.Glossary of symbols**

$A_\mu(x)$ Covariant components of the electromagnetic four potential.

$A^\mu(x)$ Contravariant components of the electromagnetic four potential.

$F_{\mu\nu}(x)$ Covariant components of the antisymmetric electromagnetic field tensor. $F_{0r} = -F_{r0}, r = 1,2,3$ are the electric field components while $F_{12}, F_{23}, F_{31}$ are the magnetic field components.

$\rho(t)$ density matrix representing a mixed state of a quantum system at time $t$.

$\rho(t,r,r') =< r|\rho(t)|r' >$ position space representation of the density matrix of a mixed state of a quantum system.

$\rho_{12..N}(t)$ Joint mixed state of $N$ particles of a quantum system.

$Tr_{23...N}\rho_{12...N}(t) = \rho_1(t)$ marginal mixed state of the first particle of an $N$ particle quantum system. $Tr_{23...N}$ denotes the partial trace operation. In the position space representation,

$$[Tr_{23...N}\rho_{12..N}](t,r_1,r_1') = \int \rho_{12...N}(t,r_1,r_2...,r_N,r_1',r_2...,r_N)d^3r_2...d^3r_N$$

or equivalently in terms of countable orthonormal bases,

$$< e_{i_1}|[Tr_{23...N}\rho_{12...N}](t)|e_{j_1} >=$$

$$\sum_{i_2,...,i_N} < e_{i_1} \otimes e_{i_2} \otimes ... \otimes e_{i_N}|\rho_{12...N}|e_{j_1} \otimes e_{2,i_2} \otimes ... \otimes e_{N,i_N} >$$

**Acknowledgements**: I am grateful to Dr.Stephen A.Langford for several constructive suggestions based on his experimental data of refractive indices that helped me improve the content and the presentation of this paper. The data shown by Dr.Langford that helped in improving the content pertain to temperature and wavelength dependence of the refractive index of liquids.

**13.References**

[1] Landau and Lifshitz, "The classical theory of fields", Butterworth and Heinemann.

[2] Steven Weinberg, "Gravitation and Cosmology:Principles and Applications of the General Theory of Relativity", Wiley.

[3] Steven Weinberg, "The quantum theory of fields, vol.I", Cambridge University Press.

[4] Harish Parthasarathy, "Developments in mathematical and conceptual physics:Concepts and applications for engineers", Springer Nature 2020.

[5] K.R.Parthasarathy, "An introduction to quantum stochastic calculus", Birkhauser, 1992.

[6] M.Green, J.Schwarz, E.Witten, "Superstring Theory", Cambridge University Press.

[7] Stephen Arthur Langford, "Cumulative distribution functions for the refractive index of liquids as a function of temperature and wavelength", private communication.

[8] A.Fetter and Walecka, "Quantum Theory of Many Particle Systems", Dover, 1971.

[9] Richard Feynman, "The Feynman lectures on physics, vol.II"

[10] P.A.M.Dirac, "The principles of quantum mechanics", Oxford.

[11] Landau and Lifshitz, "Fluid mechanics", Butterworth and Heinemann.

**Some additional remarks about refractive index computation using classical statistical mechanics**

Let the fluid consist of electric and magnetic dipoles and let $f(t,r,v,p,p',m)d^3rd^3vd^3pd^3p'd^3m$ denote the number of particles of the fluid that have positions in the volume $d^3r$, velocity in the volume $d^3v$, electric dipole moment in the volume $d^3p$, dipole rate of change in the volume $d^3p'$ and magnetic dipole moment in the volume $d^3m$. The equations of motion for these variables are

$$dr/dt = v, dv/dt = -\gamma v + Q(E(t,r) + v \times B(t,r)),$$

$$dL/dt = p \times E + m \times B,$$

$$m = QL/2m_0,$$

$$d^2\xi/dt^2 = -\gamma d\xi/dt - U'(\xi) + Q(E(t,r) + v \times B(t,r))$$

$$p = Q\xi$$

Note that $L$ is the orbital angular momentum of the charge $Q$ which is assumed to be an electron ($Q = -e$) moving in the nuclear electrostatic field of an atom and $\xi$ is the position of the electron w.r.t the nucleus of the atom of which it is a part. Thus, we can write

$$dm/dt = (Q/2m_0)(p \times E(t,r) + m \times B(t,r)),$$

$$d^2p/dt^2 = -\gamma.dp/dt - U'(p/Q) + Q^2(E(t,r) + v \times B(t,r))$$

Formally, we can formulate a classical Boltzmann equation for the rate of change of $f$ taking into account a collision term and then by solving it along with the Maxwell equations

$$divE = \rho/\epsilon_0, divB = 0,$$

$$curlE = -\partial_t B, curlB = \mu_0 J + \mu_0\epsilon_0\partial_t E$$

where

$$J(t,r) = \partial_t P(t,r) + \nabla \times M(t,r)$$

$$\rho(t,r) = -divP(t,r)$$

with

$$P(t,r) = \int pf(t,r,v,p,p',m)d^3vd^3pd^3p'd^3m,$$

$$M(t,r) = \int m.f(t,r,v,p,p',m)d^3vd^3pd^3p'd^3m$$

Note that since an atom including its electrons is neutral, we are assuming no free charges in the fluid. The charge density of the fluid comes only from polarized atoms and as is well known in classical electrodynamics, it is determined from the polarization in accordance with the above equation. Likewise, the only current density in the fluid comes from the polarization and magnetization currents of each atom, no contribution from free charges. Let us now see how these equations can be formulated from the quantum mechanical viewpoint.

[3] Let $a_n, n \geq 1$ be a sequence of commuting annihilation operators so that

$$[a_n, a_m^*] = \delta(n,m)$$

Let $\phi_n(t,r)$ be functions of time and space and assume that they satisfy

$$\sum_n \phi_n(t,r)\phi_n(s,r')^* = min(t,s)K(r-r')$$

Construct a space-time process

$$A(t,r) = \sum_n [a_n\phi_n(t,r)],$$

so that

$$A(t,r)^* = \sum_n [a_n^*\phi_n(t,r)^*]$$

Show that

$$[A(t,r), A(s,r')^*] = min(t,s)K(r-r')$$

Hence, deduce that if $d$ denotes time differential, then

$$dA(t,r).dA(t,r')^* = K(r-r')dt$$

which is the quantum noise-field theoretic generalization of the celebrated Quantum Ito formula of Hudson and Parthasarathy. Now, assume that

$$F(t,r) = A(t,r) + A(t,r)^*$$

is the $z$-component of the magnetic vector potential in space-time. Calculate the electric and magnetic fields $\mathbf{E}(t,r), \mathbf{H}(t,r)$ corresponding to this vectro potential and evaluate in a coherent state $|\phi(u)>$, the following averages

$$<\phi(u)|F(t,r)|\phi(u)>=<\phi(u)|F(t,r)F(s,r')|\phi(u)>$$

and

$$<\phi(u)|\mathbf{E}(t,r)\otimes\mathbf{E}(s,r')|\phi(u)>$$

$$<\phi(u)|\mathbf{H}(t,r)\otimes\mathbf{H}(s,r')|\phi(u)>$$

$$<\phi(u)|\mathbf{E}(t,r)\otimes\mathbf{H}(s,r')|\phi(u)>$$

# Chapter 5

# Statistics of Refractive Index and Fundamental Laws of Nature

**Some remarks about the relationship between the statistics of refractive indices of materials and the fundamental laws of nature**

Note:This appendix is a part of my research work carried out jointly with Dr.Steven A.Langford.

## 5.1 RI from a classical Newtonian angle

The refractive index of a material is usually computed using a classical formula based on writing down the equation of motion of an electron bound to its nucleus by a harmonic force under small displacements around the equilibrium position. In this equation, velocity damping is taken into account and we also add an additional forcing term coming from an external electric field acting on the electronic charge. The solution to this linearized differential equation yields a solution for the electronic displacement given in the frequency domain by a function of the frequency times the electric field at that specific frequency. The dipole moment of the electron at that frequency is then computed as a function of the frequency times the electric field. This function of the frequency is parametrized by the natural frequency of the harmonic nuclear binding force, the damping force coefficient and the effective mass of the electron. In general, owing to the anisotropy of the medium, this function will be a $3 \times 3$ matrix valued function of frequency. By using Planck's formula for the number of oscillators per unit volume at a given temperature having natural frequency in a given range, we can evaluate the average dipole moment (ie, polarization) of the electron/atom per unit volume. The ratio of this polarization to the the electric field yields the average refractive index as a function of the radiation frequency/wavelength and the temperature of the medium. The reason why we observe statistical fluctuations in the refractive index and hence require to describe the RI in terms of a CDF rather than a specific value at a given wavelength and temperature is also explained by this classical model. It is because the natural frequency of the oscillator which binds the electron has a Planckian probability distribution function which is equivalent to saying that the number of electrons per unit volume having natural binding harmonic frequency in a given range is given by the Planckian black-body law. Such a model was in fact used by Einstein to modify the Debye theory of specific heat of solids. It should be noted that using classical Newtonian physics combined with the Lorentz force equation for a charged particle moving in an electromagnetic field and the expression for the torques exerted by an electric field on an electric dipole and by a magnetic field on a magnetic dipole, we can write down the equations of motion for the electric and magnetic dipoles of an atom in an external electromagnetic field taking damping forces and the electromagnetic force of the nucleus on an electron into account. The electric dipole equation in same as the equation of motion for the displacement of the electron relative to its nucleus while the magnetic dipole equation is same as the torque equation since the magnetic dipole moment is proportional to the orbital plus spin angular momentum of the electron and the rate of change of total angular momentum equals the torque. In this generalized Newtonian formalism, the electric dipole moment and hence the polarization/permittivity will depend on both the electric and magnetic field and likewise for the magnetic dipole moment. Specifically, these equations are as follows:

$$M(r)d^2\xi(t)/dt^2 + \Gamma(r).d\xi/dt + U'(\xi(t)) = -e(E(t,r) + d\xi(t)/dt \times B(t,r))$$

$$dL(t)/dt = p(t) \times E(t,r) + m(t) \times B(t,r)$$

where

$$p(t) = -e\xi(t), m(t) = -eL(t)/2m$$

and $U(\xi)$ is the binding potential of the electron to its nucleus.

## 5.2 RI from classical statistical mechanics

There exist other procedures based on classical statistical mechanics that explain the average RI as well as the CDF of the RI of materials. For example, one can start with the energy of interaction between an electric dipole and an externally applied electric field plus the energy of interaction between a magnetic dipole and a magnetic field and then apply the Gibbs formula for calculating the probability distribution of these dipoles in a given external electromagnetic field and hence determine using this probability distribution both the average dipole moments and their mean square fluctuations. From the joint probability distributions of the electric and magnetic dipoles, we obtain the probability distribution of the permittivity and permeability which are simply the ratios of the dipole moments per unit volume to the electric and magnetic field intensities and from the probability distribution of the permittivity and permeability, we obtain the probability distribution of the RI. Another way to look at the origin of the CDF of the RI from a classical dynamical standpoint is to start with the classical Boltzmann equation for the joint probability density of the position, velocity, electric and magnetic dipole moment of an atom taking collision terms into account. Such an equation can be derived by applying the number conservation principle after taking into account the equations of motion for the charged particle's position, velocity and electric and magnetic dipole moments in an external electromagnetic field. The equations for position and velocity are obtained from the Lorentz force equation while those for the and magnetic dipole moments are obtained from the precession equations that relate the rate of change of total angular momentum to the torque on the dipoles produced by the electric and magnetic fields and noting that the magnetic dipole moment is proportional to the angular momentum. The equations of motion for the electric dipole are again obtained from the equations of motion of the displacement of the electron relative to its nucleus caused by the force on it generated by the external electromagnetic field in accordance to the Lorentz force equation and due to the internal electromagnetic field generated by the nucleus in the moving reference frame of the electron. By looking at the equilibrium solution to such a Boltzmann equation in a static electromagnetic field starting with the Gibbs density, we can calculate the CDF for the electric and magnetic dipoles and hence of the RI.

## 5.3 RI, Boltzmann equation and Cosmic expansion

The classical problem of determining the effect of cosmic expansion on the measured values of RI can also be addressed here. The fundamental equations that one will have to start with will be the geodesic equations for the particle's position and velocity taking into account terms apart from space-time curvature additional forcing terms coming from the interaction between the external electromagnetic field and the charge of the particle (ie, the general relativistic equations of motion for a particle in an electromagnetic field) and also the covariant equation of motion for the total angular momentum precessing in a gravitional field taking into account the torque generated by the electromagnetic field on the particle. The metric used in calculating these general relativistic equations of motion may be taken to be the Robertson-Walker metric for a homogeneous and isotropic expanding universe and then the resulting classical Boltzmann equation will depend upon this expansion scale factor. The solution of this equation will then yield the CDF of the dynamically varying RI of the material and if sensitive measurements are made, one may then also be able to determine the scale factor and hence Hubble's constant for the expanding universe at the current epoch from the RI CDF measurements. A more general viewpoint of this circle of ideas involves taking into account the propagation of inhomogeneities in our universe in the form of galactic matter and also in the form of electromagnetic radiation and looking at the consequences of this on the RI of a material. For example, by linearizing the Einstein-Maxwell equations around the Robertson-Walker homogeneous and isotropic solution and also taking into account the changes in the energy-momentum of the matter fluid field due to viscous and thermal effects, we obtain partial differential equations that describe the evolution of metric perturbations, velocity and density perturbations and also electromagnetic field perturbations. In particular, if we apply an initial electromagnetic field to our laboratory sample of fluid, then the perturbations to this field caused by the expansion of our universe as well the propagation of metric and matter velocity and density inhomogeneities will generate a perturbation to the electric and magnetic dipole densities, or more precisely a perturbation to the CDF of these dipole densities derived from the Gibbs distribution for the interaction energy between the electromagnetic field and the dipoles. Thus, by analyzing the CDF of our sample fluid, we obtain information about not only the expansion scale factor of our universe but also about the nature of inhomogeneities, ie, the density and velocity field of the galaxies. Another way to analyze this problem which combines the classical Boltzmann equation with the Einstein-Maxwell field equations and which can also be used to calculate the CDF of the RI of our material is as follows: The Boltzmann distribution function can be used to calculate the energy-momentum tensor of matter as a function of space-time coordinates by averaging over the velocities. This averages energy momentum tensor $T^{ab}(t,r) = \int f(t,r,v)v^a v^b d^3v$ is substituted into the linearized Einstein field equations as a perturbation. Then, we observe that the Boltzmann equation for $f(t,r,v)$ also involves the metric tensor of space time as well as the electromagnetic field in view of the equations of motion of a single particle in an an electromagnetic field with a background curved metric of space-time. Solving this coupled Einstein-Maxwell-Boltzmann equation would then in principle yield the distribution function $f(t,r,v)$ in phase space being a function of the expansion scale factor and hence any average computed using this distribution function like polarization, magnetization and RI would also be a function of the scale factor. Temperature and viscosity of the medium enters into this picture via the corrections to the energy-momentum tensor of the matter field caused by viscous stress and thermal conduction. The final solution for the Boltzmann distribution function then involves all the parameters of interest: The frequency of electromagnetic radiation which appears in the frequency domain formulation of the Maxwell equations and the temperature and viscosity. More precisely, the Boltzmann distribution function $f$ should be regarded as a function of time, spatial coordinates, velocity of the particle and the electric and magnetic dipole moments and its dynamics apart from a collision term would involve the equations of motion of all these physical quantities of the particle. Here, by particle, we mean an atom with its mobile electrons. The state of such an atom is defined by its position, velocity, its electric dipole moment due to its displaced electrons and also its magnetic dipole moment due to its electrons moving around it. The electric dipole has an equation of motion determined by that of the electron moving relative to its nucleus in the external gravitational and electromagnetic field and also in the electromagnetic field generated by the nucleus. The magnetic dipole moment of the atom is proportional to the angular momentum whose rate of change is the torque determined by the cross products of the electric and magnetic dipole moments with the electric and magnetic fields. It should be remarked at this point that one can derive almost all the equations of cosmology approximately using classical Newtonian physics. The idea is to consider a fluid ball of radius $S(t)$ containing matter having a large homogeneous component $\rho_0(t)$ and small inhomogeneous perturbation $\delta\rho(t,r)$ and a velocity field having a large comoving component $H(t)r$ where $H(t) = S'(t)/S(t)$ is Hubble's constant and a small inhomogeneous perturbation $\delta v(t,r)$ and the set up the Navier-Stokes equations for such a fluid taking into account heat transfer terms, ie, the energy equation coming from temperature gradients and viscous shear forces by making use of the first and second laws of thermodynamics. Prior to applying these perturbed fluid equations, the unperturbed Robertson-Walker equations for $S(t), \rho_0(t)$ can be derived by considering the rate of change of the total energy/mass within the sphere and equating this rate to the rate at which pressure forces do work on the sphere and next set up the energy conservation equation for a particle located at the boundary of this sphere with the potential energy coming from gravitational interaction of the particle with the matter within the sphere. The resulting heat and mass transfer equations [Landau and Lifshitz, Fluid mechanics] can be used to explain almost all cosmological phenomena including galactic formation.

## 5.4 RI and quantum mechanics

If one wishes to explain the statistics of the RI using the quantum theory taking into account gravitational interactions with the expanding universe, with the cosmic microwave background radiation, as well as with the galactic matter in the form of inhomogeneities in the density, velocity and metric perturbations, then there are once again several avenues open to us. The first is to start with the Dirac equation in a curved space-time metric [Steven Weinberg, Gravitation and Cosmology] for $N$ electrons with the electrons interacting with an electromagnetic field and also with their respective nuclei and also interacting with each other via the Coulomb repulsive electrostatic field and taking bath noise interactions into consideration via Lindblad operators, derive a quantum Boltzmann equation for the marginal density matrix of a single electron. An approximate solution to such a quantum Boltzmann equation using perturbation theory can be used to compute the probability distribution of the electric and magnetic dipoles of each atom in this setup and also the probability distribution of the refractive index of the material by regarding the RI as an observable defined by the square root of the product of the permittivity and permeability which in turn are proportional to the ratios of the dipole moments to the electric and magnetic fields. It is clear from this picture that temperature dependence will arise when we take as initial conditions to the quantum Boltzmann equation, the Gibbs density with the Dirac Hamiltonian in curved space-time. When the metric of space-time is the Robertson-Walker metric, then this picture gives the

CDF of the RI as a function of the scale factor of the expanding universe while when the metric is the Schwarzchild metric of a massive gravitating body, it gives the dependence of the CDF on the mass of the body or if it is a rotating blackhole described by the Kerr metric, it gives the dependence of the CDF of the RI on both the mass and angular momentum of the blackhole.

## 5.5 RI and quantum entanglement

It is noteworthy at this point to see how quantum entanglement occurs in a fluid of atoms having electrons at two different spatial locations. Two quantum systems are said to be entangled if their total state is non-separable. It is then easy to see that measurement of an observable on the first system will after the outcome is noted cause the state of the first system to collapse to an eigenstate of the observable being measured and will also cause the state of the second system to simultaneously collapse to another state that is dependent upon the state to which the first has collapsed and also to the nature of the entanglement. From this statement it is clear that entanglement between two systems can be used to generate an efficient quantum channel for transmission of quantum information but that is another story. What is of importance to us is that the entangled state of two particles in our fluid can be quantified by deriving the quantum Boltzmann equation for the marginal density of not one particle but rather of two particles in the system of $N$ particles by taking a partial trace of the Dirac equation containing two particle interaction terms (like the electrostatic repulsion between two electrons) with respect to the remaining $N-2$ particles. An external field that interacts with two particles can also be a source of entanglement because the effect of the same field is now present in both the particles causing their states to be correlated. Thus, if our universe is expanding and we model the metric perturbations as quantum observables (as is done in quantum gravity), then, this can cause the particle's states to get entangled eventually in time even if initially they are not. This is because if $H_g$ denotes the interaction Hamiltonian of the $N$ particles with the gravitational field and if initially the $N$ particles were not entangled, ie their state was $\rho_1 \otimes \rho_2 \otimes ...\rho_N$, then finally after time $T$, the joint state of the first two particles would be

$$\rho_{12}(T) = Tr_{34...N}(exp(-iTH_g)(\rho_1 \otimes \rho_2 \otimes ...\rho_N).exp(iTH_g))$$

which is clearly a non-separable state since $H_g$ is not separable.

From our RI viewpoint, entanglement of the particle's states would cause the RI measured at two different spatial points to be correlated which means that we would get a joint non-separable CDF for the RI's at two different spatial locations. We can attempt to explain such RI correlations using a model for the external gravitational and electromagnetic fields by introducing unknown parameters into this model and optimizing over these parameters to get a correlation match between experiment and theory. The optimal parameters can then be used to predict something hitherto unknown about the large scale structure of space time which affects the electromagnetic and radiation fields.

## 5.6 Spin, Polarization, Quantum electrodynamics and RI

The spin of an electron interacts with an external magnetic field causing a Zeeman splitting of its energy levels. If we have several particles arranged in a lattice each one having a spin of its own, then, adjacent spins will interact with each other producing an interaction Hamiltonian $H_I = \sum_{i=1}^{n-1} g(i)(\sigma_i, \sigma_{i+1})$ and furthermore each spin will interact with the external magnetic field with a Hamiltonian being given by $H_B = \sum_i h(i)(\sigma_i, B(r_i))$. The spin magnetic moment of the $i^{th}$ particle is $m_i = ge\sigma_i/2m$ and hence according to the principle of quantum statistics the average magnetic moment of the $i^{th}$ particle has the form

$$< m_i >= Tr(m_i.exp(-\beta(H_I + H_B))), \beta = 1/kT$$

If we adopt the Schrodinger picture for the electron then we have to add an extra Pauli spin term to the Hamiltonian which is simply the spin-magnetic field Zeeman interaction or if we have a system of several electrons, then to the sum of each electron's Hamiltonian we have to add the above Ising term plus the Zeeman term ie a total of $H_I + H_B$ and then solve Schrodinger's equation or more precisely formulate the quantum Boltzmann equation taking this modified Hamiltonian into consideration.

The Dirac equation however is better and more accurate to use than the Schrodinger equation since it takes into account relativistic speeds and moreover, one does not have to introduce artifically, the spin-magnetic field interaction for the electron. When one tries to convert the Dirac equation for the electron in an external electromagnetic field into a Klein-Gordon like equation by premultiplying the Dirac operator with its conjugate, then apart from the standard Klein-Gordon term there are additional terms that describe the interaction of the electron's spin with the electric and magnetic fields. In short, Dirac's equation states that spin is naturally a relativistic effect, it does not have to be artifically introduced into the dynamics as one does in the Schrodinger equation. One can use the Dirac equation for a system of $N$ particles interacting with each other via self generated electromagnetic fields and via spins and also having interactions with external electromagnetic and gravitational fields to study their entanglement. In this model for $N$ Dirac particles, we can also introduce bath noise via Lindblad operators. The resulting two particle quantum Boltzmann equation will describe the evolution of the joint two particle state from which correlations can be extracted. For example, given a two particle mixed state we can try to approximate it by a pure entangled state by minimizing the fidelity between the mixed state and the pure entangled state. Of course then, the mixed state will carry non-zero entropy while the pure entangled state will not carry any entropy. A better scheme would be to extend the Hilbert space of the two particle system and purify their joint mixed state by using the enlarged Hilbert space.

An interesting aspect of quantum field theory appears at this stage and is very much relevant to the RI problem as follows. Usually the electromagnetic field of photons and the Dirac field of electrons and positrons are quantized in a vacuum background. One could consider quantizing these two fields in a background gravitational field described by a space-time metric and also when the medium is not vacuum but rather has a inhomogeneous and anisotropic permittivity and permeability tensor that depends upon frequency. Such a model for the permittivity and permeability may be obtained for example using classical methods described earlier based on the equations of motion of electric and magnetic dipoles or using quantum mechanics. Now, the idea is to write down the Maxwell equations for the electromagnetic four potential taking into account the space-time curvature as well as the permittivity and permeability tensor and also to write down the Dirac equation for electrons and positrons in the presence of the electromagnetic four potential taking into account the spinor connection of the gravitational field. Then, we quantize the resultant field equations. This is equivalent to quantizing the electromagnetic field in curved space-time and in the presence of sources in the form of polarization and magnetiization charge and current densities. First we observe that the Hamiltonian of the resulting quantized Maxwell and Dirac field contains coupling terms of the field with the permittivity and permeability tensor fields and to the metric field. We then apply perturbation theory to this Hamiltonian by expressing it as the Hamiltonian of the free Maxwell and Dirac field in vacuum plus interaction terms involving coupling between the Dirac field and the Maxwell field (as appears in conventional quantum electrodynamics) plus other interaction terms between the Maxwell field, the Dirac field and the gravitational field and the electric and magnetic susceptibility tensors. The permittivity and permeability tensors in this formalism are expressed as the sum of an identity tensor and a weak susceptibility tensor. The gravitational coupling arises in the Maxwell field in the form of covariant derivatives in curved space-time or alternately while constructing the scalar action functional of the Maxwell field by raising and lowering indices using the metric tensor and using the four dimensional invariant volume element in curved space time during the integration process. The gravitational coupling appears in the Dirac field action owing to the requirement of having to introduce a connection to maintain local Lorentz and diffeomorphism invariance of the Dirac action. Now these additional perturbation terms in the free field Hamiltonian of the Dirac and the Maxwell field will appear in the Dyson series expansion of the resulting evolution operator and hence will contribute to the Feynman diagrams involved in the computing of the matrix elements of the scattering matrix between electron-positron and photon states. This means that scattering, absorption and emission processes that are calculated conventionally in quantum electrodynamics using the Feynman diagrammatic method will acquire correction terms involving the background permittivity and permeability tensor fields and one can hope to estimate the permittivity, permeability and hence the RI sensitively using scattering, emission and absorption experiments carried out in an accelerator. Of course, one could go a step further by quantizing even the gravitational field by setting up the Hamiltonian of the gravitational field upto a certain degree in the metric perturbations and expanding the solution to the linearized Einstein field equations for the metric perturbations in terms of superpositions o plane waves with coefficients being the graviton creation and annihilation operator fields in wavennumber/momentum space. The terms in the Einstein gravitational Hamiltonian that are greater than second degree in the metric perturbations will have to be regarded as perturbation/self-interaction terms whose effect on the scattering processes of electrons, positrons, photons and gravitons can be formally computed perturbatively using the Feynman diagrams or using the operator theoretic formalism of Dyson and Schwinger. The final outcome would be that the scattering cross sections for these four particles will depend upon the permittivity and permeability of the medium. In this quantum gravity formalism, gravity does not appear as a background classical field, rather it appears in the form of an elementary particle, the graviton which participates in the quantum scattering processes.

The next point that we take up here is the aspect of polarization. It is well known that a photon is a spin one particle that cannot have zero spin component along any direction. Thus, essentially, a photon has two polarization states, left circular and right circular corresponding the spin projections of $\pm 1$ in units of $h/2\pi$. Any other polarization state like linear and elliptic can be obtained by linearly combining these two mutually orthogonal polarization states. Now when we take a bunch of photons distributed over space, we can talk of a joint state for all these photons and ask what is the correlation between the polarizations of these photons in this state. The answer to this question may be posed in the form of the spin interaction described above. It is all a special instance of the following more general problem. Suppose $X(r)$ is an observable at the spatial location $r$, If the $X(r)'s$ at different $r's$ commute, then it is meaningful to talk of a joint state in which these different observables have specific values and then there would be no correlations in such a state since the values of $X(r)'s$ are determined in such a joint state. However, we can choose a state in which these $X(r)'s$ are not all diagonal, ie, they do not have definite values and hence in such a state, we can calculate their correlations. From a quantum field theoretic viewpoint, the photon four vector field has the form

$$A^{\mu}(x) = \int (\sqrt{2|K|})^{-1}[a(K,s)e^{\mu}(K,s)exp(-ik.x) + a(K,s)^{*}e^{\mu *}(K,s)exp(ik.x)]$$

where the sum is over $s = 1, 2$ corresponding to the photon having two independent degrees/states of polarization. One way to understand why the photon has only two independent states of polarization is to consider the Coulomb gauge in which the magnetic vector potential $A$ satisfies $divA = 0$ or equivalently in the spatial Fourier domain, $K.A = 0$ and the other constraint is obtained by the fact that in the Coulomb gauge, the electric scalar potential becomes a matter field since it satisfies Coulomb's law, or equivalently, Poisson's equation with a specified charge density. These two constraints reduce the four degrees of freedom of the four potential to just two degrees. The four vector $e^{\mu}(K,s)$ determines the polarization of a photon having wave vector $K$ and spin projection $s$. In the Coulomb gauge, the polarization vectors $e^r(K,s), r=1,2,3$ satisfy $K_r e^r(K,s)=0, s=1,2$ and since there are only two linearly independent vectors in three dimensional momentum space that are orthogonal to a given vector, we obtain the result that s takes just two values. We can choose the three vectors $(e^r(K,s))_{r=1}^3, s=1,2$ so that they form an orthonormal basis for the two dimensional vector space of all vectors that are orthogonal to the three wave vector $K$. Thus, in the Coulomb gauge, these polarization vectors satisfy the completeness condition:

$$\sum_{s=1,2} e^r(K,s)\bar{e}^m(K,s)+K^rK^m/K^2=\delta^{rm}, r,m=1,2,3$$

Now the photon creation and annihilation operators satisfy the canonical Boson commutation relations

$$[a(K,s),a(K',s')^*]=\delta_{s,s'}\delta^3(K-K')$$

and if for example, we are interested in calculating the higher order correlations between photon four potentials at spatial locations $r_1,...,r_N$ at times $t_1,...,t_N$ having polarizations along given directions $n_{1,\mu},...,n_{N,\mu}$ in a given state $\rho$, we would evaluate

$$Tr(\rho.n_{1,\mu_1}A^{\mu_1}(t_1,r_1)...n_{N,\mu_N}A^{\mu_N}(t_N,r_N))$$

For example, we can take $\rho$ to be a joint coherent state for the photon field and then evaluate these multiple correlations. Such a coherent state can be expressed abstractly as

$$|z>=\sum_{\{n(K,s)\}} \Pi_{K,s}z(K,s)^{n(K,s)}a(K,s)^{*n(K,s)}|0>/\Pi_{K,s}n(K,s)!$$

after appropriate discretization. Then the photon correlations can be evaluated using the Bosonic commutation rules combined with

$$a(K,s)|z>=z(K,s)|z>$$

ie the coherent state $|z>$ is a state in which the positive frequency components of the electromagnetic field assume definite values. More generally, suppose we solve the Maxwell field equations in a medium having inhomogenous and anisotropic permittivity and permeability and refractive index and express the solution in the form

$$A^r(t,\mathbf{r})=\sum_k a(k,s)u^r(k,s,t,\mathbf{r})+a(k,s)^*u^r(k,s,t,\mathbf{r})^*], r=1,2,3$$

where the sum is over $s=1,2$ and the $a(k,s),a(k,s)^*$ are the standard photon annihilation and creation operator fields in momentum space. The right circularly polarized component in the xy plane of this field at a given point in space at a given frequency $\omega$ is then determined by taking the scalar product of this vector potential with the rotating vector $\hat{x}.cos(\omega t)+\hat{y}.sin(\omega t)$ and then forming a time average. Specifically, this component equals

$$lim_{T\to\infty}\frac{1}{2T}\int_0^T (A^1(t,r)cos(\omega t)+A^2(t,r)sin(\omega t))$$

and this this is an operator valued function of the spatial coordinates. We can then in principle evaluate the multiple correlations of this polarization at different spatial points and answer questions like, "what is correlation between the right circularly polarized components of the radiation field at points $r_1,...,r_k$ in space ?" It may be noted that in the above expansion, the basis functions $u^r(k,s,t,\mathbf{r})$ are determined by the solution to the Maxwell equations in a background permittivity and permeability field and hence these functions are parameterized in particular by the refractive index field of the medium. By measuring multiple correlations of this field or of the state of a certain kind of polarization, and comparing it with the theoretically deduced one based on actual computation of the quantum moments in a given state like a coherent state making use of the Bosonic commutation relations for the $a(k,s),a(k',s')^*$ and the action of these operators on a coherent state, we can actually hope to estimate the refractive index with good accuracy.

## 5.7 RI and the principle of equivalence

As an additional feature of our connection between the RI and the fundamental laws of physics, we have a test for the principle of equivalence which is the foundation stone for Einstein's general theory of relativity. The principle of equivalence states that the gravitational mass equals the inertial mass and hence the acceleration of a particle in a gravitational field does not depend upon its mass. Therefore, gravitational fields can be cancelled out locally by moving into a freely falling frame. This means that locally any gravitational field is simply a change of the space-time coordinate system and by patching these local coordinate changes, we get a curved space-time metric. Such a curved space time metric cannot be cancelled out by a global coordinate change if and only if the gravitational field is generated by a definite quantity of matter and this happens if and only if the Riemann curvature tensor of the space time manifold does not vanish identically in any one frame and hence also in any frame. The principle of equivalence can be tested in the following way: Take a liquid of mass $m$ in a jar. Let $m_i$ denote its inertial mass and $m_g$ its gravitational mass. The principle of equivalence holds if $m_i$ and $m_g$ are proportional to each other, in other words when we can write $m_g = Km_i$ where the constant $K$ does not depend upon $m_i$. The idea is to estimate $K$ by choosing different liquids having different inertial masses $m_i$ and if we are able to show that the resulting $K's$ are the same for all the liquids, then the principle of of equivalence will be valid otherwise it will not be valid. Assume that the charge density of the liquid is constant $\rho_q$. We apply an external electromagnetic $E, B$ field on the liquid and allow to liquid to move with a uniform velocity $v$. Then the electric and magnetic fields in the rest frame of the liquid are respectively given in a non-relativistic approximation by

$$E' = E + v \times B, B' = B - v \times E/c^2$$

The After getting scattered by the liquid, the resulting electric and magnetic fields in the moving frame of the liquid will be functions of $E', B'$ and hence of $v$ and the permittivity and permeability of the liquid in accordance with the Maxwell equations. By measuring these scattered fields, we can thus estimate the permittivity and permeability of the liquid and hence also its RI. Now, if we assume that the velocity field of the fluid is a function of position and time, ie, $v = v(t, r)$, then accordingly the electric and magnetic fields as felt by a fluid particle at $(t, r)$ in its rest frame will be

$$E'(t, r) = E(t, r) + v(t, r) \times B(t, r), B'(t, r) = B(t, r) - v(t, r) \times E(t, r)/c^2$$

and the RI in this local frame will therefore depend upon $E', B'$ and hence upon $v(t, r)$. Now the Newtonian equations of motion of the fluid are given by

$$\rho_i(\partial_t v(t, r) + (v(t, r), \nabla)v(t, r)) = -\nabla p(t, r) - \rho_g \nabla \Phi(t, r) + \eta \nabla^2 v(t, r),$$

$$divv(t, r) = 0$$

where $\rho_i$ is the inertial mass density and $\rho_g$ the gravitational mass density of the liquid. $\Phi(t, r)$ is the externally applied gravitational potential. The inertial and gravitational mass densities of the fluid are also related by $\rho_g = K\rho_i$. So the idea is to first measure the RI and use it to estimate $E', B'$ and hence $v(t, r)$ and then to substitute this estimated $v(t, r)$ into the above fluid equations and thereby estimate $K$ and test whether we arrive at the same value of $K$ for all kinds of fluid. This calculation can be made more accurate by using the hydrodynamical equations in general relativity and applying it to large scale fluids like galactic matter.

## 5.8 Non-Abelian gauge theories and RI

The Yang-Mills non-Abelian gauge theory forms the basis for describing the electroweak and strong nuclear interactions. The crucial idea here is to first postulate the existence of gauge fields corresponding to a gauge group which is a subgroup of unitary matrices and then set up the Yang-Mills Lagrangian for the corresponding matter fields which describe electrons and leptons. This electron and lepton Lagrangian does not have any mass term apriori. Three linearly independent combinations can be formed out of these gauge fields out of which two acquire mass terms after coupling with a Higgs field while the third gauge field does not acquire mass and is the photon. The coupling with the Higgs field arises when one writes down the Lagrangian for the Higgs matter field in the presence of the gauge fields and notes that this Lagrangian is quadratic in the gauge covariant derivative because the Higgs field is a scalar Klein-Gordon type of matter field. When the Higgs field falls into the ground state, this quadratic form in the covariant derivative generates quadratic non-derivative terms in the gauge fields with coefficients determined by the vacuum expectation values of the Higgs field. By adjusting the gauge group coupling constants, we can determine the coefficients of these non-derivative quadratic terms so that the eigenvalues of the associated matrix yields masses for the gauge bosons in agreement with experiment. These massive gauge bosons are called the W and Z bosons and are the propagators of the weak nuclear forces just as the massless gauge boson which is the photon is the propagator of the electromagnetic forces. The electroweak theory also includes a mechanism by which the electron acquires mass. This mechanism involves adding to the masselss electron-lepton Lagrangian another term that is quadratic in the electron-lepton matter fields and linear in the Higgs scalar doublet field. This additional term also has the required gauge group symmetry but when the symmetry of this Lagrangian is broken with the Higgs field falling into the ground state, then quadratic terms in the electron-lepton matter fields generated by this additional coupling term also called the Yukawa coupling appear thereby causing the electron to acquire mass. Now the question is , where does RI enter into this picture? The obvious way to see this is to write down the guage field Lagrangians also namely the Lagrangians of the massive gauge bosons and that of the masseless photon. While writing down the Lagrangian of the photon gauge field, the background permittivity and permeability must be included. So the entire Lagrangian can be split into (a) the electron-lepton free massless matter Lagrangian, (b) the electron-lepton matter field interaction with the three gauge fields, (c) the KG Lagrangian for the scalar Higgs field interacting with the gauge fields, (d) the Yukawa coupling term invovlving non-derivative interaction of the electron-lepton matter fields with the Higgs field and (e) the massive and massless gauge field Lagrangian in a background RI medium It is then obvious that the photon gauge field Maxwell equations will involve not only the medium RI but also the electron-lepton fields and the Higgs field owing to the photon gauge field appearing in these Lagrangians. These extra or source terms for the Maxwell photon field will appear in the form of electron-lepton Dirac currents and the Higgs scalar current. A measurement of the photon field and the electron-lepton and Higgs matter fields will then enable us to infer something about the RI of the medium. More precisely, we write the electron-lepton Lagrangian as

$$L_e = \psi^* \gamma^0 (i\partial_\mu + eA_\mu^\alpha t_\alpha))\psi$$

where $t_\alpha$ are the Hermitian generators of the gauge group. The gauge field Lagrangian is

$$L_g = (-1/4)(\epsilon^{\mu\nu\rho\sigma}(x) * F_{\rho\sigma\alpha}(x))F_{\mu\nu}^\alpha(x)$$

where $\epsilon^{\mu\nu\rho\sigma}$ is the permittivity permeability tensor of the medium and $*$ denotes space-time convolution. Then, we also have the Higgs Lagrangian

$$L_H = [(\partial_\mu + ieA^\alpha_\mu t_\alpha)\phi(x)]^*.[(\partial^\mu + ieA^{\mu\alpha} t_\alpha)\phi(x)$$

and the Yukawa coupling Lagrangian

$$L_{YK} = g\psi(x)^* X(\psi(x) \otimes \phi(x))$$

where $X$ is a matrix of appropriate size so that $L_{YK}$ is real. The field equations for $\psi, A^\alpha_\mu, \phi$ are obtained by setting the variational derivatives of

$$L = L_e + L_g + L_{YK}$$

to zero w.r.t these fields. This is the first quantized picture. In the second quantized picture, we calculate the scattering matrix elements between two states either using the operator theoretic formalism or using the Feynman path integral formalism. In the operator theoretic approach, we first write down field equations for $\psi, A^\alpha_\mu, \phi$ in the absence of any interactions, neglecting non-linear terms in the gauge field equations and assuming that the medium is vacuum. This enables us to expand these linearized system of pde's in terms of the Fermion creation and annihilation operators for $\psi$, the Boson creation and annihilation operators for $A^\alpha_\mu$ and the Boson creation and annihilation operators for $\phi$. We then express the Hamiltonian of the exact nonlinear system in terms of these creations and annihilation operators and we separate out the quadratic component from the cubic and higher order components in these creation and annihilation operators and the terms involving the medium RI. We then write down the Dyson series expansion of the evolution operator for this total Hamiltonian with the unperturbed Hamiltonian being the quadratic part and the perturbation Hamiltonian being the cubic and higher order part plus the terms involving RI perturbations. Then we take an initial state say in which there are a definite number of Fermions, gauge Bosons and Higgs Bosons having specified four momenta and spins and likewise a final state and using the Dyson series expansion for the unitary evolution operator, we calculate the matrix elements of this w.r.t the initial and final states. These matrix elements give us the transition probability amplitudes for scattering from the initial to the final state. These scattering matrix elements depend upon the RI and hence by measurement of the scattering cross-sections, we can hope to get sensitive RI estimates.

## 5.9 Symphonic resonance with dark matter and the disappearance of fragments

It is well known that when a ray of light is incident normally from a medium of permittivity-permeability pair $(\epsilon_1(\omega), \mu_1(\omega))$ into a medium of permittivity-permeability pair $(\epsilon_2(\omega), \mu_2(\omega))$ at frequency $\omega$, then the coefficient of reflection at that frequency is given by

$$R(\omega) = \frac{\eta_2(\omega) - \eta_1(\omega)}{\eta_2(\omega) + \eta_1(\omega)}$$

and hence if there is a critical frequency $\omega$ at which

$$\eta_2(\omega) = \eta_1(\omega)$$

then at this critical frequency $\omega$, there will not be any reflected wave. Here the one subscript refers to the fragment and the two subscript refers to the dark matter surrounding it. We may term this as symphonic resonance. When we use quantum mechanics or quantum statistical mechanics to calculate these permittivities and permeabilities then generally, the the permittivity and permeabilty pair which will have probability distributions given the initial Gibbs state of the fragment and dark matter. In classical statistical mechanics, we can talk of the joint probability distribution of the fragment-dark matter permittivity and permeability at a given frequency and temperature and hence we can talk of the probability of a symphonic resonance occurring. On the other hand, in quantum mechanics, the fragment dipole moment and the dark matter dipole moment may be non-commuting observables in which case, we cannot talk of their joint probability distributions owing to the Heisenberg uncertainty principle. A typical calculation would proceed along the following lines: Let $\psi(t,r)$ be the wave function of a fragment atom. Its Schrodinger dynamics in an external electromagnetic field is

$$i\partial_t\psi(t,r) = -\nabla^2\psi(t,r)/2m + V(r)\psi(t,r) + e(r, E(t)) + e(L, B(t)/2m$$

where $V(r)$ is the nuclear potential and $L$ is the electron's angular momentum. Let the stationary states of the unperturbed atom be $u_n(r), n = 1, 2, ...$ with corresponding energy levels $E_n, n = 1, 2, ....$ Thus,

$$-\nabla^2 u_n(r)/2m + V(r)u_n(r) = E_n u_n(r), n = 1, 2, ...$$

If we solve the Schrodinger equation with initial state $u_n(r)$ using perturbation theory with the electronic charge $e$ as the perturbation parameter, then our perturbation series will be

$$\psi_n(t,r) = u_n(r) + \sum_{k\geq 1} e^k\psi^{(k)})_n(t,r)$$

Then treating the external electromagnetic field terms as small perturbations, we have

$$i\partial_t\psi_n^{(k)}(t,r) = H_0\psi_n^{(k)}(t,r) + ((r, E(t) + (L, B(t))/2m)\psi_n^{(k-1)}(t,r), k \geq 1$$

where

$$H_0 = -\nabla^2/2m + V(r)$$

is the unperturbed atomic Hamiltonian. We can solve this infinite system of perturbation equations sequentially to get

$$\psi_n^{(k)}(t,r)=\int_0^t U(t-s)((r,E(s))+(L,B(s))/2m)\psi_n^{(k-1)}(s,r)ds, k\geq 1, \psi_n^{(0)}(t,r) = u_n(r)exp(-iE_nt)$$

where

$$U(t)=exp(-itH_0)$$

Thus we see that if this series is truncated to the $N^{th}$ term, then the wave-function is approximated by an $N^{th}$-degree Volterra series in the electric and magnetic fields. Let now the initial state be the Gibbs mixed state rather than the pure state $u_n(r)$. Then the mixed state at time $t$ in the position representation is given by

$$\rho(t,r,r')=\sum_n p(n)\psi_n(t,r)\psi_n(t,r')^*$$

where

$$\psi_n(t,r)=\sum_{k=0}^N e^k\psi_n^{(k)}(t,r), p(k)=exp(-E_k/kT)/\sum_m exp(-E_m/kT)$$

and the average dipole moment of the atom is

$$<p(t)>=\int \rho(t,r,r)(-er)d^3r$$

If $N(r)$ denotes the number of atoms per unit volume, then the average polarization at time $t$ is

$$P(t,r)=N(r)<p(t)>$$

and this is an $N^{th}$ degree polynomial functional, ie, Volterra functional in the electromagnetic fields, ie, it can be expressed as

$$P(t,r)=P_0(r)+\sum_{k=1}^N \int H_k(t_1,...,t_k|r)(E(t-t_1),B(t-t_1))\otimes...\otimes(E(t-t_k), B(t-t_k))dt_1...dt_k$$

and likewise the magnetization $M(t,r)$ obtained by multiplying $N(r)$ with the average magnetic moment of the atom

$$<m(t)>=\int \rho(t,r,r')(-eL(r',r)/2m)d^3rd^3r'$$

is expressible as

$$M(t,r)=M_0(r)+\sum_{k=1}^N \int G_k(t_1,...,t_k|r)(E(t-t_1),B(t-t_1))\otimes...\otimes(E(t-t_k), B(t-t_k))dt_1...dt_k$$

Taking the Fourier transforms on both sides and neglecting the time independent term (which will vanish if the initial state is isotropic in the position domain)

$$\hat{P}(\omega,r)=\sum_{k=1}^{N}\hat{H}(\omega_1,...,\omega_k|r)\delta(\omega-\omega_1-...-\omega_k)(\hat{E}(\omega_1),\hat{B}(\omega_1))\otimes...\otimes(\hat{E}(\omega_k),\hat{B}(\omega_k))d\omega_1...d\omega_k$$

and likewise for the magnetization

$$\hat{M}(\omega,r)=\sum_{k=1}^{N}\hat{G}(\omega_1,...,\omega_k|r)\delta(\omega-\omega_1-...-\omega_k)(\hat{E}(\omega_1),\hat{B}(\omega_1))\otimes...\otimes(\hat{E}(\omega_k),\hat{B}(\omega_k))d\omega_1...d\omega_k$$

The frequency domain values of the average permittivity and permeability as a function of the electromagnetic field are given respectively by

$$\epsilon(\omega,E,B)=\epsilon_0(1+\delta P(\omega,r)/\delta\hat{E}(\omega))$$

$$1/\mu(\omega,E,B)=1/\mu_0-\delta M(\omega,r)/\delta B(\omega)$$

Note:

$$B=\mu_0(M+H),1=\mu_0(dM/dB+dH/dB)=\mu_0(dM/dB+1/\mu)$$

The condition for symphonic resonance becomes then the equality of $\mu/\epsilon$ for the fragment to that for the dark matter and by solving this equation, we get a relationship between the frequency, the electromagnetic field amplitudes at different frequencies and the temperature.

Now we discuss a quantum field theoretic analysis of symphonic resonance.

Consider an atom with a nucleus and an electron interacting with a quantum electromagnetic $A_\mu(x)$. The Maxwell equations taking into account the electronic current in our non-relativistic model for the atom are

$$F^{\mu\nu}_{,\nu}=\mu_0.J^\mu$$

where

$$J^\mu=(\rho,J),$$

$$\rho=\psi^*\psi,$$

$$J=(ie/2m)(\psi^*\nabla\psi-\psi\nabla\psi^*)$$

Here, we assume that the free quantum photon field is expandable as in conventional quantum electrodynamics as a superposition of photon creation and annihilation operators in momentum space. The atomic wave function satisfies Schrodinger's equation

$$[-(\nabla+ieA)^2/2m+V(r)-eA^0]\psi=i\partial_t\psi$$

When we solve this equation for $\psi$, the solution comes out in terms of the initial Gibbs state and the quantum field $A^r, A^0$. Now assume that the quantum photon field initially is in a definite state, say a coherent state. Then defining the electric dipole and magnetic dipole moment of the atom with its electron in the usual way, we see that these dipoles will have a probability distribution at time $t$ being obtained by the following procedure: First solve the Schrodinger equation for the atomic wave function at time t in terms of its initial stationary state $u_n(r)$ and the quantum electromagnetic field $A^\mu$. Denote this solution by $\psi_n(t, r)$. This will be an operator field since it depends upon the quantum electromagnetic field which is an operator field. Then the probability distribution of the electronic position at time $t$ will be given by

$$f(t, r) = \sum_n p(n, T) < \phi(z)|\psi_n(t, r)^* \psi_n(t, r)|\phi(z) >$$

where $\phi(z)$ is the coherent state of the photon field at time $t = 0$ and $p(n, T) = exp(-E_n/KT)/Z$ is the Gibbs distribution. We can calculate the average dipole moment at time $t$ using this time varying probability density and in fact, calcualate the probability density of the atomic dipole $p = -er$. For the magnetic dipole moment along a given direction $\hat{n}$, we can evaluate its moment generating function as

$$M(s) = \sum_n p(n, T) < \phi(z)| \int \psi_n(t, r)^* exp(-se\hat{n}.L/2m)\psi_n(t, r)d^3r|\phi(z) >$$

and by inverting this Laplace transform, we may derive the pdf of any given component of the atomic magnetic dipole moment. It should be noted that in practice the coherent state $|\phi(z) >$ of the photon bath may be replaced be selected so that the average value of the electric and magnetic field components in this state assume some prescribed classical values and then the probability distributions of the electric and magnetic dipoles and hence their average values in such a state of the photon field will be functions of this average electromagnetic field. From this relationship between the average electric and magnetic dipole moments and the average electromagnetic field, we can evaluate the permittivity and permeability of the atoms in the fragment and those in the dark matter surrounding it and hence derive conditions as explained above for resonance to occur leading to the disappearance of the fragments.

## 5.10 Symphonic resonance with matter waves instead of photons

Our previous analysis of symphonic resonance dealt was based upon the reflection and refraction of light at a boundary interface between two different dielectrics, one the fragment medium and two the dark matter medium. We explained how when a critical frequency is reached for a given amplitude of the electric and magnetic fields, there will not be any reflection from this interface thereby causing all the light from the fragment to pass through the dark matter as if it were transparent thereby causing total disappearance of the fragment. The same could also be said about De-Broglie matter waves instead of photons. For example, electrons or any other elementary particle issuing from the fragments are associated with matter waves described by a wave function satisfying the Schrodinger equation and when these waves are incident upon the dark matter boundary, they encounter a change in the potential thereby causing reflection and refraction of these matter waves. The analysis of the critical momentum/wave number of the incident particles at which the reflection coefficient vanishes is done as one does in standard quantum mechanical tunneling theory with the only difference now in that the spin of the electron must also be considered and the interaction of this spin with a magnetic field contributes to the potential thereby causing the potentials in the two regions to be $2 \times 2$ Hermitian matrix valued functions of the position. Suppose for example, electrons move in the fragment region under a constant $2 \times 2$ Hermitian matrix potential $\mathbf{V}_1$ and in the dark matter region under another constant $2 \times 2$ Hermitian matrix valued potential $\mathbf{V}_2$. Assume for simplicity that the motion takes place along a single dimension which we call the $x$ axis. Then the Schrodinger equation in the regions $I : x < 0, I : 0 < x < L, III : x > L$ where $III$ is the dark matter region, $I$ is the region of free space from where the incident matter wave hits the fragment and $II$ is the region of the fragment. The Schrodinger equation taking $V_1 = 0$ and $V_2 = V$ has the form

$$I : \psi''(x) + 2mE\psi(x) = 0, x < 0$$

$$II : \psi''(x) + 2m(E - V_1)\psi(x) = 0, 0 < x < L,$$

$$III : \psi''(x) + 2m(E - V_2)\psi(x) = 0, x > L$$

where $\psi(x) = [\psi_1(x), \psi_2(x)]^T$ is a two component wave function. The solution in the first region has an incident and a reflected component:

$$I : \psi(x) = exp(-ikx)u_1 + exp(ikx)u_2,$$

where

$$k = \sqrt{2mE}$$

The solutions in $II$ and and $III$ are obtained from the spectral decompostion of $V_1$ and $V_2$:

$$V_1 = v_{11}P_{11} + v_{12}P_{12}, P_{1k}^* = P_{1k} = P_{1k}^*, k = 1,2, P_{11}P_{12} = 0, P_{11} + P_{12} = I$$

$$V_2 = v_{21}P_{21} + v_{22}P_{22}, P_{2k}^* = P_{2k} = P_{2k}^*, k = 1,2, P_{21}P_{22} = 0, P_{21} + P_{22} = I$$

Then we find that $E - V_k$ has four square roots, namely

$$\pm(E - v_{k1})^{1/2}P_{k1} + \pm\sqrt{E - v_{k2}}P_{k2}, k = 1,2$$

and hence the general solution to the wave function in the regions $II$ and $III$ are respectively given by

$$II : \psi(x) = P_{11}(exp(ik_1x)v_1 + exp(-ik_1x)v_2) + P_{12}(exp(ik_2x)v_3 + exp(-ik_2x)v_4)$$

$$III : \psi(x) = P_{21}(exp(ik_3x)w_1 + exp(-ik_3x)w_2) + P_{22}(exp(ik_4x)w_3 + exp(-ik_4x)w_4)$$

where $v_k, w_k, k = 1, 2, 3, 4$ are $2 \times 1$ complex vectors (called spinors) and

$$k_1 = \sqrt{2m(E - v_{11})}, k_2 = \sqrt{2m(E - v_{12})},$$

$$k_3 = \sqrt{2m(E - v_{21})}, k_4 = \sqrt{2m(E - v_{22})}$$

In region III, we assume that there is just a transmitted wave, no reflected wave because there is no barrier at $x > L$. Hence, $w_1 = w_3 = 0$. Then the number of variable amplitude two component spinors is $1 + 4 + 2 = 7$. Out of these, one namely $u_1$ is assumed to be known since we know the incident wave. The boundary conditions are that the wave function as well as its first order spatial derivative should be continuous at $x = 0, L$. That means that we have four two vector constraints. Therefore it appears that the number of constraints is not sufficient to determine all the amplitudes. A more careful analysis shows that this is not the case because $P_{11}v_1$ and $P_{11}v_2$ are both in the range of the one dimensional projection $P_{11}$ and hence both of these together form a single two vector. Likewise, $P_{12}v_3$ and $P_{12}v_4$ together define a single two vector and likewise also $(P_{21}w_1, P_{21}w_2)$ and $(P_{22}w_3, P_{22}w_4)$. Since $w_1 = w_3 = 0$, these last two variables are effectively equivalent to just a single two vector formed out of $(P_{21}w_2, P_{22}w_4)$. Taken along with $u_2$, therefore, we have four two vector constraints and four two vector variables which means that a consistent unique solution to this problem can be obtained. Specifically, let $e_{11}$ be a unit vector in the range of the one dimensional projection $P_{11}$, $e_{12}$ in that of $P_{12}$, $e_{21}$ in that of $P_{21}$ and $e_{22}$ in that of $P_{22}$. Then, we can write the above solutions as

$$II : \psi(x) = (c_1exp(ik_1x) + c_2exp(-ik_1x))e_{11} + (c_3exp(ik_2x) + c_4exp(-ik_2x))e_{12}$$

$$III : \psi(x) = (d_1exp(ik_3x) + d_2exp(-ik_3x))e_{21} + (d_3exp(ik_4x) + d_4exp(-ik_4x))e_{22}$$

$$I : \psi(x) = u_1.exp(-ikx) + u_2.exp(ikx)$$

Here, $u_1$ is a given $2 \times 1$ complex vector and $c_1, c_2, c_3, c_4, d_1, d_2, d_3, d_4$ are eight unknown complex variables. Since however there is no reflected wave in the third region, we take $d_1 = d_3 = 0$. Taken along with the two vector $u_2$, there are in all therefore eight complex variables to be determined. The continuity of $\psi, \psi'$ at $x = 0$ furnishes us with four complex scalar equations while the continuity of the same at $x = L$ furnishes us with another four complex scalar equations and we obtain a consistent solution which can be used to determine the reflection and transmission coefficient for any given spin polarization.

## 5.11 The effect of magnetic particles in the atmosphere on the refractive index

Magnetic particles in the atmosphere produce magnetic fields and hence act as sources in the energy momentum tensor of the electromagnetic field that drives the Einstein-Maxwell field equations. Consequently the metric of space time gets modified by such perturbation terms and when we work with the Dirac equation in this magnetically modified space-time metric, the wave function also gets perturbed. Consequently, the probability distributions of the different components of the electric and magnetic dipole moments get perturbed and hence so does the refractive index of the material under consideration which we are analyzing using the Dirac equation with initial Gibbs state.

## 5.12 Quantum filtering theory applied to improve our RI measurements

The unitary evolution of our system containing the RI liquid coupled to a Fermion and Boson bath has the form

$$dU(t) = (-(iH+P)dt + L_1 dA(t) + L_2 dA(t)^* + L_3 dJ(t) + L_4 dJ(t)^*)U(t)$$

where $A(t), A(t)^*$ are the boson bath noise annihilation and creation processes while $J(t) = \int_0^t (-1)^{\Lambda(s)} dA(s)$ is the Fermionic bath annihilation process and $J(t)^* = \int_0^t (-1)^{\Lambda(s)} dA(s)^*$ is the Fermionic bath creation process. These processes satisfy the canonical commutation and anticommutation relations:

$$[A(t), A(s)^*] = min(t,s),$$

$$[A(s), dJ(t)]_+ = 0, s < t, [dA(s), dJ(t)]_+ = 0, s < t,$$

$$[A(s)^*, dJ(t)]_+ = 0, s < t, [dA(s)^*, dJ(t)]_+ = 0, s < t,$$

$$[dA(s), J(t)] = 0, s \geq t, [dA(s)^*, J(t)] = 0, s \geq t$$

$$dA(t).dJ(t)^* = (-1)^{\Lambda(t)} dt, dJ(t).dJ(t)^* = dt,$$

$$[A(s), dJ(t)^*]_+ = 0, s < t, [J(s), J(t)^*]_+ = min(t,s),$$

$$[A(s), J(t)]_+ = 0, s \leq t,$$

$$[dA(s), dJ(t)^*]_+ = 0, s < t, [dA(s), J(t)^*]_+ = (-1)^{\Lambda(s)} ds, s \leq t$$

$$[A(s), J(t)^*]_+ = \int_0^s (-1)^{\Lambda(u)} du, s \leq t$$

Now we "bosonise" the HP qsde by replacing the Fermionic noise differentials $dJ(t), dJ(t)^*$ with $\theta(t) dJ(t)$ and $\theta(t)^* dJ(t)^*$ respectively where $\theta(t), \theta(t)^*$ are Majorana Fermionic parameters satisfying the anticommutation rules

$$\theta(t)\theta(s) + \theta(s)\theta(t) = 0, \theta(t)\theta(s)^* + \theta(s)^*\theta(t) = 0$$

We also assume that these Fermionic parameters anticommute with $A(.), A(.)^*$:

$$\theta(t)A(s) + A(s)\theta(t) = 0,$$

$$\theta(t)A(s)^* + A(s)^*\theta(t) = 0,$$

and therefore taking adjoints,

$$\theta(t)^*A(s) + A(s)\theta(t)^* = 0,$$

$$\theta(t)^*A(s)^* + A(s)^*\theta(t)^* = 0$$

These relations ensure that since $A(s)$ anticommutes with $dJ(t), dJ(t)$ for $s \leq t$, $A(s)$ will commute with $\theta(t)dJ(t), \theta(t)^*dJ(t)^*$ for $s \leq t$. Now we take our input measurement process differential as

$$dY_i(t) = c_1 dA(t) + \bar{c}_1 dA(t)^* + c_2\theta(t)dJ(t) + \bar{c}_2\theta(t)^*dJ(t)^*$$

and construct the output measurement process as

$$Y_o(t) = U(t)^*Y_i(t)U(t)$$

The HP qsde therefore reads

$$dU(t) = (-(iH+P)dt + L_1 dA(t) + L_2 dA(t)^* + L_3\theta(t)dJ(t) + L_4\theta(t)^*dJ(t)^*)U(t)$$

To make $U(t)$ a unitary evolution operator, we require

$$0 = d(U(t)^*U(t)) = dU(t)^*.U(t) + U(t)^*dU(t) + dU(t)^*.dU(t)$$

This gives us

# Chapter 6

# Miscellaneous remarks on the content of the previous chapters

## 6.1 Quantum Boltzmann equation in the presence of Lindblad noise with time varying interactions

$$i\rho'(t) = [\sum_a H_a, \rho(t)] + [\sum_{a<b} V_{ab}, \rho(t)] + \theta(\rho(t))$$

where

$$\theta(\rho) = (-1/2)(L^*L\rho + \rho L^*L - 2L\rho L^*)$$

Assume $L$ to have the separable form

$$L = L_0^{\otimes N}$$

where $N$ is the number of particles. We get on taking partial trace over $2, 3, ..., N$,

$$i\rho_1'(t) = [H_1, \rho_1(t)] + (N-1)Tr_2[V_{12}, \rho_{12}(t)],$$

$$i\rho_{12}'(t) = [H_1 + H_2 + V_{12}, \rho_{12}(t)] + (N-2)Tr_3[V_{13} + V_{23}, \rho_{123}]$$

$$\rho_{12} = \rho_1 \otimes \rho_1 + g_{12}$$

Then,

$$i(\rho_1' \otimes \rho_1 + \rho_1 \otimes \rho_1' + g_{12}') =$$

$$[H_1, \rho_1] \otimes \rho_1 + \rho_1 \otimes [H_1, \rho_1] + (N-1)Tr_2[V_{12}, \rho_1 \otimes \rho_1] \otimes \rho_1$$

$$+(N-1)\rho_1 \otimes Tr_2[V_{12}, \rho_1 \otimes \rho_2]$$

$$= [H_1 + H_2 + V_{12}, \rho_1 \otimes \rho_1 + g_{12}] + (N-2)Tr_3[V_{13} + V_{23}, \rho_1 \otimes \rho_1 \otimes \rho_1]$$

approximately. Further approximation gives after making appropriate cancellations,

$$ig'_{12}(t) = [H_1 + H_2, g_{12}(t)] + [V_{12}(t), \rho_1(t) \times \rho_1(t)]$$

and this has the solution

$$g_{12}(t) = -i \int exp(-i(t-s)ad(H_1 + H_2))([V_{12}(s), \rho_1(s) \otimes \rho_1(s)])ds$$

For our RI problem, we take $H$ as the Dirac Hamiltonian and choose a system observable $X$ as a component of the atomic electric or magnetic dipole moment. Then $j_t(X) = U(t)^* XU(t)$ becomes the evolution of $X$ in the Heisenberg dynamics after time $t$. We estimate $j_t(X)$ based on our non-demolition measurements as

$$\pi_t(X) = \mathbb{E}[j_t(X)|\eta_o(t)]$$

where

$$\eta_o(t) = \sigma(Y_o(s) : s \leq t)$$

and hence obtain reliable real time estimates of the evolving refractive index. Quantum filtering therefore connects measurable values (ie estimates) of the RI to the true RI.

## 6.2 Construction of irreducible representations of a semisimple Lie algebra

Let $\mathfrak{g}$ be a semisimple Lie algebra generated by the standard $3l$ elements $(H_i, X_i, Y_i : i = 1, 2, ..., l)$. They satisfy the commutation relations

$$[H_i, X_j] = a(j,i)X_j, [H_i, Y_j] = -a(j,i)Y_j, [X_i, Y_j] = \delta(i,j)H_i$$

where $a(j,i)$ are the Cartan integers. This can be achieved by choosing a simple system $S = (\alpha_i, i = 1, 2, ..., l)$ of positive roots and defining $H_i = \bar{H}_{\alpha_i}, X_i = X_{\alpha_i}, Y_i = X_{-\alpha_i}$ and then

$$a(j,i) = \alpha_j(H_i) = \alpha_j(\bar{H}_{\alpha_i}) = 2 < \alpha_j, \alpha_i > / < \alpha_i, \alpha_i >$$

Let $\mathcal{U}$ denote the universal enveloping algebra of $\mathfrak{g}$. Let $\pi$ denote a representation of $\mathcal{U}$ in a vector space $V$. Then, $\pi$ is said to be a representation with weights if

$$V = \bigoplus_{\mu \in \mathfrak{h}^*} V_\mu$$

where

$$V_\mu = \{v \in V : \pi(H)v = \mu(H)v \forall H \in \mathfrak{h}\}$$

and

$$H = span(H_i : 1 \leq i \leq l)$$

Suppose $\pi$ is a representation with weights and let $U$ be a $\pi(\mathfrak{h})$-invariant subspace of $V$. Then, we claim that

$$U = \bigoplus_{\mu \in \mathfrak{h}^*} U \cap V_\mu$$

In fact, if $u \in U$, then we can write

$$u = \sum_{\mu \in \mathfrak{h}^*} u_\mu, u_\mu \in V_\mu$$

Now the rhs is a finite sum so we can write

$$u = \sum_{k=1}^{p} u_k, u_k \in V_{\mu_k}, k = 1, 2, ..., p, \mu_k \in \mathfrak{h}^*$$

Since the complement in $\mathfrak{h}$ of the set $\{H : \mu_k(H) = \mu_j(H) some k \neq j\}$ is dense in $\mathfrak{h}$, it follows that there exists an $H \in \mathfrak{h}$ such that $\mu_k(H), k = 1, 2, ..., p$ are all distinct. Therefore, we can choose polynomials $p_j, j = 1, 2, ..., p$ in a single complex variable such that $p_j(\mu_k(H)) = \delta_{jk}$ and then it follows from the $\pi(H)$-invariance of $U$ that

$$p_j(\pi(H))u \in U,$$

$$p_j(\pi(H))u_k = p_j(\mu_k(H))u_k = \delta_{jk} u_k$$

and therefore,

$$u_j = p_j(\pi(H)) \sum_k u_k = p_j(\pi(H))u \in U, j = 1, 2, ..., p$$

proving that

$$u_j \in U \cap V_{\mu_j}, j = 1, 2, ..., p$$

It follows that

$$U = \bigoplus_{\mu \in \mathfrak{h}^*} (U \cap V_\mu)$$

Theorem: Let $\mathcal{U}$ be the universal enveloping algebra of a Lie algebra $\mathfrak{g}$ and let $\pi$ be a representation of $\mathcal{U}$ in a vector space $V$. Let $\lambda \in \mathfrak{h}^*$ and let $v$ be a cyclic vector of $\pi$, ie, $\pi(\mathcal{U})v = V$ and assume that $v \in V_\lambda$, ie, $\pi(H)v = \lambda(H)v \forall H \in \mathfrak{h}$ and $\pi(X_i)v = 0, i = 1, 2, ..., l$. Then (a) $\pi$ is a representation with weights, (b) $v$ is a highest weight vector of $\pi$, ie, $\lambda$ is the highest weight of $\pi$, (c) $dim V_\lambda = 1$ and (c) $\pi(\mathcal{N}^-)v = V$ where $\mathcal{N}^-$ is the universal enveloping algebra of the Lie algebra $\mathfrak{n}^-$ generated by $Y_i, i = 1, 2, ..., l$. Further, if $\pi$ is irreducible, then

$$V_\lambda = \mathbb{C}.v = \{w \in V : \pi(X_i)w = 0, 1 \leq i \leq l\}$$

Proof: Let

$$\bar{V} = span\{\pi(Y_{i_1})...\pi(Y_{i_k})v : i_1, ..., i_k = 1, 2, ..., l, k = 0, 1, 2, ...\}$$

Then it is easy to see that $\bar{V}$ is $\pi(\mathcal{U})$-invariant by using the commutation relations

$$[\pi(X_i), \pi(Y_j)] = \delta_{ij}\pi(H_i), [\pi(H_i), \pi(Y_j)] = -a(j,i)\pi(Y_j)$$

Further, $v \in \bar{V}$ and hence by cyclicity of $\pi$ it follows that $\bar{V} = V$. This proves that

$$V = \pi(\mathcal{N}^-)v$$

Further, it is easily seen that

$$v(i_1, ..., i_k) = \pi(Y_{i_1}...Y_{i_k})v \in V_{\lambda-\alpha_{i_1}-...-\alpha_{i_k}}$$

and hence by cyclicity of $v$, it follows that if $\mu \in \mathfrak{h}^*$, then

$$V_\mu = span\{v(i_1, ..., i_k) : \alpha_{i_1} + ...\alpha_{i_k} = \lambda - \mu\}$$

Thus,

$$V = \pi(\mathcal{N}^-)v = \bigoplus_{\mu \in \mathfrak{h}^*} V_\mu$$

In other words, we have proved that $\pi$ is a representation with weights, $\lambda$ and that $dimV_\lambda = 1$, ie, $V_\lambda = \mathbb{C}.v$, $\lambda$ is the highest weight of $\pi$ and $v$ is a highest weight vector for $\pi$. Note that the fact that $dimV_\lambda = 1$ follows immediately from $\pi(\mathcal{N}^-)v = V$.

Further, suppose that $\pi$ is irreducible and

$$U = \{u \in V : \pi(X_i)u = 0, 1 \leq i \leq l\}$$

Then it is easily seen that $U$ is $\pi(\mathfrak{h})$-invariant and hence

$$U = \bigoplus_{\mu \in \mathfrak{h}^*} U \cap V_\mu$$

Then, if for some $\mu \in \mathfrak{h}^*$, $U \cap V_\mu \neq 0$ then we can select a nonzero $u \in U \cap V_\mu$, ie, $u \in V_\mu$ and $\pi(X_i)u = 0, 1 \leq i \leq l$. Since further $\pi$ is now assumed to be irreducible, it follows by the same argument as above, with $v$ replaced by $u$, that $u$ is a highest weight vector for $\pi$ and therefore $\mu = \lambda$. This is because, all the weights of $\pi$ as we saw above were of the form $\lambda - \sum_{i=1}^{l} m_i\alpha_i$ with the $m_i's$ being non-negative integers. This proves that $U = V_\lambda = \mathbb{C}.v$.

Let $\pi$ be a representation of a Lie algebra $\mathfrak{g}$ or equivalently, of its universal enveloping algebra $\mathcal{U}$ in $V$ and let $v$ be a cyclic vector of $\pi$. Define the map

$$T : \mathcal{U} \to V$$

by

$$T(a) = \pi(a)v, a \in \mathcal{U}$$

Let $K$ denote the kernel of $T$, ie,

$$K = \{a \in \mathcal{U} : T(a)v = 0\}$$

Then, $K$ is an ideal of $\mathcal{U}$. Then let $T_1$ denote the map induced by $T$ from $\mathcal{U}/K \to V$, ie,

$$T_1(a+I) = T(a) = \pi(a)v, a \in \mathcal{U}$$

This map is well defined because $T$ vanishes on $K$. Further, it easily follows that $T_1$ is a vector space isomorphism. Thus,

$$\pi_1(a) = T_1^{-1} o\pi(a)oT_1, a \in \mathcal{U}$$

is a representation of $\mathcal{U}$ in $\mathcal{U}/K$ that is equivalent to $\pi$. Therefore, $\pi$ is irreducible iff $K$ is a maximal ideal of $\mathcal{U}$. More generally, suppose $U$ is a $\pi$-invariant subspace of $V$ not containing $v$ and define

$$M(U) = \{a \in \mathcal{U} : \pi(a)v \in U\}$$

Then, $M(U)$ is a proper ideal in $\mathcal{U}$ containing $K$. Proper because $\pi(1)v = v \notin U$ and therefore, $1 \notin K(U)$. Conversely, suppose $M$ is any proper ideal in $\mathcal{U}$ containing $K$ and define

$$U = \pi(M)v = \{\pi(a)v : a \in M\}$$

Then, $U$ is a $\pi$-invariant subspace of $V$ not containing $v$ and further,

$$M(U) = M(\pi(I)v) = M$$

To see this last part, suppose first that $a \in M(U)$. Then, $\pi(a)v \in U$ by definition and hence $\pi(a)v = \pi(b)v$ for some $b \in M$. Therefore, $\pi(a-b)v = 0$ and hence, $a - b \in K \subset M$. Therefore, $a \in M + M = M$. Conversely, suppose $a \in M$. Then, $\pi(a)v \in U$ and hence $a \in M(U)$. This proves the claim. Thus, the mapping $U \to M(U)$ is a bijection between all $\pi$-invariant subspaces of $V$ not containing $v$ (or equivalently, all proper $\pi$-invariant subspaces of $V$, since cyclicity of $v$ implies that a $\pi$-invariant subspace is proper iff it does not contain $v$) and all proper maximal ideals of $\mathcal{U}$ containing $K$.

Now we come to another result: Let $\lambda \in \mathfrak{h}^*$. Define

## 6.3 Questions on Matrix theory

[1] Write down explicitly a formula for $B = \sqrt{A^*A}$ where $A$ is an arbitrary $2 \times 2$ square matrix.

[2] If $\mathfrak{g}$ is a semisimple Lie algebra and $X \in \mathfrak{g}$, then consider the polynomial

$$p_X(t) = det(t - ad(X))$$

Let $l(X)$ denote the least power of $t$ appearing with a non-zero coefficient in $p_X(t)$. Choose $X \in \mathfrak{g}$ so that $l(X)$ is a minimum. Then, define

$$\mathfrak{h}_X = \{Y \in \mathfrak{g} : ad(X)^m(Y) = 0 forsomem \geq 1\}$$

Prove the following statements:
[a] $\mathfrak{h}_X$ is a Lie sub-algebra of $\mathfrak{g}$.
[b] $\mathfrak{h}_X$ is its own normalizer in $\mathfrak{g}$.
[c] $\mathfrak{h}_X$ is maximal nilpotent.

[3] Let $\mathfrak{g} = \mathfrak{sl}(2, \mathbb{C})$, ie, the Lie algebra of all $2 \times 2$ complex matrices having zero trace. Show that this is indeed a Lie algebra. Construct a basis $\{H, X, Y\}$ for $\mathfrak{g}$ satisfying the commutation relations

$$[H, X] = 2X, [H, Y] = -2Y, [X, Y] = H$$

Using the idea of weight vectors, construct all the finite dimensional irreducible representations of $\mathfrak{g}$. Show that

$$\omega = 4XY + H^2 - H$$

regarded as an element of the universal enveloping algebra $\mathcal{U}$ of $\mathfrak{g}$ is in the centre of $\mathcal{U}$, ie, it commutes with every element of $\mathcal{U}$.

[4] If $\mathcal{U}$ is the universal enveloping algebra of a Lie algebra $\mathfrak{g}$ and $\pi$ is a representation of $\}$, explain how $\pi$ can be extended to a representation of $\mathcal{U}$. Further, if $\pi$ is an irreducible representation, then prove that $\pi(z) = c(z)I$ for any $z$ in the centre of $\mathcal{U}$.

[5] Draw the Dynkin diagram for the Lie algebras $\mathfrak{so}(4, \mathbb{C})$ and $\mathfrak{sl}(3, \mathbb{C})$. Here, $\int o(n, \mathbb{C})$ is the set of all $n \times n$ complex matrices that are antisymmetric and have zero trace while $\int l(n, \mathbb{C})$ is the set of all $n \times n$ complex matrices having zero trace. Before solving this problem, identify the Cartan subalgebra and the root vectors in the root space decomposition.

[6] Write down the primary decomposition of an $(n + m) \times (n + m)$ matrix consisting of exactly two Jordan blocks, one of size $n \times n$ having eigenvalue $\lambda$ and another of size $m \times m$ having eigenvalue $\mu$. Consider both the cases $\lambda = \mu$ and $\lambda \neq \mu$.

[7] Given three linearly independent column vectors $a, b, c$ in $\mathbb{R}^4$, write down the QR decomposition of the $4 \times 3$ matrix $A = [a, b, c]$ and explain how this QR decomposition can be used to solve the least squares problem

$$min(\| y - Ax \|^2 : x \in \mathbb{R}^3)$$

## 6.4 Questions on Antenna Theory

[1] If a Dirac field and an electromagnetic field both are present within a rectangular cavity resonator, of dimensions $a, b, d$, then expand the both the magnetic vector potential and the Dirac wave field in terms of the basis functions

$$u_{mnp}(x,y,z) = (2\sqrt{2})/\sqrt{abd}.sin(m\pi x/a)sin(n\pi y/b)sin(p\pi z/d)$$

with coefficients depending upon time and write down a perturbative algorithm to solve for these coefficients. Your perturbation theory should be based on regarding the interaction between the electromagnetic field and the Dirac field to be small with the electronic charge being the perturbation parameter.

[2] In the previous problem, from the differential equations satisfied by the the time dependent coefficients appearing in the expansion of the Dirac wave field and the electromagnetic field, and their approximate solutions, calcuate the surface current density induced on the perfectly conducting walls of the cavity due to the magnetic field and also the volume current density within the cavity produced by the Dirac field. Hence calculate the far field radiation pattern outside the cavity produced by these two currents using the retarded potential.

[3] What is a Fermionic coherent state ? For a single Fermion, write down its explicit form in terms of a linear combination of particle number states $|0>$ and $|1>$ in terms of a Grassmannian parameter $\gamma$ which anticommutes with itself, with $\gamma^*$ and with $a, a^*$.

[4] Consider a situation in which there is one photon at frequency $\omega$ with creation- annihilation pair, $c^*, c$ and one electron-positron pair with annihilation-creation pair $a, a^*, b, b^*$ at a different frequency $\omega'$. The photon field is then

$$A^\mu(t,r) = e^\mu(t,r)c + e^\mu(t,r)^* c^*$$

and the electron field is

$$\psi(t,r) = a.u(t,r) + b^* v(t,r)$$

write down explicitly the formula for the electric and magnetic fields within the cavity upto first order perturbation terms by considering the electronic current density to be $J^\mu(t,r) = \psi(t,r)^* \alpha^\mu \psi(t,r)$ and the interaction energy between the photon and the Fermions to be $e(\alpha, A(t,r)) - eA^0(t,r)$. Also write down upto first order perturbation theory the Dirac wave function within the cavity. Hence calculate the mean and correlations of the far field radiation pattern produced by the surface current and the volume current in the cavity when the photon and the Fermions are in the joint coherent state

$$|z, \gamma_1, \gamma_2>$$

where

$$c|z>= z|z>, a|\gamma_1, \gamma_2>= \gamma_1|\gamma_1, \gamma_2>, b|\gamma_1, \gamma_2>= \gamma_2|\gamma_1, \gamma_2>$$

## 6.5 Superstrings and the Refractive Index of materials

[1] Consider the super-Yang-Mills action

$$S[A,\chi] = \int [(-1/4)Tr(F_{\mu\nu}F^{\mu\nu}) + \bar{\chi}\gamma.D\chi]d^Dx$$

$$= \int [(-1/4)F^a_{\mu\nu}F^{\mu\nu a} + \chi^{aT}\gamma^5\epsilon\gamma^\mu D_\mu\chi^a]d^D$$

where

$$ieF_{\mu\nu} = [D_\mu, D_\nu] = [\partial_\mu + ieA_\mu, \partial_\nu + ieA_\nu]$$

or equivalently,

$$F_{\mu\nu} = A_{\nu,\mu} - A_{\mu,\nu} + ie[A_\mu, A_\nu]$$

or equivalently in terms of the gauge group coordinates,

$$F^a_{\mu\nu} = A^a_{\nu,\mu} - A^a_{\mu,\nu} + eC(abc)A^b_\mu A^c_\nu$$

where $C(abc)$ are the structure constants for the chosen basis of the gauge group Lie algebra. Note that the covariant derivative $D_\mu$ acts on the spinor $\chi$ in the adjoint representation in contrast to what happens in conventional non-supersymmetric Yang-Mills theory. There, the gauge covariant derivative acts on the Dirac wave function spinor in the spinor representation, ie, $\psi \rightarrow D_\mu\psi$ or equivalently

$$\psi \rightarrow \partial_\mu\psi + ieA_\mu\psi$$

where

$$A_\mu\psi = A^a_\mu\tau_a\psi$$

with the $\tau_a$ being Hermitian basis elements for the gauge Lie algebra. Here,

$$D_\mu\chi = \partial_\mu\chi + ie[A_\mu, \chi]$$

or in terms of gauge group coordinates,

$$(D_\mu\chi)^a = \partial_\mu\chi^a + eC(abc)A^b_\mu\chi^c$$

since

$$\chi = \chi^a\tau_a, [\tau_a, \tau_b] = -iC(cab)\tau_c$$

The supersymmetry transformations which leave the above action invariant are

$$\delta A^{\mu a} = \bar{\eta}\gamma^\mu\chi^a = \eta^T\gamma^5\epsilon\gamma^\mu\chi^a$$

$$\delta\chi^a = F^a_{\mu\nu}\gamma^{\mu\nu}\eta$$

where $\eta$ is a Majorana Fermionic parameter and $\gamma^{\mu\nu} = [\gamma^\mu, \gamma^\nu$. The cubic term in the Fermionic fields $\chi^a$ that arise from the variation of the above action is

$$\chi^{aT}\gamma^5\epsilon\gamma^\mu\delta(D_\mu)\chi^a = \chi^{aT}\gamma^5\epsilon\gamma^\mu C(abc)(\delta A^b_\mu)\chi^c$$

$$= C(abc)\chi^{aT}\gamma^5\epsilon\gamma^\mu\chi^c(\eta^T\gamma^5\epsilon\gamma_\mu\chi^b)$$

and for supersymmetry, we require that this term should vanish. This happens only when the space-time dimension assumes a critical value. When we control the Fermionic field $\chi$ using supercurrents, the Yang-Mills field $A^a_\mu$ will get altered in view of the field equations and when a Dirac particle moves in this controlled Yang-Mills field, its corresponding wave function will get altered. Consequently, the statistics of the electric and magnetic dipole moment of the electron-nucleus pair in this altered Dirac wave function will get altered and hence the RI will undergo a change. Another place to see how one can formulate super RI problems for gauge fields interacting with superstrings is by noting that the string-gauge field Lagrangian has the form

$$L = (1/2)\partial_\alpha X^\mu\partial^\alpha X_\mu + (1/2)\psi^T\rho^0\rho^\alpha\partial_\alpha\psi + B_{\mu\nu}(X)\epsilon_{\alpha\beta}\partial^\mu X^\alpha.\partial^\nu X^\beta + c.H_{\mu\nu\rho}H^{\mu\nu\rho}$$

where

$$H = dB$$

ie

$$H_{\mu\nu\rho} = B_{\mu\nu,\rho} + B_{\nu\rho,\mu} + B_{\rho\mu,\nu}$$

and by $\psi^T\rho^0\rho^\alpha\partial_\alpha\psi$, we actually mean $\psi^{\mu T}\rho^0\rho^\alpha\partial_\alpha\psi_\mu$. Thus, $\psi$ actually has two indices, one a four vector index $\mu$ and two a spinor index $\alpha$. $B_{\mu\nu}(X)$ is the string-theoretic analog of the electromagnetic four potential $A_\mu$ in point particle field theory and $H_{\mu\nu\rho}(X)$ is the string theoretic analogue of the electromagnetic antisymmetric field tensor $F_{\mu\nu}$ in point particle field theory. $((\epsilon_{\alpha\beta}))$ is the standard antisymmetric $2x2$-matrix. . It should be noted that just as in point particle field theory, there is an interaction term between the electromagnetic four potential and the Dirac field, here, in the string theoretic context also we could include such an interaction term

$$B_{\mu\nu}\psi^T\gamma^5\epsilon\gamma^{\mu\nu}\psi, \gamma_{\mu\nu} = [\gamma_\mu, \gamma_\nu]$$

provided that $\psi$ has an additional four bi-spinor index. However, one cannot achieve supersymmetry using such a term. It would be more plausible to represent the interaction between $\psi = ((\psi^{\mu\alpha}))$ and $B_{\mu\nu}$ by incorporating a term such as

$$\psi^{\sigma T}(\tau,\sigma)\rho^0[\rho^\alpha,\rho^\beta]\psi_\sigma(\tau,\sigma)B_{\mu\nu}(X(\tau,\sigma))X^\mu_{,\alpha}(\tau,\sigma)X^\nu_{,\beta}(\tau,\sigma)$$

or equivalently,

$$\psi^{\sigma T}\rho^0\psi_\sigma\epsilon^{\alpha\beta}B_{\mu\nu}(X)X^\mu_{,\alpha}X^\nu_{,\beta}$$

or even a term such as

$$B_{\mu\nu}(X)\psi^{\mu T}\psi^\nu$$

Note that $\rho^0$ is proportional to $\epsilon$ which is proportional to $\sigma_2$.

The corresponding string theoretic Maxwell equations that arise by varying the potentials $B_{\mu\nu}$ then have the form

$$\Box B_{\mu\nu} = J_{\mu\nu}$$

where the currents $J_{\mu\nu}$ comprise of a superposition of terms such as $\epsilon^{\alpha\beta}\partial_\alpha X_\mu \partial_\beta X_\nu$, $\psi_\mu^T \psi_\nu$ and $\psi^{\sigma T}\rho^0\psi_\sigma \epsilon^{\alpha\beta} X^\mu_{,\alpha} X^\nu_{,\beta}$. Note that since the $\psi^\mu$ anticommute, it follows that all these currents are antisymmetric w.r.t the interchange of $\mu$ and $\nu$ as it should be since $B_{\mu\nu}$ is antisymmetric.

Note:The supersymmetry transformations that leave the action invariant in the absence of an external gauge field are

$$\delta X^\mu = \bar{\epsilon}\psi^\mu, \delta\psi^\mu = \rho^\alpha \epsilon \partial_\alpha X^\mu$$

where $\bar{\epsilon} = \epsilon^T \rho^0$.

Another way to see how the string dynamics in a string potential can affect the RI of materials is to assume that the string carries a linear charge density $\rho$. The dynamics of the string taking into account its interaction with the string gauge field $B_{\mu\nu}(X)$ is given by

$$\begin{aligned}\partial^\alpha \partial_\alpha X_\mu &= (\delta/\delta X^\mu)(B_{\rho\sigma}(X)\epsilon^{\alpha\beta} X^\rho_{,\alpha} X^\sigma_{,\beta}) \\ &= B_{\rho\sigma,\mu}(X)\epsilon^{\alpha\beta} X^\rho_{,\alpha} X^\sigma_{,\beta} \\ &\quad -2\partial_\alpha(B_{\mu\sigma}(X)\epsilon^{\alpha\beta} X^\sigma_{,\beta})\end{aligned}$$

Note:The last term on the right can be expanded as

$$\begin{aligned}&\partial_\alpha(B_{\mu\sigma}(X)\epsilon^{\alpha\beta} X^\sigma_{,\beta}) \\ &= B_{\mu\sigma,\rho}(X)\epsilon^{\alpha\beta} X^\rho_{,\alpha} X^\sigma_{,\beta} \\ &\quad +B_{\mu\sigma}(X)\epsilon^{\alpha\beta} X^\sigma_{,\alpha\beta}\end{aligned}$$

Remark:This equation of motion of the Bosonic string in an external gauge field should be compared with the geodesic equation of motion of a point particle in a curved space time with the particle world line $x^\mu(\tau)$ being replace by the string world sheet $X^\mu(\tau, \sigma)$ and the metric $g_{\mu\nu}(x)$ of space-time being replaced by the gauge potential $B_{\mu\nu}(X)$. Now the free Bosonic string in quantum mechanics has the expansion

$$X^\mu(\tau, \sigma) = -i \sum_{n \neq 0} (\alpha^\mu(n)/n) exp(in(\tau - \sigma))$$

where

$$[\alpha^\mu(n), \alpha^\nu(m)] = \eta^{\mu\nu}\delta[n+m]$$

We regard this as the unperturbed solution to the string field. To proceed further, we require to calculate the Hamiltonian of the Bosonic string field interacting with the gauge field in terms of the string gauge field $B_{\mu\nu}(X)$. The Hamiltonian of the free string part is

$$H = (1/2)(X^\mu_{,\tau} X_{\mu,\tau} + X^\mu_{,\sigma} X_{\mu,\sigma})$$

and this evaluates to the sums of the Hamiltonians of independent harmonic oscillators. To proceed further we must first compute the perturbation to this string Hamiltonian due the gauge field. The perturbation $\delta X^\mu$ to the string field satisfies

$$\Box\delta X^\mu =$$

$$\delta[B_{\rho\sigma,\mu}(X)\epsilon^{\alpha\beta}X^\rho_{,\alpha}X^\sigma_{,\beta}$$

$$-2B_{\mu\sigma,\rho}(X)\epsilon^{\alpha\beta}X^\rho_{,\alpha}X^\sigma_{,\beta}$$

$$-2B_{\mu\sigma}(X)\epsilon^{\alpha\beta}X^\sigma_{,\alpha\beta}]$$

We can express this as

$$\Box\delta X^\mu = \delta[H_{\rho\sigma\mu}(X)\epsilon^{\alpha\beta}X^\rho_{,\alpha}X^\sigma_{,\beta}]$$

$$-2\delta[B_{\mu\sigma}(X)\epsilon^{\alpha\beta}X^\sigma_{,\alpha\beta}]$$

In principle, we can solve this linear pde for $\delta X^\mu$ and express it as a nonlinear combination of the string creation-annihilation operators $\alpha^\mu(n)$. If the string carries a linear charge density, we can then calculate the statistical moments of the electromagnetic field produced by this charge in any given string state, say in a coherent state of the string. The effect of this electromagnetic field on the wave function of a particle can then be computed and this would yield the probability distribution function of the atomic electric and magnetic dipole moments.

# Chapter 7

# Applications of Large deviation theory to engineering problems

**Some problems of interest**

[A] Diffusion process theory.

[1] Construction of the stochastic integral w.r.t Brownian motion.

[2] Properties of the stochastic integral.

[3] Construction of the quantum stochastic integral w.r.t creation, annihilation and conservation processes in the Hudson-Parthasarathy formalism.

[4] Expressing stochastic differential equations as special cases of Evans-Hudson flows.

[5] Existence and Uniqueness theorems for classical and quantum stochastic differential equations.

$$dj_t(X) = j_t(\theta_b^a(X))d\Lambda_a^b$$

or in integral form

$$j_t(X) = X + \int_0^t j_s(\theta_b^a(X))d\Lambda_a^b(s)$$

Solve by the Picard iteration method:

$$j_t^n(X) = X + \int_0^t j_s^{n-1}(\theta_b^a(X))d\Lambda_a^b(s), n = 1, 2, ...$$

with initial condition

$$j_s^0(X) = X$$

$X$ is a system space operator. This equation should be viewed as defining the operators $j_t^n(X), n = 1, 2, ...$ in Boson Fock space for any given system space operator $X$. Thus, iterating this equation gives us

$$j_t^n(X) = X + \theta_{b1}^{a1}(X)\Lambda_{a1}^{b1}(t) + \theta_{b2}^{a2}(\theta_{b1}^{a1}(X)) \int_{0<s_2<s_1<t} d\Lambda_{a2}^{b2}(s_2)d\Lambda_{a1}^{b1}(s_1) + ...+$$

$$+\theta_{b1}^{a1}(X)\theta_{b2}^{a2}(X)...\theta_{bn}^{an}(X) \int_{0<s_n<...<s_1<t} d\Lambda_{an}^{bn}(s_n)...d\Lambda_{a1}^{b1}(s_1)$$

Then,

$$(j_t^n(X) - j_t^{n-1}(X))|fe(u) >= \int_0^t (j_s^{n-1}(\theta_b^a(X)) - j_s^{n-2}(\theta_b^a(X))d\Lambda_s^b(s)|fe(u) >$$

By using this equation along with Gromwall's lemma, the existence of a solution to the qsde can be deduced.

[5] Large deviation principle for Gaussian and Poisson processes.

[6] Large deviation principle for quantum Gaussian processes and quantum Poisson processes in a coherent state.

[7] Large deviation principle for non-independent random variables based on the Gartner-Ellis theorem.

[8] Large deviation principle based design of controllers of plant dynamics based on feedback from state observers and the desired trajectory.

[10] Large deviations in plasma physics within a cavity resonator.

Let $C$ denote a cavity resonator and accordingly assume that the modal expansions of the electric and magnetic fields are

$$\mathbf{E}(\omega, \mathbf{r}) = \sum_n [c_1(n,\omega)\psi_{1n}(\mathbf{r}), c_2(n,\omega)\psi_{2n}(\mathbf{r}), c_3(n,\omega)\psi_{3n}(\mathbf{r})]^T$$

$$\mathbf{B}(\omega, \mathbf{r}) = \sum_n [d_1(n,\omega)\phi_{1n}(\mathbf{r}), d_2(n,\omega)\phi_{2n}(\mathbf{r}), d_3(n,\omega)\phi_{3n}(\mathbf{r})]^T$$

where $\psi_{kn}, \phi_{kn}, k = 1,2,3$ are basis functions within the cavity satisfying the required boundary conditions, namely, the tangential components of the electric field and the normal components of the magnetic field vanish on the boundary. The plasma within the cavity generates a current density $\mathbf{J}(\omega, \mathbf{r})$ and a charge density $\rho(\omega, \mathbf{r})$ as

$$\mathbf{J}(\omega, \mathbf{r}) = Q \int f_1(\omega, \mathbf{r}, \mathbf{v})\mathbf{v}d^3v,$$

$$\rho(\omega, \mathbf{r}) = Q \int f_1(\omega, \mathbf{r}, \mathbf{v})d^3v$$

where $f_1$ is the time Fourier transform of the perturbation to the Boltzmann distribution function. $f_1$ satisfies the first order perturbed Boltzmann equation with the relaxation time approximation used for the collision term:

$$\partial_t f_1(t, \mathbf{r}, \mathbf{v}) + (\mathbf{v}, \nabla_r)f_1(t, \mathbf{r}, \mathbf{v}) +$$

$$Q(\mathbf{E}(t, \mathbf{r}) + \mathbf{v} \times \mathbf{B}(t, \mathbf{r}), \nabla_v)f_0(\mathbf{r}, \mathbf{v}) - Q(\nabla\Phi(\mathbf{r}), \nabla_v)f_1(t, \mathbf{r}, \mathbf{v})$$

$$+f_1(t,\mathbf{r},\mathbf{v})/\tau(\mathbf{v})=0$$

where

$$f_0(\mathbf{r},\mathbf{v})=C(\beta).exp(-\beta(-Q\nabla\Phi(\mathbf{r})+mv^2/2))$$

is the equilibrium Gibbs distribution function corresponding to a static electrostatic potential $\Phi(\mathbf{r})$ that is assumed to be present even prior to the generation of the electric and magnetic fields within the cavity. It satisfies the equilibrium/unperturbed Boltzmann equation

$$(\mathbf{v},\nabla_r)f_0(\mathbf{r},\mathbf{v})-Q(\nabla\Phi(\mathbf{r}),\nabla_v)f_0(\mathbf{r},\mathbf{v})=0$$

Taking the time-Fourier transform of the above first order perturbed Boltzmann equation gives us

$$i\omega.f_1(\omega,\mathbf{r},\mathbf{v})+(\mathbf{v},\nabla_r)f_1(\omega,\mathbf{r},\mathbf{v})+$$
$$Q((\mathbf{E}(\omega,\mathbf{r})+\mathbf{v}\times\mathbf{B}(\omega,\mathbf{r}),\nabla_v)f_0(\mathbf{r},\mathbf{v})$$
$$-Q(\nabla\Phi(\mathbf{r}),\nabla_v)f_1(\omega,\mathbf{r},\mathbf{v})+f_1(\omega,\mathbf{v},\mathbf{v})/\tau(\mathbf{v})=0$$

The Maxwell equations within the cavity are

$$curl\mathbf{E}(\omega,\mathbf{r})=-i\omega\mathbf{B}(\omega,\mathbf{r}),$$

$$curl\mathbf{B}(\omega,\mathbf{r})=i\omega\mu\epsilon\mathbf{E}(\omega,\mathbf{r})+\mu\mathbf{J}(\omega,\mathbf{r}),$$
$$div\mathbf{E}(\omega,\mathbf{r})=\rho(\omega,\mathbf{r}),div\mathbf{B}(\omega,\mathbf{r})=0$$

From these equations, we get after some manipulation

$$-\nabla^2\mathbf{E}+\nabla(\rho/\epsilon)=\omega^2\mu\epsilon\mathbf{E}-i\mu\omega\mathbf{J}$$

$$-\nabla^2\mathbf{B}=\omega^2\mu\epsilon\mathbf{B}+\mu\nabla\times\mathbf{J}$$

or equivalently,

$$(\nabla^2+\omega^2\mu\epsilon)\mathbf{E}=\nabla(\rho/\epsilon)+i\omega\mu\mathbf{J},$$
$$(\nabla^2+\omega^2\mu\epsilon)\mathbf{B}=-\mu\nabla\times\mathbf{J}$$

We also expand the Boltzmann distribution function using basis functions:

$$f_1(\omega,\mathbf{r},\mathbf{v})=\sum_n L(n,\omega)\chi_n(\mathbf{r},\mathbf{v})$$

The basis functions $\chi_n(\mathbf{r},\mathbf{v})$ are now functions of the position and velocity which vanish on the boundary of the cavity. Sometimes, it is more convenient to absorb the dependence upon $\mathbf{v}$ into the coefficients $L_n$ and assume that the $\chi_n's$ are functions of only the position variable:

$$f_1(\omega,\mathbf{r},\mathbf{v})=\sum_n L_n(\omega,\mathbf{v})\chi_n(\mathbf{r})$$

The Large deviation principle for this problem can now be formulated as follows: Let $\delta\mathbf{J}(\omega,\mathbf{r})$ be a small perturbation to the cavity current density coming from an external probe. We assume that this small perturbation is a Gaussian random field. Then, we must determine change in the distribution function perturbations $f_1$ caused by this perturbation to the current density. Correspondingly, the far field electromagnetic radiation pattern would get perturbed and we wish to restore equilibrium, ie, we wish to give an error feedback between the desired antenna pattern and the actual antenna pattern transformed into a corrector current source into the cavity with the controller feedback coefficients so chosen that the probability of the far field error pattern deviating from zero by an amount greater than a prescribed threshold level over a given frequency-space region is as small as possible.

The perturbed Boltzmann equation in the frequency-velocity domain:

$$i\omega.\sum_n L_n(\omega,v)\chi_n(r)+\sum_n L_n(\omega,v)(\mathbf{v},\nabla_r)\chi_n(r)$$

$$+Q(\sum_n C(n,\omega)\psi_n(r),\nabla_v)f_0(\mathbf{r},\mathbf{v})$$

$$-Q\sum_n \chi_n(r)(\nabla\Phi(\mathbf{r}),\nabla_v)L_n(\omega,v)+\sum_n L_n(\omega,v)\chi_n(r)/\tau(v)=0$$

Note that the cancellation $(v\times B,\nabla_v)f_0(r,v)=0$ has been used here. Using the assumed orthonormality of the $\chi_n's$, this gives us

$$(i\omega+1/\tau(v))L_n(\omega,v)+\sum_m L_m(\omega,v)<\chi_n,(v,\nabla_r)\chi_m>$$

$$+Q\sum_m C(m,\omega)<\chi_n,(\psi_m,\nabla_v)f_0>(v)$$

$$-Q\sum_m(<\chi_m,\chi_n.\nabla\Phi>,\nabla_v)L_m(\omega,v)=0$$

[11] Large deviation principle in general control theory for dynamical equations described by partial differential equations with random noise.

The dynamical system is given by

$$\partial_t\phi(t,x)=L_t\phi(t,x)+G(\phi)w(t,x)$$

where $L_t$ is a linear/nonlinear integro-partial differential-operator acting on the function $\phi(t,x)$ of time and space and $\phi(t,x)$ is a vector valued function of the time and space coordinates. $w(t,x)$ is random noise. Further $G(\phi)$ has a kernel $G(\phi)(x,y)$ defined by the equation

$$G(\phi)w(t,x)=\int G(\phi)(x,y)w(t,y)dy$$

For example, for a fluid with three velocity field $v(t,x)$ and density field $\rho(t,x)$ and pressure $p(t,x)$ given by the equation of state $p = f(\rho)$, we define our state vector $\phi = [v^T, \rho]^T \in \mathbb{R}^4$ and our dynamical system is

$$\partial_t v = -(v, \nabla)v - \rho^{-1} f'(\rho)\nabla\rho + eta/\rho)\nabla^2 v + w/\rho$$

$$\partial\rho = -\nabla.(\rho v) + g$$

where $g(t,x)$ is the rate of matter generation per unit volume. The desired state $\phi_d$ is assumed to satisfy the noiseless equation

$$\partial_t \phi_d = L_t \phi_d(t,x)$$

The state observer $\hat{\phi}(t,x)$ based on the noisy output measurements

$$dZ(t,x) = h(\phi(t,.))(x)dt + dv(t,x)$$

is given by

$$d\hat{\phi}(t,x) = L_t\hat{\phi}(t,x)dt + K_t(dZ(t,x) - H(\hat{\phi}(t,.))(x)dt)$$

where $K_t$ is the Kalman gain linear operator with kernel $K_t(x,y)$ so that

$$K_t(dZ(t,x) - h(\hat{\phi}(t,.))(x)dt) \int K_t(x,y)(dZ(t,y) - H(\hat{\phi}(t,.))(y))dy$$

If we are using the infinite dimensional EKF as our state observer, then with $P_t$ denoting the observer error correlation kernel $P_t(x,y)$, we have

$$K_t = P_t H_t^T R_t^{-1}$$

where $R_t(x,y)dt = \mathbb{E}(dv(t,x).dv(t,y)^T)$ is the measurement noise correlation and $H_t$ has the kernel $H_t(x,y)$ defined by the variational Jacobian kernel of $h$ at the state estimate:

$$H_t(x,y) = (\delta h(\phi)(x)/\delta\phi(y))|_{\phi=\hat{\phi}(t,.)}$$

and further the infinite dimensional Riccati equation for $P_t$ is given by

$$dP_t/dt = M_t P_t + P_t M_t^T - P_t H_t^T R_t^{-1} P_t + G_t Q_t G_t^T$$

where

$$Q_t(x,y) = \mathbb{E}(w(t,x)w(t,y)^T), M_t(x,y) = (\delta L_t(\phi)(x)/\delta\phi(y))|_{\phi=\hat{\phi}(t,.)}$$

$$G_t(x,y) = G(\hat{\phi}(t,.))(x,y), G_t(\phi)w(t,x) = \int G_t(\hat{\phi}(t,.))(x,y)w(t,y)dy$$

After linear feedback control based on observed output tracking error $\phi_d - \hat{\phi}$, our state dynamics gets modified to

$$d\hat{\phi}(t,x) = L_t\hat{\phi}(t,x) + G_t(\phi)w(t,x) + K_c(h(\phi_d) - h(\hat{\phi}))(t,x)$$

where

$$K_c(h(\phi_d)-h(\hat{\phi}))(t,x)=\int K_c(x,y)(h(\phi_d)(t,y)-h(\hat{\phi})(t,y))dy$$

$$=\int K_c(x,y)(h(\phi_d(t,.))-h(\hat{\phi}(t,.))(y)dy$$

[12] The large deviation principle for the Boltzmann distribution function: Consider a system of $N$ identical particles where $N$ is a very large integer. The Boltzmann distribution function $f(t,r_n,v_n,n=1,2,...,N)$ evolves with time as

$$\partial_t f+\sum_{n=1}^N(v_n,\nabla_{r_n})f+\sum_{n=1}^N(F_n(r,v),\nabla_{v_n})f=0$$

where

$$F_n(r,v)=F_{ext}(r_n,v_n)+\sum_{k=1,k\neq n}^N F_{int}(r_n,v_n|r_k,v_k)$$

is the total force acting on the $n^{th}$ particle where $F_{ext}$ denotes the force on it due to the external fields and $F_{int}(r_n,v_n|r_k,v_k)$ is the force on it due to the $k^{th}$ particle. This equation is simply the Liouville equation that arises from the conservation of probability for the dynamical system

$$dr_n/dt=v_n, dv_n/dt=F_{ext}(r_n,v_n)+\sum_{k\neq n}F_{int}(r_n,v_n|r_k,v_k)=F_n(r,v)$$

The initial conditions $r_n(0),v_n(0),n=1,2,...,N$ are assumed to be random. We could add stochastic and damping force terms to this system so that they read

$$dr_n=v_ndt, dv_n=-\gamma v_ndt+F_n(r,v)dt+\sigma.dB_n(t)$$

where $B_n(.),n=1,2,...,N$ are independent Brownian motion processes. Then for each $n$, $(r_n(t),v_n(t):t\in\mathbb{R})$ will be a stationary process provided the initial conditions are chosen appropriately and further for different $n's$, these processes will will have the same statistics. We can now ask the following question: Suppose that we write down the one particle empirical distribution function based on particle and time averages as

$$\hat{f}(r,v|N,T)=\frac{1}{NT}\sum_{n=1}^N\int_0^T\delta_{(r_n(t),v_n(t))}dt$$

Then by the ergodic theorem as $N,T\to\infty$, this will converge a.s. to a random density $\tilde{f}(r,v|\omega)$. Further, under appropriate conditions, this limiting density will be nonrandom and will equal the true one particle stationary density. The question is can we determine the rate functional for this convergence ? Note that the one particle density $f_1(t, r_1, v_1)$ satisfies under the conditions of identicality of the particles,

$$\partial_t f_1(t, r_1, v_1) + (v_1, \nabla_{v_1}) f_1(t, r_1, v_1) + (F_{ext}(r_1, v_1), \nabla_{v_1}) f_1(t, r_1, v_1)$$

$$+(N-1) \int (F_{int}(r_1, v_1|r_2, v_2), \nabla_{v_1}) f_{12}(t, r_1, v_1, r_2, v_2) d^3r_2 d^3v_2 = 0$$

in the absence of stochastic terms. Here, $f_{12}$ is the two particle marginal density function and we are also assuming that

$$div_{v_2} F_{int}(r_2, v_2|r_1, v_1) = 0$$

Under the molecular chaos approximation,

$$f_{12}(t, r_1, v_1, r_2, v_2) \approx f_1(t, r_1, v_1) f_1(t, r_2, v_2)$$

and with this approximation it follows that the one particle density $f_1$ follows approximately a quadratic non-linear evolution equation. When damping and stochastic forcing terms are included and molecular chaos assumption is made, we get the following version of the Boltzmann equation

$$\partial_t f_1(t, r_1, v_1) + (v_1, \nabla_{v_1}) f_1(t, r_1, v_1) + (F_{ext}(r_1, v_1), \nabla_{v_1}) f_1(t, r_1, v_1)$$

$$+(N-1) \int f_1(t, r_2, v_2)(F_{int}(r_1, v_1|r_2, v_2), \nabla_{v_1}) f_1(t, r_1, v_1) d^3r_2 d^3v_2$$

$$-\gamma(v_1, \nabla_{v_1}) f_1(t, r_1, v_1) - (1/2)\sigma^2 \nabla_{v_1}^2 f_1(t, r_1, v_1) = 0$$

This can be termed as the Fokker-Planck-Boltzmann equation.

Evaluation of the rate function for the empirical density of Brownian motion:

$$\hat{\mu}(E|T) = T^{-1} \int_0^T \delta_{B(t)} dt$$

$$\int f(x) d\hat{\mu}(x|T) = T^{-1} \int_0^T f(B(t)) dt = X(T, f)$$

say. Then

$$\mathbb{E}exp(T.X(T, f)|B(0) = x) = \mathbb{E}[exp(\int_0^T f(B(t))dt)|B(0) = x] = u(T, x)$$

say. Then,

$$u(T, x) = (1 + f(x)dT)\mathbb{E}[u(T - dT, B(dT))|B(0) = x]$$

Thus,

$$\partial_T u(T, x) = (1/2)\partial_x^2 u(T, x) + f(x) u(T, x)$$

Let $\lambda(f)$ denote the maximum eigenvalue of the operator

$$L_f = (1/2)\partial_x^2 + f(x)$$

Then

$$lim_{T\to\infty}log(u(T,x)) = \lambda(f)$$

Thus, the rate functional of the time average emprical density of Brownian motion is given by the Legendre transform of $\lambda(f)$, ie,

$$I(\mu) = sup_f(\int f d\mu - \lambda(f))$$

[12] The moment generating function for a mixture of quantum Gaussian and quantum Poisson observables. First let $[a, a^*] = 1$ and define

$$X = a + a^* + ka^*a$$

where $k$ is a real number. We wish to calculate the moment generating function of $X$ in a coherent state, ie,

$$< \phi(u)|exp(tX)|\phi(u) >, |\phi(u) >= exp(- \| u \|^2 /2)|e(u) >$$

Write,

$$exp(tX) = exp(tka^*a).F(t)$$

Then,

$$exp(tka^*a)(ka^*aF(t) + F'(t)) = (a + a^* + ka^*a)exp(tka^*a)F(t)$$

so that

$$exp(tka^*a)F'(t) = (a + a^*)exp(tka^*a)F(t)$$

or equivalently,

$$F'(t) = exp(-tk.ad(a^*a))(a + a^*)F(t)$$

Now,

$$< n|exp(-tkad(a^*a))(a)|m >=< n|exp(-tka^*a)a.exp(tka^*a)|m >$$

$$= exp(tk(m - n)) < n|a|m >= exp(tk(m - n))\sqrt{m}.\delta[n - m + 1]$$

and in particular,

$$< n|exp(-tka^*a)a|n >= 0$$

The coherent state is

$$|\phi(u) >= exp(-|u|^2/2)\sum_n u^n|n > /\sqrt{n!}$$

Then,

$$< \phi(u)|exp(-tk.ad(a^*a))(a)|\phi(u) >=$$

$$exp(-|u|^2)\sum_{n,m}(\bar{u}^n u^m/\sqrt{n!m!}) < n|exp(-tk.ad(a^*a))(a)|m >$$

$$= exp(-|u|^2)\sum_{n,m}(\bar{u}^n u^m/\sqrt{n!m!}).exp(tk(m-n))\sqrt{m}\delta[n-m+1]$$

$$= exp(-|u|^2+tk)\sum_n(\bar{u}^n u^{n+1}/n!) = u.exp(tk)$$

Likewise,

$$< \phi(u)|exp(-tkad(a^*a))(a^*)|\phi(u) >= \bar{u}.exp(tk)$$

and hence

$$< \phi(u)|exp(-tk.ad(a^*a))(a+a^*)|\phi(u) >= 2Re(u).exp(tk)$$

More generally,

$$< \phi(u)|exp(-tkad(a^*a))(a)|\phi(v) >=$$

$$exp(-|u|^2/2-|v|^2/2)\sum_{n,m}(\bar{u}^n v^m/\sqrt{n!m!}) < n|exp(-tk.ad(a^*a))(a)|m >$$

$$= exp(-|u|^2/2-|v|^2/2)\sum_{n,m}(\bar{u}^n v^m/\sqrt{n!m!})exp(ik(m-n))\sqrt{m}\delta[n-m+1]$$

$$= exp(-|u|^2/2-|v|^2/2)\sum_n(\bar{u}^n v^{n+1}/n!)$$

$$= v.exp(-|u|^2/2-|v|^2/2+\bar{u}v)$$

and likewise,

$$< \phi(u)|exp(-tk.ad(a^*a))(a^*)|\phi(v) >=$$

$$exp(-|u|^2/2-|v|^2/2+\bar{u}v)\bar{u}$$

so that

$$< \phi(u)|exp(-tk.ad(a^*a))(a+a^*)|\phi(v) >= exp(-|u|^2/2-|v|^2/2+\bar{u}v)(\bar{u}+v)$$

Now we use the Glauber-Sudarshan P-representation

$$I = \int \phi(u) >< \phi(u)|dud\bar{u}/\pi$$

to get

$$< \phi(u)|F'(t)|\phi(v) >= d/dt(< \phi(u)|F(t)|\phi(v) >=$$

$$\pi^{-1}\int < \phi(u)|exp(-tk.ad(a^*a))(a+a^*)|\phi(w) >< \phi(w)|F(t)|\phi(v) > d^2w$$

The large deviation problem for quantum independent increment processes: Consider the process

$$X(t) = A_t(u) + A_t(u)^* + \Lambda_t(H), t \geq 0$$

where $u \in \mathcal{H}$ and $H$ is a self-adjoint operator in $\mathcal{H}$ that commutes with the spectral family of time flow. The process $X(t)$ is commutative since

$$[dA_t(u), dA_s(u)] = [dA_t(u), dA_s(u)^*] = [dA_t(u), d\Lambda_s(H)]$$

$$= [d\Lambda_s(H), d\Lambda_s(H)] = 0, \forall s = t$$

Therefore, in a coherent state $|\phi(v)>$ of the bath, the process $X(t)$ is a classical independent increment process. The problem is to determine the rate function of the empirical probability distribution of this process in this coherent state.

[13] Given a noisy quantum system described by the qsde

$$dU(t) = (-(iH + \delta^2 LL^*/2)dt + \delta.LdA(t) - \delta.L^*dA(t)^*)U(t)$$

calculate upto $O(\delta^2)$ the probability of the system state making a transition from the energy eigenstate $|n>$ of the system Hamiltonian $H$ to another energy eigenstate $|m>$ of the same with the bath being in a coherent state $|\phi(u)>$. Clearly as the noise parameter $\delta \to 0$, this transition probability converges to zero if $m \neq n$. The question is that at what rate does this probability converge to zero ? Secondly, let $X$ be a system space self-adjoint operator. Calculate the probability distribution $P_{t,\delta}$ of $X(t) = U(t)^*XU(t)$ in the coherent state $|\phi(u)>$ of the bath with the system being in the state $|f>$. Does this probability distribution have a large deviation rate function ? If so, how then to evaluate it. The third question is that given the sample average $Z(T) = (1/T)\int_0^T X(t)dt$, calculate the large deviation rate function of the probability distribution of $Z(T)$ in the state $|f\phi(u)>$ as $T \to \infty$.

[14] $[a, a^*] = 1$. Write

$$b = a + c$$

where $c$ is a scalar. Then

$$b^*b = a^*a + \bar{c}a + ca^*, [b, b^*] = 1$$

Let $|\phi(u)>$ be the normalized coherent state for $a$ and $|\psi(u)>$ that for $b$. Then,

$$a|\phi(u)>= u|\phi(u)>, b|\psi(u)>= u|\psi(u)>$$

That gives

$$a|\psi(u)>= (u - c)|\psi(u)>$$

and hence

$$|\psi(u)>= |\phi(u - c)>, |\phi(u)>= |\psi(u + c)>$$

Then,

$$<\phi(u)|exp(\beta b^*b)|\phi(u)>=<\psi(u+c)|exp(\beta b^*b)|\psi(u+c)>$$

$$=<\phi(u+c)|exp(\beta a^*a)|\phi(u+c)>$$

$$= \sum_{n\geq 0} |<n|\phi(u + c)>|^2 exp(\beta n)$$

where

$$<n|\phi(u + c)>= exp(-|u + c|^2/2)(u + c)^n/\sqrt{n!}$$

Thus,

$$<\phi(u)|exp(\beta b^*b)|\phi(u)>= exp(|u + c|^2.(exp(\beta) - 1))$$

Using this we can calculate the moment generating function of the observable $a^*a + ca + \bar{c}a^*$ by completing the squares. More generally, let $[a_n, a_m^*] = \delta_{n,m}, [a_n, a_m] = 0 = [a_n^*, a_m^*]$. Then consider

$$b_n = a_n + c_n, c_n \in \mathbb{C}$$

Let $|\phi(u)>$ be a normalized coherent state for $\{a_n\}$, ie,

$$a_n|\phi(u)>= u_n|\phi(u)>, n \geq 1, <\phi(u)|\phi(u)>= 1$$

and $|\psi(u)>$ that for $\{b_n\}$:

$$b_n|\psi(u)>= u_n|\psi(u)>, n \geq 1, <\psi(u)|\psi(u)>= 1$$

Then

$$a_n|\psi(u)>= (u_n - c_n)|\psi(u)>$$

and therefore,

$$|\psi(u)>= |\phi(u-c)>, |\phi(u)>= |\psi(u+c)>$$

Thus,

$$<\phi(u)|exp(\sum_n \beta(n)b_n^*b_n)|\phi(u)>$$

$$=<\psi(u+c)|exp(\sum_n \beta(n)b_n^*b_n)|\psi(u+c)>=$$

$$<\phi(u+c)|exp(\sum_n \beta(n)a_n^*a_n)|\phi(u+c)>$$

$$= exp(\sum_n(|u_n + c_n|^2(exp(\beta(n)) - 1)))$$

This formula immediately yields

$$<\phi(u)|exp(\sum_n \beta(n)(a_n^*a_n + c_n a_n^* + \bar{c}_n a_n))|\phi(u)>=$$

$$exp(\sum_n(|u_n + c_n|^2(exp(\beta(n)) - 1) - \beta(n)|c_n|^2))$$

Writing

$$c_n = \gamma(n)/\beta(n)$$

gives us

$$<\phi(u)|exp(\sum_n(\beta(n)a_n^*a_n + \gamma(n)a_n^* + \bar{\gamma}(n)a_n)|\phi(u)>=$$

$$exp(\sum_n(|u_n + \gamma(n)/\beta(n)|^2(exp(\beta(n)) - 1) - |\gamma(n)|^2/\beta(n)))$$

Taking the limit of this as $\beta(n) \to 0$ and using

$$lim_{\beta(n)\to 0}|u_n + \gamma(n)/\beta(n)|^2(\beta(n) + \beta(n)^2/2) - |\gamma(n)|^2/\beta(n)$$

$$= \bar{\gamma}(n)u_n + \gamma(n)\bar{u}_n + |\gamma(n)|^2/2$$

gives us the well known result

$$< \phi(u)|exp(\sum_n(\gamma(n)a_n^* + \bar{\gamma}(n)^* a_n))|\phi(u) >=$$

$$=< \phi(u)|\Pi_n exp(\gamma(n)a_n^* + \bar{\gamma}(n)a_n)|\phi(u) >=$$

$$exp(\sum_n(\bar{\gamma}(n)u_n + \gamma(n)\bar{u}_n + |\gamma(n)|^2/2))$$

which is the standard formula for the moment generating function of quantum Gaussian observables. In other words, in the coherent state $|\phi(u)>$, the observables $\gamma(n)a_n^* + \bar{\gamma}(n)a_n, n = 1, 2, ...$ are independent Gaussian random variables with means $\gamma(n)\bar{u}_n + \bar{\gamma}(n)u_n$ and variances $|\gamma(n)|^2$.

[15] Feedback controllers of the Hudson-Parthasarathy QSDE. Let $X_d(t)$ be the desired process and $j_t(X)$ satisfy the Evans-Hudson flow equation

$$dj_t(X) = j_t(\theta_0(X))dt + j_t(\theta_1(X))dA(t) + j_t(\theta_2(X))dA(t)^*$$

Let $\eta_0(t)$ be the output non-demolition measurement algebra at time $t$ and assume that we have a Belavkin filter for

$$\pi_t(X) = \mathbb{E}[j_t(X)|\eta_0(t)]$$

satisfying the classical sde

$$d\pi_t(X) = F_t(X)dt + G_t(X)dY_o(t)$$

where $F_t(X)$ and $G_t(X)$ are both $\eta_o(t)$-measurable. The control input given to the Evans-Hudson flow is a linear feedback term to make $j_t(.)$ satisfy the "controlled Evans-Hudson flow" equation

$$dj_t(X) = j_t(\theta_0(X))dt + j_t(\theta_1(X))dA(t) + j_t(\theta_2(X))dA(t)^*$$

$$+K_c(X_d(t) - \pi_t(X))dt$$

We assume that $k_t(X) = X_d(t)$ satisfies the noiseless Evans-Hudson flow equation

$$dk_t(X) = k_t(\theta_0(X))dt$$

Then the problem is to calculate the mean square tracking and observer error energy at time $t$:

$$\mathbb{E}[(j_t(X) - k_t(X))^2], \mathbb{E}[(j_t(X) - \pi_t(X))^2]$$

[16] Large deviation problems in superstring theory.
For Bosonic fields on the two $\mathbb{R}_+ \times [0, \pi]$, we have the action

$$S[\phi] = \int \partial_+\phi.\partial_-\phi.d^2\sigma$$

and hence the field equations

$$\partial_+\partial_-\phi = 0$$

This has two linearly independent solutions $\phi_+$ and $\phi_-$. $\phi_+$ is a function of $\sigma^+$ only and $\phi_-$ is a function of $\sigma^-$ only. Obviously

$$\partial_-\phi_+ = 0, \partial_+\phi_- = 0$$

and hence if we evaluate the action at the solution $\phi_+ + \phi_-$, we get

$$S = \int (\partial_+\phi_+.\partial_-\phi_-)d^2\sigma$$

We calculate the various components of the energy-momentum tensor

$$T_{--} = T_-^+ = (\delta L/\delta\partial_+\phi)(\partial_-\phi) = \partial_-\phi.\partial_-\phi$$

$$= \partial_-\phi_-.\partial_-\phi_-$$

Note that the metric is

$$d\tau^2 - d\sigma^2 = d\sigma^+.d\sigma^-, \sigma^+ = \tau + \sigma, \tau^- = \tau - \sigma$$

Hence,

$$h_{++} = h_{--} = 0, h_{+-} = h_{-+} = 1$$

Likewise,

$$T_{++} = T_+^- = (\delta L/\delta\partial_-\phi).\partial_+\phi$$

$$= \partial_+\phi.\partial_+\phi = \partial_+\phi_+.\partial_+\phi_+$$

Further,

$$T_{-+} = T_+^+ = (\delta L/\delta\partial_+\phi).\partial_+\phi - L$$

$$= \partial_-\phi.\partial_+\phi - L = 0$$

and likewise,

$$T_{+-} = T_-^- = 0$$

Green's function for the Fermionic string:

$$\partial_- < T(\phi_+(\sigma).\psi_+(\sigma')) >= \delta^2(\sigma - \sigma')$$

because the equal time Fermion anticommutation relations are

$$[\phi(\sigma^0, \sigma^1), \psi(\sigma^0, \sigma'^1)]_+ = \delta(\sigma^1 - \sigma'^1)$$

and

$$\partial_-\psi_+(\sigma) = 0, \partial_0\theta(\sigma^0 - \sigma'^0) = \delta(\sigma^0 - \sigma'^0)$$

Therefore,

$$< T(\phi_+(\sigma).\psi_+(\sigma')) >= \int exp(ik.\sigma)d^2k/k^-$$

$$= \int k^+.exp(ik.(\sigma - \sigma'))d^2k/k^2, k^2 = k^+.k^- = (k^0)^2 - (k^1)^2$$

Here,

$$k.\sigma = k^0\sigma^0 + k^1\sigma^1 = (1/2((k^0+k^1)(\sigma^0+\sigma^1)+(k^0-k^1)(\sigma^0-\sigma^1)) = (k^+\sigma^+ - k^-\sigma^-)$$

Hence

$$< T(\phi_+(\sigma).\psi_+(\sigma')) >= \partial_+ \int exp(ik.(\sigma - \sigma'))d^2k/k^2$$

Now,

$$\int exp(ik.\sigma)d^2k/k^2 = \int exp(i(k^0\sigma^0 + k^1\sigma^1))d^2k/((k^0)^2 - (k^1)^2)$$

$$= \int exp(i(k^0\sigma^0 + k^1(i\sigma^1)))d^2k/((k^0)^2 + (k^1)^2)$$

$$= ln((\sigma^0)^2 + (i\sigma^1)^2) = ln(\sigma^+\sigma^-)$$

and hence,

$$< T(\phi_+(\sigma).\psi_+(\sigma')) >= \partial_+(ln((\sigma^+ - \sigma'^+)(\sigma^- - \sigma'^-))) = 1/(\sigma^+ - \sigma'^+)$$

Now, we evaluate the two point function for the right moving part of the energy-momentum tensor: Using the fact that

$$\partial_- T_{++} = 0$$

because of

$$\partial_-\phi_+ = 0, \partial_-\psi_+ = 0, T_{++} = \phi_+.\partial_+.\psi_+,$$

we get

$$\partial_- < T(T_{++}(\sigma).T_{++}(\sigma')) >=$$

$$= \delta(\sigma^0 - \sigma'^0) < [T_{++}(\sigma^0, \sigma^1), T_{++}(\sigma^0, \sigma'^1)] >$$

On the other hand,

$$=< T(T_{++}(\sigma)T_{++}(\sigma')) >$$

$$=< T(\phi_+(\sigma).\partial'_+\psi_+(\sigma')) >^2$$

$$= (\partial'_+ < T(\phi_+(\sigma).\psi_+(\sigma')) >)^2$$

$$= K/(\sigma^+ - \sigma'^+)^4 = K(\partial^3/\partial\sigma^{+3})1/(\sigma^+ - \sigma'^+)$$

So

$$\partial_- < T(T_{++}(\sigma)T_{++}(\sigma')) >= K(\partial_+^3\partial_-(1/(\sigma^+ - \sigma^{'+}))$$

$$= K\partial_+^3\partial_-\partial_+ln(\sigma^+ - \sigma^{'+})$$

$$= K\partial_+^3\partial_-\partial_+ln((\sigma - \sigma')^2) = K\partial_+^3\partial^2 ln(\sigma - \sigma')^2) = K\partial_+^3\delta^2(\sigma - \sigma')$$

So in the momentum domain,

$$p_- < T_{++}(p)T_{++}(-p) >= Kp_+^3$$

or equivalently,

$$< T_{++}(p)T_{++}(-p) >= Kp_+^3/p_-$$

A comparison of the above two formulas, shows that there is a quantum mechanical anomaly which implies that the energy-momentum tensor is not conserved. If it were, then we would have $\partial_- T_{++}(\sigma) = 0$ and this would imply that

$$\partial_- < T(T_{++}(\sigma).T_{++}(\sigma')) >=$$

$$= \delta(\sigma^0 - \sigma^{'0}) < [T_{++}(\sigma^0, \sigma^1), T_{++}(\sigma^0, \sigma^{'1})] >$$

while another computation of this two point function based on a loop Feynman diagram would imply that

$$< T(T_{++}(\sigma).T_{++}(\sigma')) >= \int (Kp^{+3}/p^-)exp(p.(\sigma - \sigma'))d^2p$$

or equivalently, that

$$\partial_- < T(T_{++}(\sigma).T_{++}(\sigma')) >= \int K(p^+)^3.exp(ip.(\sigma - \sigma'))d^2p$$

$$= K\delta'''(\sigma^+ - \sigma^{'+})$$

which would not vanish if $\sigma^0 \neq \sigma^{'0}$.

Anomaly for the Bosonic string: The energy-momentum tensor for the right moving component is

$$T_{++}(\sigma) = \partial_+\phi_+.\partial_+\phi_+$$

The Bosonic string Lagrangian is

$$L = (\partial_0\phi)^2 - (\partial_1\phi)^2 = \partial_+\phi.\partial_-\phi$$

and writing $\phi = \phi_+\phi_-$, we can see that the above expression for $T_{++}$ is correct. Now, the Bosonic two point function can be computed as follows:

$$\partial_0 < T(\phi(\sigma)\phi(\sigma')) >= \delta(\sigma^0 - \sigma^{'0}) < [\phi(\sigma), \phi(\sigma')] >$$

$$+\theta(\sigma^0 - \sigma^{'0}) < \partial_0\phi(\sigma).\phi(\sigma') >$$

$$+\theta(\sigma^{'0} - \sigma^0) < \phi(\sigma')\partial_0\phi(\sigma) >$$

$$= \theta(\sigma^0 - \sigma^{'0}) < \partial_0\phi(\sigma).\phi(\sigma') >$$

$$+\theta(\sigma^{'0} - \sigma^0) < \phi(\sigma')\partial_0\phi(\sigma) >$$

where we have used the commutation relation

$$[\phi(\sigma), \phi(\sigma')] = 0$$

Another time differentiation using $\partial_0^2\phi_+ = \partial_1^2\phi_+$ gives us

$$\partial^2 < T(\phi(\sigma)\phi(\sigma')) >= \delta(\sigma^0 - \sigma^{'0}) < \partial_0\phi(\sigma), \phi(\sigma')] >$$

$$= -\delta^2(\sigma - \sigma')$$

in view of the equal time Bosonic commutation relations

$$[\phi(\sigma^0, \sigma^1), \partial_0\phi(\sigma^0, \sigma'1)] = \delta(\sigma^1 - \sigma'^1)$$

Note that

$$\partial^2 = \partial_+.\partial_- = \partial_-.\partial_+ = \partial_0^2 - \partial_1^2$$

Thus,

$$< T(\phi(\sigma)\phi(\sigma')) >= K.ln((\sigma - \sigma')^+(\sigma - \sigma')^-) = K.ln(\sigma - \sigma')^2)$$

where

$$\sigma^2 = \sigma^+\sigma^- = (\sigma^0)^2 - (\sigma^1)^2$$

The general solution to the Bosonic field equations

$$\partial^2\phi = 0$$

is

$$\phi = \phi_+ + \phi_-$$

where $\phi_+$ is only a function of $\sigma^+$ while $\phi_-$ is only a function of $\sigma^-$. Then since the forward string commutes with the backward string and any creation or annihilation operator acting on the vacuum state is zero,

$$< T(\phi(\sigma)\phi(\sigma')) >=< T(\phi_+(\sigma^+).\phi_+(\sigma^{'+})) > + << T(\phi_-(\sigma^-)\phi_-(\sigma'-) >$$

$$= K.(ln(\sigma - \sigma')^+) + ln((\sigma - \sigma')^-))$$

and hence in particular,

$$< T(\phi_+(\sigma^+).\phi_+(\sigma^{'+})) >= K.ln((\sigma - \sigma')^+),$$

$$< T(\phi_-(\sigma^-).\phi_-(\sigma^{'-})) >= K.ln((\sigma - \sigma')^-),$$

Note:

$$(\sigma - \sigma')^+ = \sigma^+ - \sigma^{'+}, (\sigma - \sigma')^- = \sigma^-\sigma^{'-}$$

Now the right moving part of the energy-momentum tensor is

$$T_{++} = T_{++}(\sigma^+) = \partial_+\phi_+.\partial^+\phi_+$$

and hence

$$< T(T_{++}(\sigma^+).T_{++}(\sigma^{'+})) >=$$
$$[< T(\partial_+\phi_+(\sigma^+).\partial'_+\phi_+(\sigma^{'+})) >]^2$$
$$= [\partial_+.\partial'_+ < T(\phi^+(\sigma^+).\phi^+(\sigma^{'+})) >]^2$$
$$= K[\partial_+\partial'_+ ln(\sigma^+ - \sigma^{'+})]^2$$
$$= K/(\sigma^+ - \sigma^{'+})^4$$

and in the same way, we get

$$\delta(\sigma^0 - \sigma^{'0}) < [T_{++}(\sigma^+), T_{++}(\sigma^{'+})] >$$
$$= \partial_- < T(T_{++}(\sigma^+).T_{++}(\sigma^{'+})) >=$$
$$K.\partial_-(\sigma^+ - \sigma^{'+})^{-4} = K.\delta'''(\sigma^+ - \sigma^{'+})$$

which in the momentum domain, translates to

$$< T_{++}(p)T_{++}(-p) >= K(p^+)^3/p^-$$

A more clear way to understand these anomalies is to observe that if no anomaly were present then the two point function $< T(T_{++}(\sigma^+).T_{++}(\sigma^{'+})) >$ would be a function of only $\sigma^+, \sigma^{'+}$ and hence if we apply $\partial_-$ to it, we would get zero. The same is true in the Fermionic case.

Note: The Fermionic Lagrangian is

$$L = \phi_+.\partial_-\psi_+ + \phi_-.\partial_+\psi_-$$

which results in the field equations

$$\partial_-\psi_+ = 0, \partial_+\psi_- = 0, \partial_-\phi_+ = 0, \partial_+\phi_- = 0$$

The components of the energy-momentum tensor are

$$T_{++} = T_+^- = (\delta L/\delta\partial_-\psi_+).\partial_+\psi_+ = \phi_+.\partial_+\psi_+$$

and

$$T_{--} = T_-^+ = (\delta L/\delta\partial_+\psi_-).\partial_-\psi_- = \phi_-.\partial_-\psi_-$$

the other components $T_+ = T_{-+} = 0$. Now we wish to derive the Virasoro commutation/anticommutation relations for the Fourier components of the energy-momentum tensor.

The large deviation problem: Take the Fermionic string Lagrangian

$$L = \psi_+\partial_-\psi_+ + \psi_-\partial_+\psi_-$$

and perturb it by a small random mass term say of the form

$$\delta L = \delta m_1.\psi_+\psi_+ + \delta m_2\psi_-.\psi_-$$

or more generally,

$$\delta L = \psi_+.\delta M_1.\psi_+ + \psi_-\delta M_2.\psi_-$$

where $\delta M_k, k = 1,2$ are small random skewsymmetric matrices. Then evaluate the rate function of the pair of fields $(\psi_+, \psi_-)$. Also evaluate the rate function of the energy momentum tensor perturbations caused by these mass perturbations. Finally, evaluate the rate function of the perturbations in the Fourier series components of the energy-momentum tensor.

Note: The perturbed Fermionic fields satisfy the field equations (derived from the Lagrangian)

$$\partial_-\psi_+ + \delta M_1.\psi_+ = 0 = \partial_+\psi_- + \delta M_2.\psi_- = 0$$

Now expanding in a Fourier series

$$\psi_+ = \sum_n S_n^+ exp(in\sigma^+) + \sum_n \delta S_n^+(\sigma^+, \sigma^-) exp(in\sigma^+)$$

gives us using first order perturbation theory the following approximate equation:

$$\partial_-\delta S_n^+ + \delta M_1.S_n^+ = 0$$

This translates in the momentum domain to

$$p^-\delta S_n^+(p) + \delta M_1 S_n^+ = 0$$

or equivalently,

$$\delta S_n^+(p) = -(p^-)^{-1}\delta M_1.S_n^+$$

Energy-momentum tensor for the massive Fermionic string:

$$L = \psi_+\partial_-\psi_+ + \psi_-\partial_+\psi_- - \psi_+.\delta M_1.\delta\psi_+ - \delta\psi_-.\delta M_2.\delta\psi_-$$

The various components of the energy-momentum tensor are

$$T_{--} = T_-^+ = (\partial L/\partial\partial_+\psi_-).\partial_-\psi_- =$$

$$\psi_-.\partial_-\psi_-,$$

$$T_{++} = \psi_+.\partial_+.\psi_+,$$

$$= T_-^- = T_{+-} = (\partial L/\delta\partial_-\psi_-)\partial_-\psi_- - L = -L$$

$$T_+^+ = T_{-+} = -L$$

Note that when the field equations are satisfied,

$$T_{+-} = T_{-+} = \psi_+.\delta M_1.\psi_+ + \psi_-.\delta M_2.\psi_-$$

Fourier component evaluation of the Fermionic energy-momentum tensor.

$$T_{++} = \psi_+ \partial_+ \psi_+ = \sum_{n,m} S_n^+ exp(in\sigma^+)(im) S_m^+ .exp(im\sigma^+)$$

$$= \sum_n G_n^+ exp(in\sigma^+)$$

where

$$G_n^+ = i \sum_n m S_{n-m}^+ S_m^+$$

Equal time Fermionic anticommutation relations imply

$$[\psi_+(\sigma), \psi_+(\sigma')]_+ = \delta(\sigma^+ - \sigma'^+)$$

$$[\psi_-(\sigma), \psi_-(\sigma')]_+ = \delta(\sigma^- - \sigma'^-)$$

and hence

$$[S_n^+, S_m^+]_+ = \delta[n+m], [S_n^-, S_m^-]_+ = \delta[n+m]$$

Problem: Using these anticommutation relations, deduce the commutation relations for $G_n^+$.

[17] Large deviations in stochastic optimal control: Consider the sde

$$dx(t) = c(t, x(t))dt + \sigma.dB(t)$$

where $x(t), B(t) \in \mathbb{R}^n$ with $B(.)$ being standard vector valued Brownian motion. Consider the cost function

$$V(T, c, \sigma) = \mathbb{E} \int_0^T L(s, c(s, x(s))ds$$

Calculate the drift $c$ so that $V$ is a minimum. Then if $\sigma \to 0$, calculate the rate function of $V$. To solve this problem define

$$V(t, T, x, c) = \mathbb{E}[\int_t^T L(s, c(s, x(s))ds|x(t) = x]$$

Let

$$V^*(t, T, x) = inf_{c(s,x(s)), t \le s \le T} V(t, T, x, c)$$

Then an easy computation shows that the optimal $c$ that minimizes $V$ is given by the solution to

$$V^*(t, T, x) = min_{c(t,x)}(L(t, c(t, x))dt + \mathbb{E}[V^*(t+dt, T, x(t+dt))|x(t) = x])$$

from which we immediately get

$$\frac{\partial V^*(t,T,x)}{\partial t} + min_u(L(t,u) + u^T \frac{\partial V^*(t,T,x)}{\partial x}) + \frac{\sigma^2}{2} \Delta V^*(t,T,x) = 0, 0 \le t \le T$$

with the final condition

$$V^*(T,T,x)=0$$

The optimal value of $c(t,x)$ is given by the solution to $u$ in this equation. It is clear from the nature of the problem that if $L$ does not depend upon time explicitly, ie $L(t,c)=L(c)$, then $V^*(t,T,x)=V^*(0,T-t,x)$ and hence denoting this quantity by $V^*(T-t,x)$, we get the Stochastic Bellman-Hamilton-Jacobi equation

$$\partial_t V^*(t,x)=\frac{\sigma^2}{2}\Delta V^*(t,x)+min_u(L(u)+u^T\frac{\partial V^*(t,x)}{\partial x}),0\leq t\leq T$$

with initial condition

$$V^*(0,x)=0$$

Now,

$$\int_0^t L(c(s/\delta,x(s/\delta))ds=\delta\int_0^{t/\delta} L(c(s,x(s))ds$$

which by the ergodic theorem converges as $\delta\to 0$ to

$$t\int L(c(x))dP_c(x)$$

provided that $c(t,x)=c(x)$ is independent of tiem and ergodicity of the process $x(t)$ assumed with $P_c$ being a stationary probability measure for $x(t)$. More generally, we can consider random cost functions $L(c,\omega)$ and ergodicity would then yield

$$\int_0^t L(c(s/\delta,x(s/\delta),\omega),\tau_{x(s/\delta)}(\omega))ds=$$

$$\delta.\int_0^{t/\delta} L(c(s,x(s),\omega),\tau_{x(s)}\omega)ds$$

which assuming that $c(t,x,\omega)=b(\tau_x\omega)$ (where $\tau_x, x\in\mathbb{R}^d$ are spatial shifts) and ergodicity of shifts converges as $\delta\to 0$ to

$$t\int L(b(\omega),\omega)dP(\omega)$$

where $P$ is a stationary measure for the process.

Remark: More precisely, the ergodicity of the process $x(t)$ implies that if $P_\omega(dx)$ is an invariant measure for it with $c(t,x,\omega)=b(\tau_x\omega)$, then

$$lim_{\delta\to 0}\int_0^t L(c(s/\delta,x(s/\delta),\omega),\tau_{x(s/\delta)}(\omega))ds$$

$$=t.\int L(b(\tau_x\omega),\tau_x\omega)dP_\omega(x)$$

[18] Random walk in a random environment: Let the state space of the random walk be $\mathbb{Z}$ to start with. Let $X(n) \in \mathbb{Z}, n = 0, 1, 2, ...$ be the random walk. Let $p(x), x \in \mathbb{Z}$ be a stationary process taking values in $[0, 1]$. We may write this as $p(x, \omega) = p(\tau_x \omega), x \in \mathbb{Z}$ where $\tau_x$ is the spatial shift by $x$. By stationary, we mean that there is a probability distribution $P$ on $\Omega$ such that $Po\tau_x^{-1} = P, x \in \mathbb{Z}$. Now a random walk in this random spatially stationary environment can be constructed as follows: For a given $\omega \in \Omega, X(n+1, \omega) = X(n, \omega)+1$ with probability $p(X(n, \omega))$ and $= X(n, \omega)-1$ with a probability $1 - p(X(n, \omega))$. Note that the quantity $p(X(n, \omega))$ has two degrees of randomness, one the randomness inherent in the function $p(x), x \in \mathbb{Z}^d$ which is a stationary process on $\mathbb{Z}$, and two, the randomness in the process $X(n, \omega)$. It should be noted that the randomness in the process has a part coming from that in the stationary function $p(.)$. More generally, we can consider a generalized random walk to be such that given the environment and given its position $X_n$ at time $n$, the process jumps to $X_n + m$ at time $n + 1$ with a probability $p(m, X_n)$, ie, $p(m, X_n, \omega)$ where $m \in \mathbb{Z}$. We assume the environment to be stationary, ie, $(p(m, x))_{m \in \mathbb{Z}}, x \in \mathbb{Z}$ is a finite or infinite vector valued stationary process on $\mathbb{Z}$. The probability distribution of this process is given by

$$\mathbb{E}(f_n(X_n)...f_1(X_1)|X_0 = x) =$$

$$\sum_{m_1,...,m_n} f_n(x + m_1 + ... + m_n) f_{n-1}(x + m_1 + .. + m_{n-1})...f_1(x + m_1)$$

$$\times \mathbb{E}[p(m_n, x + m_1 + ... + m_{n-1}).p(m_{n-1}, x + m_1 + ... + m_{n-2})...p(m_1, x)]$$

In particular, if we assume that $p(m, x) = p(m)$ is independent of $x$ but is random, then the above expression simplifies to

$$\mathbb{E}(f_n(X_n)...f_1(X_1)|X_0 = x) =$$

$$\sum_{m_1,...,m_n} f_n(x + m_1 + ... + m_n) f_{n-1}(x + m_1 + ... + m_{n-1})...f_1(x + m_1)$$

$$\times \mathbb{E}[p(m_n)p(m_{n-1})...p(m_1)]$$

Remark: It is clear that $X_n$ is not a Markov process, for if it were, then the above expectation would equal

$$\mathbb{E}(...\mathbb{E}(\mathbb{E}(\mathbb{E}(f_n(X_n)|X_{n-1})f_{n-1}(X_{n-1})|X_{n-2})f(X_{n-2})|X_{n-3})...f(X_1)|X_0 = x)$$

which is not the case.

# Chapter 8

# Large deviations for filtering in a mixture of Boson-Fermion noise

[19] Filtering in Fermionic noise:

$$dJ = (-1)^{\Lambda} dA, dJ^* = (-1)^{\Lambda} dA^*$$

$$dJdJ^* = dt, dJdA^* = (-1)^{\Lambda} dt, dAdJ^* = (-1)^{\Lambda} dt,$$

$$A(s)dJ(t) = -dJ(t)A(s), s \le t$$

HP qsde

$$dU = (-(iH + P)dt + L_1 dA + L_2 dA^* + L_3 dJ + L_4 dJ^*)U$$

$$dU^*U + U^*dU + dU^*dU = d(U^*U) = 0$$

gives

$$0 = U^*((iH - P)dt + L_1^* dA^* + L_2^* dA + L_3^* dJ^* + L_4^* dJ$$

$$-(iH + P)dt + L_1 dA + L_2 dA^* + L_3 dJ + L_4 dJ^*)U$$

$$+U^*(L_1^d A^* + L_2^* dA + L_3^* dJ^* + L_4^* dJ)(L_1 dA + L_2 dA^*$$

$$+ L_3 dJ + L_4 dJ^*)U$$

$$= U^*(-2Pdt + (L_1^* + L_2)dA^* + (L_2^* + L_1)dA + (L_3^* + L_4)dJ^*$$

$$+(L_3 + L_4^*)dJ + (L_2^* L_2 + L_4^* L_4)dt +$$

$$L_2^* L_4 (-1)^{\Lambda} dt + L_4^* L_2 (-1)^{\Lambda} dt)U$$

This gives

$$P = (1/2)(L_2^* L_2 + L_4^* L_4 + (-1)^{\Lambda}(L_2^* L_4 + L_4^* L_2)),$$

$$L_1^* + L_2 = 0, L_3^* + L_4 = 0$$

as the condition for unitary evolution.

$$j_t(X) = U(t)^* X U(t)$$

$$dj_t(X) = dU(t)^*XU(t) + U(t)^*XdU(t) + dU(t)^*XdU(t)$$
$$= U^*([(iH-P)dt + L_1^*dA^* + L_2^*dA + L_3^*dJ^* + L_4^*dJ]X + X[-(iH+P)dt$$
$$+L_1dA + L_2dA^* + L_3dJ + L_4dJ^*])U$$
$$+U^*(L_1^*dA^* + L_2^*dA + L_3^*dJ^* + L_4^*dJ)X(L_1dA + L_2dA^*$$
$$+ L_3dJ + L_4dJ^*)U\ L_2 + L_4^*XL_4)dt +$$
$$= U^*(i[H,X]dt - (PX+XP)dt + (L_1^*X + XL_2)dA^* + (L_2^*X + XL_1)dA$$
$$+(L_3^*X + XL_4)dJ^* + (XL_3 + L_4^*X)dJ + (L_2^*X$$
$$+L_2^*XL_4(-1)^\Lambda dt + L_4^*XL_2(-1)^\Lambda dt)U$$

These equations can be cast in the form

$$dj_t(X) = j_t(\theta_0(X) + \theta_1(X)(-1)^\Lambda)dt + j_t(\theta_2(X))dA + j_3(\theta_3(X))dA^*$$

Assume that the input measurement process is

$$dY_i(t) = c_1dA(t) + \bar{c}_1dA(t)^* + c_2\theta(t)dJ(t) + \bar{c}_2\theta(t)^*dJ(t)^*$$

where the $\theta(t)'s$ are anticommuting Fermionic parameters. These parameters are also assumed to anticommute wtih the processes $A(s), A(s)^*, s \le t$. We shall check whether $Y_i(t), t \ge 0$ forms an Abelian family of operators. For this we require first that $[\theta(t)dJ(t), \theta(s)dJ(s)] = 0$ for all $s,t$. Now $dJ(t)$ anticommutes with $dJ(s)$ and $\theta(t)$ anticommutes with $\theta(s)$. If we assume that $\theta(t)$ anticommutes with $dJ(s)$, then this commutator will be zero. Of course it will also be zero if $\theta(t)$ commutes with $dJ(s)$ but we choose the latter. For $Y_i$ to form an Abelian family, we also require that $[dA(t), \theta(s)dJ(s)] = 0$ for all $s,t$. Now, for $t \le s$, $dA(t)$ anticommutes with $dJ(s)$ so we require that $\theta(s)$ anticommute with $dA(t)$ for $t < s$. For $t > s$, $dA(t)$ commutes with $dJ(s)$, so we require that $\theta(s)$ commutes with $dA(t)$ for $t > s$. In summary, for $Y_i(.)$ to form an Abelian family, we require that $\theta(s)$ and $\theta(s)^*$ commute with $dA(t), dA(t)^*$ for $t \ge s$, anticommute with the same for $t < s$ and anticommute with $dJ(t), dJ(t)^*$ for all $t$.

Then we have

$$\theta(t)dJ(t)A(s) = -\theta(t)A(s)dJ(t) = A(s)\theta(t)dJ(t), s \le t$$

The output measurement process is

$$Y_o(t) = U(t)^*Y_i(t)U(t)$$

and for it to be non demolition, we require

$$d_T(U(T)^*Y_i(t)U(T)) = 0, T \ge t$$

ie,

$$dU(T)^*Y_i(t)U(T) + U(T)^*Y_i(t)dU(T) + dU(T)^*Y_i(t)dU(T) = 0, T > t$$

or equivalently,

$$U(T)^*(-(iH+P(T))dT + L_1^*dA^*(T) + L_2^*dA(T) + L_3*dJ^*(T) + L_4^*dJ(T))Y_i(t)U(T)$$

$$+U(T)^*Y_i(t)(-(iH+P(T))dT+L_1dA(T)+L_2dA(T)^*+L_3dJ(T)+L_4dJ(T)^*)U(T)$$

$$+U(T)^*(L_2^*dA(T)+L_4^*dJ(T))Y_i(t)(L_2dA(T)^*+L_4dJ(T)^*U(T)=0, T\geq t$$

For this, we first of all require that $P(T)$ should commute with $Y_i(t)$ and this amounts to $(-1)^{\Lambda(T)}$ to commute with $Y_i(t)$ for $T \geq t$. But actually the component $A(t)$ in $Y_i(t)$ anticommutes with $(-1)^{\Lambda(T)}$. So to rectify this, we incorporate the anticommuting Fermionic parameters $\theta(t)$ in $dJ(t)$ and likewise $\theta(t)^*$ in $dJ(t)^*$ in the HP qsde (This is the second reason for incorporating the Fermionic parameters, earlier to achieve Abelian input measurements we had incorporated Fermionic parameters in the measurement model, now we also incorporate it in the HP qsde to ensure the non-demolition property of the output measurement process). Then the HP equation then becomes

$$dU = (-(iH+P)dt + L_1dA + L_2dA^* + L_3\theta.dJ + L_4\theta^*.dJ^*)U$$

where $\theta = \theta(t)$ etc. Then, the last condition for unitary of $U(t)$ becomes

$$P = (1/2)(L_2^*L_2 + |\theta(t)|^2L_4^*L_4) + (-1)^{\Lambda(t)}(\theta(t)^*L_2^*L_4 + \theta(t)L_4^*L_2))$$

since now

$$\theta(t).dJ(t).\theta(t)^*.dJ(t)^* = -\theta(t)\theta(t)^*.dJ(t).dJ(t)^*$$

$$= -\theta(t)\theta(t)^*dt = \theta(t)^*\theta(t)dt = |\theta(t)|^2dt$$

Note that we have used

$$\theta(t)dJ(t).dA(t)^* = \theta(t)(-1)^{\Lambda(t)}dt,$$

$$dA(t)\theta(t)dJ(t)^* = \theta(t)dA(t)dJ(t)^* = \theta(t)(-1)^{\Lambda(t)}dt$$

since $\theta(t)$ commutes with $dA(t)$.

The evolution equation for $j_t(X)$ will accordingly get modified to

$$dj_t(X) =$$

$$= U^*([(iH-P)dt+L_1^*dA^*+L_2^*dA+\theta^*L_3^*dJ^*+\theta L_4^*dJ]X+X[-(iH+P)dt$$

$$+L_1dA+L_2dA^*+\theta L_3dJ+\theta^*L_4dJ^*])U$$

$$+U^*(L_1^*dA^*+L_2^*dA+\theta^*L_3^*dJ^*+\theta L_4^*dJ)X(L_1dA+L_2dA^*+\theta L_3dJ+\theta^*L_4dJ^*)U$$

$$= U^*(i[H,X]dt-(PX+XP)dt+(L_1^*X+XL_2)dA^*+(L_2^*X+XL_1)dA$$

$$+\theta^*(L_3^*X+XL_4)dJ^*+\theta(XL_3+L_4^*X)dJ+$$

$$(L_2^*XL_2 + |\theta|^2L_4^*XL_4)dt+$$

$$+\theta^*L_2^*XL_4(-1)^{\Lambda}dt + \theta L_4^*XL_2(-1)^{\Lambda}dt)U$$

The non-demolition condition after incorporation of the Fermionic parameters becomes

$$U(T)^*(-(iH+P(T))dT$$

$$+L_1^*dA^*(T)+L_2^*dA(T)+\theta(T)^*L_3*dJ^*(T)+\theta(T)L_4^*dJ(T))Y_i(t)U(T)$$

$$+U(T)^*Y_i(t)(-(iH+P(T))dT+L_1dA(T)+L_2dA(T)^*$$

$$+\theta(T)L_3dJ(T)+\theta(T)^*L_4dJ(T)^*)U(T)$$

$$+U(T)^*(L_2^*dA(T)+\theta(T)L_4^*dJ(T))Y_i(t)(L_2dA(T)^*+\theta(T)^*L_4dJ(T)^*U(T)=0, T\geq t$$

and the second condition for non-demolition that we require is that $Y_i(t)$ should commute with $\theta(T)^*dJ(T)^*$ and with $\theta(T)dJ(T)$ which is true since $\theta(T)(-1)^{\Lambda(T)}$ commutes with $dA(t), dA(t)^*, \theta(t)(-1)^{\Lambda(t)}dA(t)$ and with $\theta(t)^*(-1)^{\Lambda(t)}dA(t)^*$ for $T \geq t$.

Note: $\theta(T)dJ(T)$ commutes with $dA(t)$ for $T > t$ since $\theta(T)$ anticommutes with $dA(t)$ and $dJ(T)$ also anticommutes with $dA(t)$. $\theta(T)dJ(T)$ commutes with $\theta(t)dJ(t)$ since $\theta(T)$ and $dJ(T)$ both anticommute with each of $\theta(t)$ and $dJ(t)$. It should be noted that $\theta(t)$ anticommutes with $dJ(T) = (-1)^{\Lambda(T)}dA(T)$ and commutes with $dA(T)$ for $T \geq t$ implies that $\theta(t)$ anticommutes with $(-1)^{\Lambda(T)}$ for $T \geq t$. Also, $\theta(T)$ anticommutes with $dJ(t) = (-1)^{\Lambda(t)}dA(t)$ and with $dA(t)$ for $T > t$ implies that $\theta(T)$ commutes with $(-1)^{\Lambda(t)}$ for $T > t$. Further, $dA(T)$ commutes with $dJ(t) = (-1)^{\Lambda(t)}dA(t)$ for $T > t$ and also commutes with $\theta(t)dJ(t)$ for $T > t$ implies that $dA(T)$ commutes with $\theta(t)$ for $T > t$. Thus the entire picture is self-consistent.

The Boson-Fermion quantum filter: Let

$$\eta_o(t) = \sigma(Y_o(s) : s \leq t), Y_o(t) = U(t)^*Y_i(t)U(t)$$
$$= U(T)^*Y_i(t)U(T), T \geq t$$

Assume that the filter equations for

$$\pi_t(X) = \mathbb{E}(j_t * (X)|\eta_o(t))$$

are given by

$$d\pi_t(X) = F_t(X)dt + G_t(X)dY_o(t)$$

wher $F_t(X), G_t(X)$ are $\eta_o(t)$-measurable. Note that

$$dY_o(t) = d(U(t)^*Y_i(t)U(t))$$
$$= U(t)^*dY_i(t)U(t) + dU(t)^*Y_i(t)U(t) + U(t)^*Y_i(t)dU(t) + dU(t)^*Y_i(t)dU(t)$$

Note that we are assuming that each of $\theta(t), \theta(t)^*$ anticommutes with each of $dA(s), dA(s)^*$ for $s < t$ and commutes with $dA(s), dA(s)^*$ for $s \geq t$. Now

$$U(t)^*dY_i(t)U(t) = U(t)^*(c_1dA(t) + \bar{c}_1dA(t)^* + c_2\theta(t)dJ(t) + \bar{c}_2\theta(t)^*dJ(t)^*)U(t)$$

Now, $dA(t), dA(t)^*$ both commute with $dA(s), dA(s)^*$ and also with $\theta(s)dJ(s), \theta(s)^*dJ(s)^*$ for $s < t$ and therefore, $U(t)^*$ commutes with $dA(t), dA(t)^*$. Likewise, $\theta(t)dJ(t), \theta(t)^*dJ(t)^*$ both commute with $dA(s), dA(s)^*$ and also with $\theta(s)dJ(s), \theta(s)^*dJ(s)^*$ for $s < t$. Thus, $dY_i(t)$ commutes with each of $\{dA(s), dA(s)^*, \theta(s)dJ(s), \theta(s)^*dJ(s)^*\}$ for $t > s$ and hence $U(t)$ commutes with $dY_i(t)$. This implies

$$U(t)^*dY_i(t)U(t) = dY_i(t).U(t)^*U(t) = dY_i(t)$$

by the unitarity of $U(t)$.

Note: It is imperative that $\theta(s), \theta(s)^*$ commute with $dA(t), dA(t)^*$ for $t \geq s$ because of the following reason: For $T > t$,

$$d_TU(T)^*Y_i(t)U(T) = dU(T)^*Y_i(t)U(T) + U(T)^*Y_i(t)dU(T) + dU(T)^*Y_i(t)dU(T)$$

For non-demolition, we require this to be zero. Since $U(T)$ is unitary, this condition boils down to requiring that $Y_i(t)$ and hence $dA(s), dA(s)^*, \theta(s)dJ(s), \theta(s)^*dJ(s)^*$ all commute with $dA(T), dA(T)^*, \theta(T)dJ(T), \theta(T)^*dJ(T)^*$ for all $s < T$. Then for example since $dA(s)$ is required to commute with $\theta(T)dJ(T)$ while $dA(s)$ anticommutes with $dJ(T)$, it follows that $dA(s)$ must anticommute with $\theta(T)$ for $T > s$. Likewise, since $\theta(s)dJ(s)$ is required to commute with $dA(T)$ for $T > s$ and since $dJ(s)$ commutes with $dA(T)$, we require that $\theta(s)$ commute with $dA(T)$ for $T > s$.

We also note that since $\theta(s)dJ(s)$ commutes with $dA(t)$ for $t > s$ and since $dJ(s)$ commutes with $dA(t)$ for $t > s$, it follows that $\theta(s)$ must commute with $dA(t)$ for $t > s$. Since $\theta(s)$ is assumed to anticommute with $dJ(t) = (-1)^{\Lambda(t)}dA(t)$ for all $t$, and $\theta(s)$ commutes with $dA(t)$ for $t > s$, it follows that $\theta(s)$ must necessarily anticommute with $(-1)^{\Lambda(t)}$ for $t > s$. The same holds for $\theta(s)^*$. Further, $\theta(t)$ is assumed to anticommute with $dJ(s) = (-1)^{\Lambda(s)}dA(s)$ for all $s$ and in particular for $t > s$ and $\theta(t)$ anticommutes with $dA(s)$ for $t > s$, it follows that $\theta(t)$ must necessarily commute with $(-1)^{\Lambda(s)}$ for $t > s$. Note that $\theta(t)$ must anticommute with $dA(s)$ for $t > s$ because we require that $\theta(t)dJ(t)$ commute with $dA(s)$ for $t > s$ while $dJ(t)$ anticommutes with $dA(s)$ for $t > s$.

Remark: Suppose that $\theta(t)$ is assumed to commute with $dJ(s)$ for all $t, s$. Then, $\theta(t)dJ(t)$ would anticommute with $dJ(s)$ for $t > s$ since $dJ(t)$ anticommutes with $dJ(s)$ for all $t, s$. But then,

$$\theta(t)dJ(t).\theta(s).dJ(s) = \theta(t)\theta(s)dJ(t)dJ(s) =$$

$$\theta(s)\theta(t)dJ(s)dJ(t) = \theta(s)dJ(s)\theta(t)dJ(t)$$

as required by the non-demolition condition. However, the non-demolition condition also requires that $\theta(t)dJ(t)$ also commute with $dA(s)$ for all $t, s$. This would imply that $\theta(t)$ anticommute with $dA(s)$ for $t > s$ (since $dJ(t)$ anticommutes with $dA(s)$ for $t > s$) and commute with $dA(s)$ for $s > t$ (since $dJ(t)$ commutes with $dA(s)$ for $s > t$). Likewise, $\theta(t)$ would anticommute with $dA(s)^*$ for $t > s$ and commute with $dA(s)^*$ for $s > t$. Then, $\theta(t)$ would commute with $\Lambda(s) = \int_0^s dA(u)^*dA(u)/du$ and hence with $(-1)^{\Lambda(s)}$ for $t > s$. We could thus develop a Boson-Fermion filtering theory based on this assumption also. This could be seen in another way also as follows. Since $\theta(t)$ is assumed to commute with $dJ(s) = (-1)^{\Lambda(s)}dA(s)$ for all $t, s$ and since we've seen that $\theta(t)$ commutes with $dA(s)$ for $s > t$, it follows that $\theta(t)$ must commute with $(-1)^{\Lambda(s)}$ for $s > t$ (since $\theta(t)$ is assumed to commute with $dJ(s)$). On the other hand, since $\theta(t)$ anticommutes with $dA(s)$ for $t > s$, it must follow that $\theta(t)$ anticommutes with $(-1)^{\Lambda(s)}$ for $t > s$. There is thus a contradiction involved here.

This contradiction is resolved if we assume that $\theta(t)$ anticommutes (rather than commutes) with $dJ(s)$ for all $s$. For we then get that

$$\theta(t)dJ(t).\theta(s)dJ(s) = -\theta(t)\theta(s)dJ(t)dJ(s) = -\theta(s)\theta(t)dJ(s)dJ(t)$$

$$= \theta(s)dJ(s)\theta(t)dJ(t)$$

for all $t, s$ as required by the non-demolition property. Further, the non-demolition property requires that $\theta(t)dJ(t)$ commute with $dA(s)$ for all $t, s$ and therefore that $\theta(t)$ anticommute with $dA(s)$ for $t > s$ and commute with $dA(s)$ for $s > t$. Likewise, $\theta(t)$ should then anticommute with $dA(s)^*$ for $t > s$ and commute with $dA(s)^*$ for $s > t$. This then implies that $\theta(t)$ commutes with $\Lambda(s) = \int_0^s dA(u)^* dA(u)/du$ for $t > s$. On the other hand, since $\theta(t)$ is assumed to anticommute with $dJ(s) = (-1)^{\Lambda(s)} dA(s)$ for all $t, s$ while $\theta(t)$ anticommutes with $dA(s)$ for $t > s$ and commutes with $dA(s)$ for $s > t$, it would follow that $\theta(t)$ anticommutes with $(-1)^{\Lambda(s)}$ for $s > t$ and commutes with the same for $t > s$. The earlier contradiction is therefore resolved here.

The quantum Boson-Fermion filter:

$$d\pi_t(X) = F_t(X)dt + G_t(X)dY_o(t)$$

$$dC(t) = f(t)C(t)dY_o(t), C(0) = 1$$

$$\mathbb{E}[(j_t(X) - \pi_t(X))C(t)] = 0$$

$$dj_t(X) = j_t(\theta_0(X) + \theta_1(X)(-1)^{\Lambda(t)})dt$$

$$+j_t(\theta_2(X))dA(t) + j_t(\theta_3(X))dA(t)^*$$

$$+j_t(\theta_4(X))\theta(t)dJ(t) + j_t(\theta_5(X))\theta(t)^* dJ(t)^*$$

where $\theta_0(X)$ contains terms involving $|\theta(t)|^2$ while $\theta_1(X)$ contains terms linear in $\theta(t), \theta(t)^*$. Now

$$\mathbb{E}[(dj_t(X) - d\pi_t(X))C(t)] + \mathbb{E}[(j_t(X) - \pi_t(X))dC(t)] + \mathbb{E}[(dj_t(X) - d\pi_t(X))dC(t)] = 0$$

From this equation and the arbitrariness of the complex valued function $f(t)$, we infer that

$$\mathbb{E}[(dj_t(X) - d\pi_t(X))C(t)] = 0,$$

$$\mathbb{E}[(j_t(X) - \pi_t(X))C(t)dY_o(t)] + \mathbb{E}[(dj_t(X) - d\pi_t(X))C(t)dY_o(t)] = 0$$

Therefore

$$\mathbb{E}[(dj_t(X) - d\pi_t(X))|\eta_o(t)] = 0,$$

$$\mathbb{E}[(j_t(X) - \pi_t(X))dY_o(t)|\eta_o(t)] + \mathbb{E}[(dj_t(X) - d\pi_t(X))dY_o(t)|\eta_o(t)] = 0$$

We define in addition

$$\nu_t(X) = \mathbb{E}(j_t(X)(-1)^{\Lambda(t)}|\eta_o(t))$$

Then,

$$d(j_t(X).(-1)^{\Lambda(t)}) = dj_t(X).(-1)^{\Lambda(t)} - 2j_t(X)(-1)^{\Lambda(t)}d\Lambda(t)$$

$$-2dj_t(X)d\Lambda(t).(-1)^{\Lambda(t)}$$

$$= [j_t(\theta_0(X) + \theta_1(X)(-1)^{\Lambda(t)})dt + j_t(\theta_2(X))dA(t) + j_t(\theta_3(X))dA(t)^*$$

$$+j_t(\theta_4(X))\theta(t)dJ(t)+j_t(\theta_5(X))\theta(t)^*dJ(t)^*](-1)^{\Lambda(t)}$$
$$-2j_t(X)(-1)^{\Lambda(t)}d\Lambda(t)$$
$$-2[j_t(\theta_0(X)+\theta_1(X)(-1)^{\Lambda(t)})dt$$
$$+j_t(\theta_2(X))dA(t)+j_t(\theta_3(X))dA(t)^*$$
$$+j_t(\theta_4(X))\theta(t)dJ(t)+j_t(\theta_5(X))\theta(t)^*dJ(t)^*]d\Lambda(t)(-1)^{\Lambda(t)}$$

We then observe that by quantum Ito's formula,

$$dJ(t)d\Lambda(t)=(-1)^{\Lambda(t)}dA(t)=dJ(t), dA(t)d\Lambda(t)=dA(t),$$
$$dJ(t)^*d\Lambda(t)=0, dA(t)^*d\Lambda(t)=0,$$
$$dJ(t)(-1)^{\Lambda(t)}=dA(t), dJ(t)^*(-1)^{\Lambda(t)}=dA(t)^*$$
$$dJ(t)=(-1)^{\Lambda(t)}dA(t), dJ(t)^*=(-1)^{\Lambda(t)}dA(t)^*$$

and thus the above simplifies to

$$d(j_t(X).(-1)^{\Lambda(t)})=$$
$$[j_t(\theta_1(X))dJ(t)+j_t(\theta_2(X))dJ(t)^*$$
$$+j_t(\theta_2(X))\theta(t)dA(t)+j_t(\theta_3(X))\theta(t)^*dA(t)^*]$$
$$-2j_t(X)(-1)^{\Lambda(t)}d\Lambda(t)$$
$$-2[j_t(\theta_2(X))dJ(t)+j_t(\theta_4(X))\theta(t)dA(t)]$$

Observe that

$$dj_t(X)=j_t(\theta_0(X))dt+j_t(\theta_1(X))dA(t)+j_t(\theta_2(X))dA(t)^*$$

where

$$\theta_0(X)=\theta_{00}(X)+\theta_{01}(X)(-1)^{\Lambda(t)},$$
$$\theta_1(X)=\theta_{10}(X)+\theta_{11}(X)(-1)^{\Lambda(t)}$$
$$\theta_2(X)=\theta_{20}(X)+\theta_{21}(X)(-1)^{\Lambda(t)}$$

where $\theta_{ab}(X)$ are system space operators containing the Fermionic parameters $\theta(t), \theta(t)^*$. We can also write

$$dY_i(t)=(c_1+c_2\theta(t)(-1)^{\Lambda(t)})dA(t)+(\bar{c}_1+\bar{c}_2\theta(t)^*(-1)^{\Lambda(t)})dA(t)^*$$
$$=\lambda(t)dA(t)+\lambda(t)^*dA(t)^*$$

with

$$\lambda(t)=c_1+c_2\theta(t)(-1)^{\Lambda(t)}$$

Then

$$dY_o(t)=j_t(M)dt+\lambda_1.dA+\lambda_2dA^*,$$
$$M=M_0+(\theta M_{10}+\theta^*M_{11})(-1)^{\Lambda}=M_0+M_1(-1)^{\Lambda}$$

$$\lambda_1 = \lambda_{10} + \tilde{\lambda}_{11}\theta(-1)^\Lambda, = \lambda_{10} + \lambda_{11}(-1)^{\Lambda(t)}$$
$$\lambda_2 = \lambda_{20} + \tilde{\lambda}_{21}\theta^*(-1)^\Lambda = \lambda_{20} + \lambda_{21}(-1)^{\Lambda(t)}$$

Note that the Boson-Fermion HP qsde can be expressed as

$$dU(t) = (-(iH + P)dt + L_1 dA + L_2 dA^*)U(t)$$

where

$$L_1 = L_{10} + L_{11}\theta(-1)^\Lambda,$$
$$L_2 = L_{20} + L_{21}\theta^*(-1)^\Lambda$$
$$P = P_0 + P_1(-1)^\Lambda$$

where

$$P_0 = P_{00} + \theta P_{01}\theta + P_{02}\theta^*$$
$$P_1 = P_{10} + P_{11}|\theta|^2$$

with $L_{ab}, P_{ab}$ all being system space operators not involving the Fermionic parameters $\theta, \theta^*$. We define for a system space operator $X$, the following conditional expectations:

$$\pi_t(X) = \mathbb{E}[j_t(X)|\eta_o(t)], \nu_t(X) = \mathbb{E}[j_t(X)(-1)^{\Lambda(t)}|\eta_o(t)],$$
$$\rho_t(X) = \mathbb{E}[j_t(X.(-1)^{\Lambda(t)})|\eta_o(t)],$$
$$\sigma_t(X) = \mathbb{E}[j_t(X.(-1)^{\Lambda(t)})(-1)^{\Lambda(t)}|\eta_o(t)]$$

and we derive differential equations for these.

$$\mathbb{E}[dj_t(X)|\eta_o(t)] = \mathbb{E}[j_t(\theta_0(X))dt + j_t(\theta_1(X))dA(t) + j_t(\theta_2(X))dA(t)^*|\eta_o(t)]$$
$$= \mathbb{E}[j_t(\theta_{00}(X) + \theta_{01}(X)(-1)^{\Lambda(t)})|\eta_o(t)]dt$$
$$+u(t)dt\mathbb{E}[j_t(\theta_{10}(X) + \theta_{11}(X)(-1)^{\Lambda(t)})|\eta_o(t)]$$
$$+\bar{u}(t)dt\mathbb{E}[j_t(\theta_{20}(X) + \theta_{21}(X)(-1)^{\Lambda(t)})|\eta_o(t)]$$
$$= dt[\pi_t(\theta_{00}(X)) + \rho_t(\theta_{01}(X))]+$$
$$u(t)dt[\pi_t(\theta_{10}(X)) + \rho_t(\theta_{11}(X))]+$$
$$\bar{u}(t)dt[\pi_t(\theta_{20}(X)) + \rho_t(\theta_{21}(X))]$$
$$\mathbb{E}[j_t(X)dY_o(t)|\eta_o(t)] =$$
$$\mathbb{E}[j_t(X)(j_t(M)dt + \lambda_1.dA + \lambda_2 dA^*)|\eta_o(t)]$$
$$= dt\mathbb{E}[j_t(XM)|\eta_o(t)] + u(t)dt\mathbb{E}[j_t(X)\lambda_1|\eta_o(t)]$$
$$+\bar{u}(t)dt\mathbb{E}[j_t(X)\lambda_2|\eta_o(t)]$$

Now,

$$\mathbb{E}[j_t(XM)|\eta_o(t)] = \mathbb{E}[j_t(X(M_0 + M_1(-1)^{\Lambda(t)}))|\eta_o(t)]$$
$$= \pi_t(XM_0) + \rho_t(XM_1)$$

$$\mathbb{E}[j_t(X)\lambda_1|\eta_o(t)] = \pi_t(X)\lambda_{10} + \nu_t(X)\lambda_{11},$$

$$\mathbb{E}[j_t(X)\lambda_2|\eta_o(t)] = \pi_t(X)\lambda_{20} + \nu_t(X)\lambda_{21},$$

Thus,

$$\mathbb{E}[j_t(X)dY_o(t)|\eta_o(t)] =$$

$$[\pi_t(XM_0) + \rho_t(XM_1) + \pi_t(X)\lambda_{10} + \nu_t(X)\lambda_{11} + \pi_t(X)\lambda_{20} + \nu_t(X)\lambda_{21}]dt$$

We write our filtering equations as

$$d\pi_t(X) = F_{1t}(X)dt + G_{1t}(X)dY_o(t),$$
$$d\nu_t(X) = F_{2t}(X)dt + G_{2t}(X)dY_o(t),$$
$$d\rho_t(X) = F_{3t}(X)dt + G_{3t}(X)dY_o(t),$$
$$d\sigma_t(X) = F_{4t}(X)dt + G_{4t}(X)dY_o(t)$$

Note that all the quantities $F_{kt}(X), G_{kt}(X), k = 1, 2, 3, 4$ are in $\eta_o(t)$ and are therefore commutative. The orthogonality principle in estimation theory states that

$$\mathbb{E}[(j_t(X) - \pi_t(X))C(t)] = 0,$$

$$\mathbb{E}[(j_t(X)(-1)^{\Lambda(t)} - \nu_t(X))C(t)] = 0,$$

$$\mathbb{E}[(j_t(X.(-1)^{\Lambda(t)}) - \rho_t(X))C(t)] = 0,$$

$$\mathbb{E}[(j_t(X.(-1)^{\Lambda(t)}).(-1)^{\Lambda(t)} - \sigma_t(X))C(t)] = 0$$

where $C(t)$ is any $\eta_o(t)$ measurable observable. In particular, this is true for

$$dC(t) = f(t)C(t)dY_o(t), t \geq 0, C(0) = 1$$

By this orthogonality principle in estimation theory, we observe that on taking differentials in time, using $d(\xi.\eta) = d\xi.\eta + \xi.d\eta + d\xi.d\eta$ and choosing $f$ appropriately that

$$\mathbb{E}[dj_t(X) - d\pi_t(X)|\eta_o(t)] = 0,$$

$$\mathbb{E}[(j_t(X) - \pi_t(X))dY_o(t)|\eta_o(t)] + \mathbb{E}[dj_t(X) - d\pi_t(X))dY_o(t)|\eta_o(t)] = 0$$

$$\mathbb{E}[dj_t(X.(-1)^{\Lambda(t)}) - d\rho_t(X)|\eta_o(t)] = 0,$$

$$\mathbb{E}[(j_t(X(-1)^{\Lambda(t)}) - \rho_t(X))dY_o(t)|\eta_o(t)] + \mathbb{E}[dj_t(X.(-1)^{\Lambda(t)}) - d\rho_t(X))dY_o(t)|\eta_o(t)] = 0$$

$$\mathbb{E}[d(j_t(X)(-1)^{\Lambda(t)}) - d\nu_t(X)|\eta_o(t)] = 0,$$

$$\mathbb{E}[(j_t(X)(-1)^{\Lambda(t)}) - \nu_t(X))dY_o(t)|\eta_o(t)] + \mathbb{E}[d(j_t(X)(-1)^{\Lambda(t)}) - d\nu_t(X))dY_o(t)|\eta_o(t)] = 0$$

$$\mathbb{E}[d(j_t(X.(-1)^{\Lambda(t)})(-1)^{\Lambda(t)}) - d\nu_t(X)|\eta_o(t)] = 0,$$

$$\mathbb{E}[(j_t(X.(-1)^{\Lambda(t)})(-1)^{\Lambda(t)} - \sigma_t(X))dY_o(t)|\eta_o(t)] + \mathbb{E}[d(j_t(X(-1)^{\Lambda(t)})(-1)^{\Lambda(t)}) - d\sigma_t(X))dY_o(t)|\eta_o(t)] = 0$$

Now

$$\mathbb{E}[dj_t(X)|\eta_o(t)] =$$

$$dt[\pi_t(\theta_{00}(X)) + \rho_t(\theta_{01}(X)) +$$

$$u(t)(\pi_t(\theta_{10}(X)) + \rho_t(\theta_{11}(X)))+$$
$$\bar{u}(t)(\pi_t(\theta_{20}(X)) + \rho_t(\theta_{21}(X)))]$$
$$\mathbb{E}[dj_t(X).dY_o(t)|\eta_o(t)] =$$
$$\mathbb{E}[j_t(\theta_1(X))dA(t).\lambda_2.dA^*(t)|\eta_o(t)]$$
$$= dt.\mathbb{E}[j_t(\theta_{10}(X) + \theta_{11}(X)(-1)^{\Lambda(t)}).(\lambda_{20} + \lambda_{21}(-1)^{\Lambda(t)})|\eta_o(t)]$$
$$= dt.[\pi_t(\theta_{10}(X))\lambda_{20} + \nu_t(\theta_{10}(X))\lambda_{21}$$
$$+\rho_t(\theta_{11}(X))\lambda_{20} + \sigma_t(\theta_{11}(X))\lambda_{21}]$$

Next,

$$d(j_t(X)(-1)^{\Lambda(t)}) =$$
$$[j_t(\theta_0(X))dt + j_t(\theta_1(X))dA(t) + j_t(\theta_2(X))dA(t)^*](-1)^\Lambda$$
$$-2j_t(X)(-1)^{\Lambda(t)}d\Lambda(t)$$
$$-2j_t(\theta_1(X))(-1)^{\Lambda(t)}dA(t)d\Lambda(t)$$

Thus, using $dA.d\Lambda = dA$, we get

$$\mathbb{E}[d(j_t(X)(-1)^{\Lambda(t)})|\eta_o(t)] =$$

$$dt[\nu_t(\theta_{00}(X))+\sigma_t(\theta_{01}(X))+u(t)(\nu_t(\theta_{10}(X))+$$
$$\sigma_t(\theta_{11}(X)))+\bar{u}(t)(\nu_t(\theta_{20}(X))+\sigma_t(\theta_{21}(X)))]$$
$$-2dt|u(t)|^2\nu_t(X) - 2u(t)dt(\nu_t(\theta_{10}(X)) + \sigma_t(\theta_{11}(X)))$$

Similarly by use of the quantum Ito formula, we can calculate

$$d(j_t(X)(-1)^{\Lambda(t)}).dY_o(t), d(j_t(X.(-1)^{\Lambda(t)}), d(j_t(X.(-1)^{\Lambda(t)}).(-1)^{\Lambda(t)})$$

in terms of $dt, dA, dA^*, d\Lambda$ and hence derive equations for $F_{kt}(.), G_{kt}(.)$. We leave this as an exercise reader. Note that this analysis immediately leads to stochastic coupled differential equations for $\pi_t, \rho_t, \nu_t, \sigma_t$. These equations have the following form:

$$d\pi_t(X) = F_1(\pi_t(\alpha_1(X)), \nu_t(\alpha_2(X)), \rho_t(\alpha_3(X)), \sigma_t(\alpha_4(X))dt$$
$$+F_2(\pi_t(\beta_1(X)), \nu_t(\beta_2(X)), \rho_t(\beta_3(X)), \sigma_t(\beta_4(X))dY_o(t)$$

and likewise for $\nu_t, \rho_t, \sigma_t$ where the maps $\alpha_k, \beta_k, k = 1, 2, 3, 4$ depend upon $u(t)$ which parametrizes the coherent state. Thus, we obtain concrete formulas for the quantum filter.

# Chapter 9

# Large deviations for classical and quantum stochastic filtering problems in general relativity

[20] Classical and quantum filtering in general relativity with applications of large deviation theory.

The matter, metric and electromagnetic fields satisfy the pde's

$$[\rho v^\mu v^\nu - pg^{\mu\nu}]_{:\nu} + S^{\mu\nu}_{:\nu} = 0$$

$$F^{\mu\nu}_{:\nu} = J^\mu = \sigma F^{\mu\nu} v_\nu$$

where $\sigma$ is the fluid conductivity and finally,

$$R^{\mu\nu} - (1/2)Fg^{\mu\nu} = K[\rho v^\mu v^\nu - pg^{\mu\nu}] + KS^{\mu\nu}$$

where $K = -8\pi G$ and $S^{\mu\nu}$ is the energy-momentum tensor of the electromagnetic field and is given by

$$S^{\mu\nu} = (-1/4)F^{\alpha\beta}F_{\alpha\beta}g^{\mu\nu} + F^{\mu\alpha}F^\nu_\alpha$$

Now we collect all these component fields into one big vector field

$$\xi(x) = [Vec((g_{\mu\nu}(x)))^T, ((Vec(g_{\mu\nu,0}(x))), (A^r(x))^T, (A^r_{,0}(x))^T, \rho(x), ((v^r(x)))^T]^T$$

It should be noted that these differential equations should be supplemented by an additional equation of state $p = F(\rho)$. The above pde's can then be cast in state variable form as

$$\partial_t \xi(x) = (L\xi)(x) + \delta.N(\xi)(x)$$

where $L$ is a linear partial differential operator of second order in the spatial variables and $N$ is a nonlinear partial differential operator in the spatial variables. It should be noted that as is usually done in the ADM action and the consequent Hamiltonian formulation of general relativity, the metric can be separated into a spatial part and a time part. We could also impose four coordinate conditions on the metric by choosing an appropriate coordinate system, so that the metric has effectively only six independent components rather than ten. The condition $g_{\mu\nu}v^\mu v^\nu = 1$ implies that the four velocity vector has just three independent components which we take as the spatial components of the velocity. Then, we add noise to this system of field equations. This added noise should be compatible to the conservation laws. For example, the Einstein Maxwell-field equations in the presence of noise gravitational noise $W^{\mu\nu}(x)$ and electromagnetic noise $W^\mu$ would read

$$R^{\mu\nu} - (1/2)Rg^{\mu\nu} = K(T^{\mu\nu} + S^{\mu\nu}) + W^{\mu\nu}$$

$$F^{\mu\nu}_{:\nu} = J^\mu + W^\mu$$

or equivalently,

$$(F^{\mu\nu}\sqrt{-g})_{,\nu} = (J^\mu + W^\mu)\sqrt{-g}$$

For consistency, we require that these random noise fields be subject to the relations

$$(T^{\mu\nu} + S^{\mu\nu})_{:\nu} = -K^{-1}W^{\mu\nu}_{:\nu},$$

$$[(J^\mu + W^\mu)\sqrt{-g}]_{,\mu} = 0,$$

or equivalently,

$$J^\mu_{:\mu} = -W^\mu_{:\mu}$$

Then, we can ask about the large deviation properties of the metric, velocity, density and electromagnetic fields.

[21] Large deviations in the Boltzmann kinetic transport equation for a two species plasma. There are two species with each particle of the $k^{th}$ species having charge $q_k, k = 1, 2$. The Boltzmann equations are

$$f_{1,t}(t,r,v) + (v,\nabla_r)f_1 + q_1(F_1(t,r,v),\nabla_v)f_1 = N_{11}(f_1,f_1) + N_{12}(f_1,f_2)(t,r,v)$$

$$f_{2,t}(t,r,v) + (v,\nabla_r)f_2 + q_2)(F_2(t,r,v),\nabla_v)f_2 = N_{21}(f_2,f_1) + N_{22}(f_2,f_2)$$

where the $N_{ab}$ are bilinear functionals of their arguments. The forces $F_1, F_2$ come from the external noisy electromagnetic fields and we assume these to be low amplitude noise. To emphasize this smallness of the noisy electromagnetic forces, we introduce a perturbation parameter $\delta$ into them and ask what is the probability that the pdfs $f_1, f_2$ will deviate from the Maxwell equilibirium density by an amount greater than a given threshold ? Another question that comes into the picture involves estimating the pdf's $f_1$ and $f_2$ empirically upon measurements of the positions and velocities of the particles in the plasma. In order to get the ldp rate function for such empirical estimates of the Boltzmann distribution function, we require the entire process statistics at finite sample times. As a preliminary calculation of this sort, we approximate the Boltzmann equation by a linear Fokker-Planck equation and then identify the drift and diffusion coefficients of a process $(r(t), v(t)), t \geq 0$ and then use the well known results for the large deviation rate function for the empirical distribution of a diffusion process. However such a calculation would not capture the nonlinearity introduced by the collision term. For example, if we look at a collision term of the sort

$$N(f,f)(t,r,v) = \int K(v,v_1,v_2)f(t,r,v+v_1)f(t,r,v_2)d^3v_1d^3v_2$$

we could approximate it by

$$\int K(v,v_1,v_2)K(v,v_1,v_2)f_0(r,v_2)(f(t,r,v)+(v_1,\nabla_v)f(t,r,v)+(1/2)Tr(v_1v_1^T\nabla_v\nabla_v^Tf(t,r,v)))d^3v_1d^3v_2$$

where $f_0$ is the equilibrium particle density function. Defining the vector

$$b(r,v) = \int K(v,v_1,v_2)f_0(r,v_2)v_1d^3v_1d^3v_2,$$

the matrix

$$a(r,v) = \int K(v,v_1,v_2)f_0(r,v_2)v_1v_1^Td^3v_1d^3v_2$$

and the scalar

$$c(r,v) = \int K(v,v_1,v_2)f_0(r,v_2)d^3v_1d^3v_2$$

we can express the approximated binary collision term as

$$N(f,f)(t,r,v) \approx c(r,v)$$
$$f(t,r,v)+b(r,v)^T\nabla_vf(t,r,v)+(1/2)Tr(a(r,v)\nabla_v\nabla_v^Tf(t,r,v))$$

which is precisely the Fokker-Planck approximation when $c = 0$.

[22] Large deviations in quantum mechanics.

Consider Schrodinger's equation for the mixed state $\rho(t)$ under a small random potential $\epsilon.V(t)$:

$$i\rho'(t) = [H_0 + \epsilon V(t), \rho(t)]$$

We can express the solution as

$$\rho(t) = U(t)\rho(0)U(t)^*$$

where

$$U(t) = U_0(t)W(t), U_0(t) = exp(-itH_0), W(t) = T\{exp(-i\epsilon\int_0^t \tilde{V}(s)ds)\},$$

$$\tilde{V}(t) = U_0(t)V(t)U_0(t)^*$$

with $T()$ being the time ordering operator. The reason for this is that the above Schrodinger equation can be expressed as

$$i\rho'(t) = (ad(H_0) + \epsilon.ad(V(t)))(\rho(t))$$

and so defining

$$T_0(t) = exp(-it.ad(H_0)) = Ad(U_0(t))$$

we can express the solution as

$$\rho(t) = T(t)(\rho(0))$$

where

$$T(t) = T_0(t)T_1(t)$$

with

$$T_1(t) = T\{exp(-i\epsilon \int_0^t S(s)ds)\}$$

where

$$S(t) = T_0(t).ad(V(t)).T_0(t)^* = Ad(U_0(t)).ad(V(t)).Ad(U_0(t))^*$$

$$= (d/ds)[Ad(U_0(t)).Ad(exp(s.V(t)).Ad(U_0(t)^*)]|_{s=0}$$

$$= (d/ds)Ad(U_0(t)exp(s.V(t)).U_0(t)^*)|_{s=0} = ad(U_0(t)V(t))U_0(t)^*)$$

and therefore, the formula follows using the idenitity

$$T\{exp(\int_0^t ad(X(s))ds)\} =$$

$$T(exp(\int_0^t (L(X(s))-R(X(s)))ds)) = T(exp(\int_0^t L(X(s))ds).exp(-\int_0^t R(X(s))ds))$$

$$= T(exp(\int_0^t L(X(s))ds)).T(exp(-\int_0^t R(X(s))ds)$$

where

$$L(X)Y = XY, R(X)Y = YX, ad(X)(Y) = XY - YX = (L(X) - R(X))Y$$

Now computing the mean value of a state at time $t$ amounts to computing the expectation of $W(t) = T(exp(-i\epsilon \int_0^t V(s)ds))$. Here, we are renaming $\tilde{V}(t)$ as $V(t)$. It is not possible to get closed form expressions for this expectation. However, using the Dyson series expansion

$$W(T, \epsilon) = I + \sum_{n\geq 1}(-i\epsilon)^n \int_{0<t_n<...<t_1<t} V(t_1)...V(t_n)dt_1...dt_n$$

and hence assuming

$$\mathbb{E}(V(t_1)...V(t_n)) = M_{V,n}(t_1, ..., t_n)$$

we have

$$W(T,\epsilon)=I+\sum_{n\geq 1}(-i\epsilon)^n.\int_{0<t_n<...<t_1<T}M_{V,n}(t_1,...,t_n)dt_1...dt_n$$

Note that $M_{V,n}(t_1,...,t_n)$ is a non-random operator valued function of $t_1,...,t_n$. If $\psi(0)>$ is an initial wave function (ie, pure state), then the wave function at time $T$ will in the interaction picture be given by

$$|\psi(T)>=W(T,\epsilon)|\psi(0)>$$

and the probability of making a transition to another state $|\phi>$ after time $T$ will be given by

$$P(T)=|<\phi|\psi(T)>|^2$$

This probability will also be a random function of $T$ because $V(.)$ is a random process. We can ask the question that if $\phi$ is orthogonal to $\psi(0)$ then as $\epsilon\to 0$, at what rate does this probability converge to zero ? The answer to this question will tell us that if noise is small, then there is a very small probability of the system making a transition from one stationary state of the unperturbed system to another and such a transition corresponds to a sudden rare spike in the system which we would like to control by altering some of the system parameters.

LDP applied to the quantum Boltzmann equation:

$$i\rho'(t)=[H,\rho(t)]+(N-1)Tr_2[V(t),\rho(t)\otimes\rho(t)]$$

where $H$ is a self-adjoint operator in the Hilbert space $\mathcal{H}$ while $V$ is a self-adjoint operator in $\mathcal{H}\otimes\mathcal{H}$. Assuming $V=V(t)$ to be a random interacting potential, we can calculate the average value $<X>(t)=Tr(\rho(t)X)$ of an observable at time $t$ and then ask that if $V$ is small amplitude noise, then what will be the rate function for the stochastic process $<X(t)>,t\geq 0$. For this, we must evaluate the moment generating functional of $<X>(t),t\geq 0$. Introducing a small perturbation parameter $\delta$ into the nonlinear part we write the quantum Boltzmann equation as

$$i\rho'(t)=[H,\rho(t)]+\delta.Tr_2[V(t),\rho(t)\otimes\rho(t)]$$

and expand

$$\rho(t)=\sum_{n\geq 0}\delta^n\rho_n(t)$$

Then equating equal powers of $\delta$ gives us the sequence of equations

$$i\rho_0'(t)=[H,\rho_0(t)],$$

$$i\rho_n'(t)=[H,\rho_n(t)]+\sum_{k=0}^{n-1}Tr_2[V(t),\rho_k(t)\otimes\rho_{n-1-k}(t)],n\geq 1$$

which can be iteratively solved as

$$\rho_0(t) = U_0(t)\rho(0)U_0(t)^*, U_0(t) = exp(-itH_0),$$

$$\rho_n(t) = -i\int_0^t U_0(t-s)(\sum_{k=0}^{n-1} Tr_2[V(t), \rho_k(s) \otimes \rho_{n-1-k}(s)])U_0(t-s)^* ds, n \geq 1$$

Exercise: Calculate the first few approximants $\rho_n(t), n = 1, 2, ...$ and evaluate their statistical means. Using the first few approximants, evaluate the moment generating function of $Tr(\rho(t)X)$ upto a finite power of $\delta$ assuming $V(t)$ to be a zero mean stationary operator valued Gaussian process with specified autocorrelation

$$R_{VV}(t-s) = \mathbb{E}[V(t) \otimes V(s)]$$

[23] Large deviations in the Schrodinger channel and the Dirac channel for transmission of information bearing sequences. Let $\{I[n] : n \in \mathbb{Z}_+\}$ be the information bearing sequence to be transmitted. First we consider the first quantized version of the Cq communication problem. The information bearing sequence $I$ is transformed into a classical current source $x(t, I)$ which is fed into an antenna thereby inducing a surface current density $J_s(t, r, I)$ on the surface of the antenna. The antenna then radiates out a classical electromagnetic field described by a magnetic vector potential $A(t, r, I)$ and an electric scalar potential $\Phi(t, r, I)$. This em field propagates over a classical channel and at the receiver end there is a atom consisting of a nucleus with charge $Ze$ and an electron on which this field is incident. The wave function of the electron accordingly satisfies Schrodinger's equation/Dirac's equation with interaction terms in the Hamiltonian arising due to the electromagnetic field at the receiver end. The wave function evolves and we measure the quantum average of an observable or a set of observables at the receiver end in this evolving state and from noisy measurements of these quantum averages, we construct a maximum likelihood estimate of the information bearing sequence $I$. The question is that if the measurement noise in the quantum averaged measurements is small, then can we calculate a large deviation rate function for the error in the information bearing sequence estimate ? Further, we know well that if the receiver is classical without any noise, then we can exactly decode $I$ from the received electromagnetic field. Now when our receiver is quantum there will be an inherent error in decoding using the maximum likelihood estimator based on using a POVM or a PVM at a finite sequence of times on the quantum receiver with the probabilities computed using the collapse postulate following each measurement with the outcome noted. How can we choose our POVM or PVM so that this decoding error is as small as possibl ? By treating Planck's constant as a small parameter which converges to zero, is it possible to obtain a large deviation rate function for the estimated information bearing sequence ?

All this analysis is based on elementary first quantization. However, if we reqard the channel as a sea of electrons and positrons, then we should describe such a channel by means of a second quantized Dirac wave function operator field $\psi(t,r)$ expanded in the absence of electromagnetic interactions in terms of the electron and positron creation and annihilation operators. Then, for example $J^{\mu}(t,r) = -e\psi(t,r)\gamma^0\gamma^{\mu}\psi(t,r)$ would describe the four current density operator of the channel and its expected value in a given state of the electrons and positrons would give us a classical four current density field, ie, a classical charge and current density field. If we transmit an information bearing sequence via a quantum antenna over such a channel, then the transmitted electromagnetic field would be a quantum electromagnetic field containing a free part and a perturbation containing the information bearing classical sequence as parameters. The entire quantum electromagnetic field that is transmitted would be expressible in terms of the free photon creation an an annihilation operator fields in momentum space and when we describe the propagation of such a field through the channel, we would have to solve Dirac's equation perturbatively taking the electron-positron-photon interactions into account as is done in conventional quantum electrodynamics with the only difference that here the photon field contains in addition the classical information bearing sequence. The resulting Dirac field is now a Fermionic operator field containing a mixture of the electron-positron-photon operators with the information bearing sequence as parameters and when we calculate the Dirac current and the resulting scattered electromagnetic field from the channel by applying the usual retarded potential formula to this current field, this resultant scattered electromagnetic field at the receiver end will again constist of a mixture of electron-positron-photon-operators with the information bearing sequence as parameters. By measuring the mean, covariance and higher moments of this field in a given state of the photons, electrons and positrons, at the receiver end, we get information about the information bearing sequence thereby enabling us to do decoding based on these quantum moments.

The LDP problem in quantum mechanics. Let $h$ denote Planck's constant divided by $2\pi$ and consider the Klein-Gordon equation for the wave function

$$((ih\partial_t + e\Phi)^2 - c^2(-ih\nabla + eA)^2 - m^2c^4)\psi = 0$$

Substitute

$$\psi(t,r) = a(t,r)exp(i\phi(t,r)/h)$$

Then using

$$ih\partial_t\psi = (ih\partial_t a - a\partial_t\phi)exp(i\phi/h)$$

$$(ih\partial_t + e\Phi)\psi = (ih\partial_t a + e\Phi a - a\partial_t\phi)exp(i\phi/h)$$

$$(ih\partial_t + e\Phi)^2\psi = (-h^2\partial_t^2 a + ieh(\partial_t\Phi)a + ieh\Phi\partial_t a - ih\partial_t a.\partial_t\phi - iha\partial_t^2\phi +$$

$$(ih\partial_t a + e\Phi - a\partial_t\phi)(e\Phi - \partial_t\phi))exp(i\phi/h)$$

[24] Large deviations for shock waves in general relativity in one dimension. The metric of two dimensional space-time is

$$d\tau^2 = a(x)dt^2 - b(x)dx^2/c^2, x^0 = t, x^1 = x, g_{00} = a, g_{11} = -b/c^2$$

The energy-momentum tensor is

$$T^{\mu\nu} = (\rho + p/c^2)v^\mu v^\nu - (p/c^2)g^{\mu\nu}$$

where

$$g^{00} = 1/a, g^{11} = -c^2/b$$

$$g_{\mu\nu}v^\mu v^\nu = 1$$

gives

$$a.(v^0)^2 - (b/c^2)(v^1)^2 = 1$$

Note that

$$v^0 = dt/d\tau, v^1 = dx/d\tau, v = dx/dt = v^1/v^0$$

so

$$a(v^0)^2 - (b/c^2)(v^0 v)^2 = 1$$

so

$$v^0 = (a - bv^2/c^2)^{-1/2}$$

The equations of motion of the fluid are

$$T^{\mu\nu}_{:\nu} = 0$$

which yield

$$[(\rho + p/c^2)v^\nu]_{:\nu} - p_{,\nu}v^\nu/c^2 = 0,$$

$$(\rho + p/c^2)v^\nu v^\mu_{:\nu} + p_{,\nu}v^\nu v^\mu/c^2 - g^{\mu\nu}p_{,\nu} = 0$$

These equations can be put into the form using

$$\alpha(x) = \sqrt{-g} = \sqrt{a(x)b(x)/c^2}$$

as

$$[(\rho + p/c^2)v^\nu \alpha]_{,\nu} - p_{,\nu}v^\nu \alpha/c^2 = 0,$$

$$(\rho + p/c^2)v^\nu(v^1_{,\nu} + \Gamma^1_{\rho\nu}v^\rho)$$

$$+p_{,\nu}v^\nu v^1/c^2 - g^{11}p_{,1} = 0$$

We next observe that

$$\Gamma^1_{00} = -g^{11}g_{00,1}/2 = c^2 a_{,1}/2b,$$

$$\Gamma^1_{01} = \Gamma^1_{10} = g^{11}\Gamma_{110} = g^{11}g_{11,0}/2 = a_{,1}/2a$$

$$\Gamma^1_{11} = g^{11}g_{11,1}/2 = b_{,1}/2b$$

These equations for $\rho(t,x), v(t,x)$ assuming an equation of state $p = f(\rho)$ can be expressed as

$$v_{,t} = F_1(\rho, v, v_{,x}, \rho_{,x}), \rho_{,t} = F_2(\rho, v, v_{,x}, \rho_{,x})$$

Here, viscosity has been ignored. We now consider two regions of space : I:$x \leq L$ and II:$x > L$. In both of these regions, the above differential equations are satisfied and at the boundary, we assume that there is a discontinuity given by

$$v(t, L+0) - v(t, L-0) = d(t), \rho(t, L+0) - \rho(t, L-0) = e(t)$$

The objective is then to solve for the fluid velocity and density fields in these two regions.

The large deviation problem: Assume that the metric of space-time undergoes small random fluctuations of the form $a(x) \to a_0(x) + \delta.a_1(x), b(x) \to b_0(x) + \delta.b_1(x)$ where $\delta$ is a small random parameter converging to zero and $a_1(.)$ and $b_1(.)$ are zero mean Gaussian random fields. The problem is then to study the probability distribution of the density and viscosity in the two regions as $\delta \to 0$ and derive a LDP rate function for the same.

PS: I am indebted to my colleague Professor J.K.Singh for suggesting to work on this problem.

[25] Boltzmann's equation for a plasma enclosed in a cavity:

$$\partial_t f(t,r,v) + (v, \nabla_r) f(t,r,v) + (q/m)(E(t,r) + v \times B(t,r)) f(t,r,v)$$
$$= (f_0(r,v) - f(t,r,v))/\tau(v)$$

$$f = f_0(r,v) + \delta f(t,r,v), E(t,r) = -\nabla\Phi(r) + \delta E(t,r)$$
$$(v, \nabla_r) f_0 - (q/m)(\nabla\Phi(r), \nabla_v) f_0(r,v) = 0$$

gives

$$f_0(r,v) = C(\beta).exp(-\beta(mv^2/2 + q\Phi(r)))$$

Then,

$$\partial_t \delta f(t,r,v) + (v, \nabla_r)\delta f$$
$$+ (q/m)(-\nabla\Phi(r) + \delta E(t,r) + v \times B(t,r), \nabla_v)\delta f(t,r,v) + (q/m)(\delta E(t,r), \nabla_v) f_0(r,v)$$
$$= -\delta f(t,r,v)/\tau(v)$$

Expansion in terms of basis functions:

$$\delta f(t,r,v) = \sum_n \delta f_n(t,v)\psi_n(r)$$

$$curl\delta E = -\partial_t B, curl B = \mu J + \mu\epsilon\partial_t \delta E, div B = 0, div\delta E = \rho/\epsilon$$

$$\rho(t,r) = q\int \delta f(t,r,v)dv, J(t,r) = q\int v.\delta f(t,r,v)dv, \delta f = f - f_0$$

In the frequency domain,

$$i\omega \sum_n f_n(\omega, v)\psi_n(r) + \sum_n f_n(\omega, v)(v, \nabla_r \psi_n(r))$$

$$+(q/m)\sum_n(-\nabla\Phi(r),\nabla_v f_n(\omega,v))\psi_n(r)+(q/m)(\delta E(\omega,r),\nabla_v)f_0(r,v)$$

$$+(q/m)\sum_n[\int(\delta E(\omega-\omega',r)+v\times B(\omega-\omega',r),\nabla_v f_n(\omega',v))d\omega']\psi_n(r)$$

$$=-\sum_n f_n(\omega,v)\psi_n(r)/\tau(v)---(1)$$

The solution to the Maxwell equations in the frequency domain is

$$\delta E(\omega,r)=\int G_{11}(\omega,r,r')J(\omega,r')dr'+\int G_{12}(\omega,r,r')\rho(\omega,r')dr'---(2),$$

$$B(\omega,r)=\int G_{21}(\omega,r,r')J(\omega,r')dr'+\int G_{22}(\omega,r,r')\rho(\omega,r')dr'---(3)$$

where

$$J(\omega,r)=q\sum_n(\int vf_n(\omega,v)dv)\psi_n(r),---(4)$$

$$\rho(\omega,r)=q\sum_n(\int f_n(\omega,v)dv)\psi_n(r)---(5)$$

Substituting (4) and (5) into (2) and (3) gives $\delta E(\omega,r),B(\omega,r)$ in terms of the functions $f_n(\omega,v)$. Then substituting these expressions for $\delta E(\omega,r)$ and $B(\omega,r)$ into (1), multiplying the resultant equation by $\psi_m(r)$ and then integrating over $r$ gives us a system of non-linear integro-differential equations for the functions $f_n(\omega,v)$. In these equations, the differential operators act only on the $v$ variable. All these equations are exact, ie no perturbation expansion is being assumed here.

The large deviation problem: When the electric potential $\Phi(r)$ is a small zero mean Gaussian random field, then obtain the rate function for the functions $f_n(\omega,v)$ and hence for the Boltzmann distribution function.

[26] S-parameters in a rectangular cavity. Dimensions are $a,b,d$. Define

$$u_{mnp}(x,y,z)=(2\sqrt{2}/\sqrt{abd})sin(m\pi x/a)sin(n\pi y/b)cos(p\pi z/d)$$

$$v_{mnp}=(2\sqrt{2}/\sqrt{abd})cos(m\pi x/a)cos(n\pi y/b)sin(p\pi z/d)$$

Assume that the cavity carries a superposition of $TE_{mnp},TM_{mnp}$ modes. Then

$$E_z=c(mnp)u_{mnp}(x,y,z),H_z=d(mnp)v_{mnp}(x,y,z)$$

The oscillation frequency for both these modes is

$$\omega=\omega(mnp)=(\mu\epsilon)^{-1}\pi\sqrt{m^2/a^2+n^2/b^2+p^2/d^2}$$

Define

$$h^2=h(m,n)^2=\pi^2(m^2/a^2+n^2/b^2)$$

We compute

$$E_x=h^{-2}\partial_z\partial_x E_z-(j\omega\mu/h^2)\partial_y H_z$$

$$= -c(mnp)h^{-2}(m\pi/a)(p\pi/d)(2\sqrt{2}/sqrtabd)cos(m\pi x/a)sin(n\pi y/b)sin(p\pi/d)$$

$$+d(mnp)(j\omega\mu/h^2)(n\pi/b)(2\sqrt{2}/\sqrt{abd})cos(m\pi x/a)sin(n\pi y/b)sin(p\pi z/d)$$

$$= (\alpha_1(mnp)c(mnp) + j\beta_1(mnp)d(mnp))w_{mnp}(x, y, z)$$

where

$$\alpha_1(mnp) = -h(m,n)^{-2}(m\pi/a)(p\pi/d), \beta_1(mnp) = \omega(mnp)(\mu/h(m,n)^2)(n\pi/b)$$

$$w_{mnp}(x, y, z) = (2\sqrt{2}/\sqrt{abd})cos(m\pi x/a)sin(n\pi y/b)sin(p\pi z/d)$$

Likewise,

$$H_y = h^{-2}\partial_z\partial_y H_z - (j\omega\epsilon/h^2)\partial_x E_z$$

$$= -d(mnp)h^{-2}(2\sqrt{2}/\sqrt{abd})(n\pi/b)(p\pi/d)cos(m\pi x/a)sin(n\pi y/b)cos(p\pi z/d)$$

$$-(j\omega\epsilon/h^2)c(mnp)(m\pi/a)(2\sqrt{2}/\sqrt{abd}))c(mnp)cos(m\pi x/a)sin(n\pi y/b)cos(p\pi z/d)$$

$$= (\alpha_2(mnp)d(mnp) + j\beta_2(mnp)c(mnp))\xi_{mnp}(x, y, z)$$

where

$$\alpha_2(mnp) = -h(m,n)^{-2}(n\pi/b)(p\pi/d), \beta_2(mnp) = -\omega(mnp)\epsilon/h(m,n)^2(m\pi/a)$$

$$\xi_{mnp}(x, y, z) = (2\sqrt{2}/\sqrt{abd})cos(m\pi x/a)sin(n\pi y/b)cos(p\pi z/d)$$

[27] Quantum systems perturbed by Martingales derived from Brownian motion. Let $B(t)$ be standard BM and let $f(x)$ be a twice continously differentiable function with bounded first and second order derivatives. Then,

$$M(t) = f(B(t)) - (1/2)\int_0^t f''(B(s))ds$$

is a continuous Martingale.

Remark: Stroock and Varadhan had developed a Martingale characterization of diffusion processes and in the process, they developed the general theory of "Ito Processes", namely processes $\xi(t)$ for which

$$M(t) = f(\xi(t)) - \int_0^t L_s f(\xi(s))ds$$

is a martingale for all well behaved functions $f$ where

$$L_t f(x) = \mu_a(t,\omega)\partial_a f(x) + (1/2)A_{ab}(t,\omega)\partial_a\partial_b f(x0$$

where $\mu_a, A_{ab}$ are progressively measurable processes. Let then $< M > (t)$ be the quadratic variation of $M$. Then we define the stop times

$$\tau(t) = inf(s > 0 :< M > (s) \geq t)$$

Then it is a well known fact that the process $B(t) = M(\tau(t))$ is a standard Brownian motion and thus $M(t) = B(< M > (t))$. Now we can ask the question that if we consider a noisy Schrodinger equation of the form

$$\partial_t \psi(t) = (-(iH(\theta)dt + V^2 d < M > (t)/2) - iVdM(t))\psi(t)$$

(The Ito correction term has been added here to ensure unitary evolution), then how to choose the function $f$ appropriately so that by taking measurements on this system like for example, the average value of an observable, we can obtain the best possible estimates of the Hamiltonian parameters $\theta$. The large deviation problem in this context is that by assuming $V$ to be small, for example we replace $V$ by $\delta.V$ where $\delta$ is a small parameter, then what is the large deviation rate function for the wave function. In particular, using such a rate function, we an calculate the asymptotic probablilty that the process $< \psi(t)|X|\psi(t) >, 0 \leq t \leq T$ will deviate from a desired process say $d(t) =< \psi_0(t)|X|\psi_0(t) >, 0 \leq t \leq T$ by an amount greater than a threshold $\epsilon$ where $\psi_0(t)$ is the solution without noise and with $\theta = \theta_0$. How then to control $\theta$ so that this deviation probability is as small as possible?

# Chapter 10

# Quantum mechanics of the eye

## 10.1 Quantum electrodynamics of photons, electrons and positrons within the eye

Introduction: When light from an object falls on the surface of the eye, it creates an image of the object field on the retina. This process of image formation is usually described by modeling the surface of the eye lens to be spherical or more precisely of a parabolic shape so that using the standard Snell's laws of refraction, the path of a given light ray from the object can be broken into three parts, first, the part before it hits the lens, second, the part after it hits the first surface of the lens and propagates within the lens till it hits the second surface and three, the part after it emerges out of the second surface of the lens and till it hits the retinal screen. It can be shown that using Snell's laws of refraction that if $u$ is the distance of the object from the lens and $v$ is the distance of the image then $1/u + 1/v = 1/f$ where $f$ is the lens focal length. This is a classical ray description of image formation. The image of a point is formed on the retina iff all the rays emanating from that point, after passing through the lens meet at a single point on the retina. The condition for this to happen puts severe restrictions on the shape of the two surfaces of the lens. The wave description of light when it passes through the eye channel and till it hits the retina is however quite different. It involves modeling the eye channel as a cylindrical cavity resonator with the retinal disc being one surface and analyzing the wave fields within this cavity using the Maxwell equations. Specifically, using this formalism, we can calculate the electromagnetic wave field pattern on the retinal disc given its pattern at the other open disc end. This process just involves expressing the transverse components of the $E - H$ field in terms of the longitudinal components of the $E$ and $H$ fields and computing the coefficients of the longitudinal components using the specified field pattern at the mouth of the

CDRA. However, if we pass over to the quantum/nano level, even this wave field description is inaccurate. Indeed, today we know that the electromagnetic field is neither a ray nor a wave field. It is actually a quantum wave operator field. More precisely, it is an ensemble of an infinite number of quantum harmonic oscillators or in fact, a superpostion of plane or constrained plane waves with coefficients being creation and annihilation operators in a Boson Fock space. It is therefore not quite precise to talk about the wave field pattern on the retina but rather specify the quantum state of the wave field like a coherent state, or a number state and then compute the statistics of the wave operator field on the retinal surface using the action of creation and annihilation operators on coherent states or on photon number states. After we do such an analysis, we can ask the more interesting question:how does light within the cylindrical cavity resonator interact with the matter field ? Specifically, when light propagates within the cylindrical eye cavity, it interacts with electrons, atoms and molecules of the fluid medium of which the eye cavity is composed and its propagation therefore gets affected by this medium. In classical wave field theory, we would model this fluid medium as a plasma composed having a definite permittivity, permeability and conductivity and would analyze the propagation of the electromagnetic field in this plasma using the Vlasov equations which are infact the coupled Boltzmann equation for the particle distribution function of the plasma and the Maxwell electromagnetic field. The result of this analysis using linearized perturbation theory would be dispersion relation between the oscillation frequencies of the plasma and the electromagnetic field and the wave vector. However, at the quantum level the description of the interaction between light and matter is more subtle. A simplified analysis would be to model the matter as just the second quantized electron-positron field using Dirac's relativistic wave equation with the wave function being an operator field and then to include the photon interaction term in the usual way. In this way, the total Hamiltonian of the photon and matter field splits into three terms: One, the Hamiltonian of the electromagnetic field described as a quadratic form in the photon creation and annihilation operators within the cavity, two the Hamiltonian of the Dirac field with cylindrical cavity boundary conditions on the wave function described as a quadratic form in the electron-positron creation and annihilation operators and three, the Hamiltonian of the interaction between the photon and electron-positron field described as a quadratic form in the electron-positron creation-annihilation operators multiplied with a linear form in the photon creation-annihilation operators. This interaction Hamiltonian can be used to compute the amplitudes for scattering but since we are primarily interested in the statistics of the electromagnetic wave field pattern on the retinal surface, we shall describe this interaction using the Dirac current density expressed as a quadratic form in the electron-positron creation-annihilation operator fields that drives the photon field using the wave equation for the electromagnetic field in the presence of a current density. Thus, we shall by perturbation theory be able to calculate the change in the electromagnetic field pattern on the retinal surface caused by the Dirac current in term of operators and then by assuming a definite state of the photon-electron-positron field, we shall be able to compute the mean and covariance of the electromagnetic field on the retinal screen surface in this state.

## 10.2 Classical and quantum descriptions of a cylindrical dielectric resonator antenna

Let $d$ be the height of the cavity and $a$ its radius. Assume that all its walls including the top and bottom surfaces are perfect conductors. At the classical scale, using the fundamental identities resulting from the Maxwell curl equations relating the transverse components of the electromagnetic field to the longitudinal components, we have

$$E_\perp = \partial_z(1/h^2)\nabla_\perp E_z - (j\omega\mu/h^2)\nabla_\perp H_z \times \hat{z} --- (1)$$

$$H_\perp = \partial_z(1/h^2)\nabla_\perp H_z + (j\omega\epsilon/h^2)\nabla_\perp E_z \times \hat{z} --- (2)$$

Also the longitudinal components of the Maxwell curl equations give us

$$\nabla_\perp \times E_\perp = -j\omega\mu H_z\hat{z} --- (3)$$

$$\nabla_\perp \times H_\perp = j\omega\epsilon E_z\hat{z} --- (4)$$

In terms of the cylindrical coordinate components,

$$E_\rho = (-\gamma/h^2)\partial_\rho E_z - (j\omega\mu/h^2)\partial_\phi H_z$$

$$E_\phi = (-\gamma/h^2\rho)\partial_\phi E_z + (j\omega\mu/h^2)\partial_\rho H_z$$

$$H_\rho = (-\gamma/h^2)\partial_\rho H_z + (j\omega\epsilon/h^2\rho)\partial_\phi E_z$$

$$H_\phi = (-\gamma/h^2)\partial_\rho H_z - (j\omega\epsilon/h^2\rho)\partial_\phi E_z$$

where $\gamma$ is to be replaced by the operator $-\partial/\partial z$. It should be noted that the values assumed by $\gamma$ are $\pi p/d$ where $p$ is an integer in order that the tangential components of the electric field and the normal component of the magnetic field vanish on the top and bottom surfaces of the CDRA. We note that

$$E_\perp(\rho, \phi, z) = E_\rho\hat{\rho} + E_\phi\hat{\phi},$$

and likewise for $H_\perp$. Also,

$$\nabla_\perp = \hat{\rho}.\partial_\rho + (\hat{\phi}/\rho)\partial_\phi$$

We are using the abbreviation

$$\partial_\xi = \partial/\partial\xi, xi = \rho, \phi, z$$

Then using equns. (1)-(4), we get

$$(\nabla_\perp^2 + h^2)(E_z, H_z) = 0$$

We note that application of the boundary conditions, namely $E_z, E_\phi, H_\rho$ vanish at $\rho = a$ while $E_\rho E_\phi, H_z$ vanish at $\rho = 0, d$ (ie, the tangential components of the electric field and the normal components of the magnetic field vanish on all the boundaries gives us for the TM modes,

$$h^2 = h^2(E, mn) = \alpha_m[n]^2/a^2 = \omega(E, mnp)^2\epsilon\mu - (\pi p/d)^2$$

where $m, n, p$ are integers and $\alpha_m[n], n = 1, 2, ..$ are the roots of the Bessel function $J_m(x)$ and for the TE modes,

$$h^2 = h^2(H, m, n) = \beta_m[n]^2/a^2 = \omega(H, mnp)^2\epsilon\mu - (\pi p/d)^2$$

where $\beta_m[n], n = 1, 2, ...$ are the roots of $J'_m(x) = 0$. Thus, we obtain the expansions

$$E_z(t, \rho, \phi, z) = \sum_{mnp} J_m(\alpha_m[n]\rho/a)cos(p\pi z/d)Re(c(E, mnp)exp(j(m\phi - \omega(Emnp)t)))$$

$$= \sum_{mnp} Re(c(E, mnp)exp(-j\omega(Emnp)t)u_{mnp}(\mathbf{r}))$$

$$H_z(t, \rho, \phi, z) = \sum_{mnp} J_m(\beta_m[n]\rho/a)sin(p\pi z/d)Re(c(H, mnp)exp(j(m\phi - \omega(H, mnp)t))$$

$$= \sum_{mnp} Re(c(H, mnp)exp(-j\omega(Hmnp)t).v_{mnp}(\mathbf{r}))$$

From these expressions, we can easily derive the corresponding expansions for the tangential components of the electromagnetic field in the time domain:

$$E_\perp = \sum_{mnp}(1/h(Emn)^2)Re(c(Emnp)exp(-j\omega(Emnp)t)\partial_z u_{mnp}(\mathbf{r}))$$

$$-\sum_{mnp}(\mu/h(Hmn)^2)Re((j\omega(Hmnp)c(Hmnp)exp(-j\omega(Hmnp)t))\nabla_\perp v_{mnp}(\mathbf{r})\times\hat{z})$$

and likewise,

$$H_\perp = \sum_{mnp}(1/h(Hmn)^2)Re(c(Hmnp)exp(-j\omega(Hmnp)t)\partial_z v_{mnp}(\mathbf{r}))$$

$$+\sum_{mnp}(\epsilon/h(Hmn)^2)Re((j\omega(Emnp)c(Emnp)exp(-j\omega(Emnp)t))\nabla_\perp u_{mnp}(\mathbf{r})\times\hat{z})$$

Note that the characteristic frequencies of oscillation of the TM modes are

$$\omega(Emnp) = (\mu\epsilon)^{-1/2}(\alpha_m[n]^2/a^2 + (\pi p/d)^2)^{1/2},$$

and those for the TE mode are

$$\omega(Hmnp) = (\mu\epsilon)^{-1/2}(\beta_m[n]^2/a^2 + (\pi p/d)^2)^{1/2},$$

Thus, we can write

$$\mathbf{E}(t,\rho,\phi,z) = \mathbf{E}(t,\mathbf{r}) = \sum_{mnp} Re(c(Emnp)exp(-j\omega(Emnp)t)\psi_{Emnp}(\mathbf{r}))$$

$$+\sum_{mnp} Re(c(Hmnp)exp(-j\omega(Hmnp)t)\chi_{Emnp}(\mathbf{r}))$$

where now $\psi_{Emnp}(\mathbf{r})$ and $\chi_{Emnp}(\mathbf{r})$ are $\mathbb{C}^3$-vector valued complex functions of the position variable. The first summation above is the contribution to the total electric field coming from the $TM$ components while the second summation is the contribution to the total electric field coming from the $TE$ components. Likewise,

$$\mathbf{H}(t,\rho,\phi,z) = \mathbf{H}(t,\mathbf{r}) = \sum_{mnp} Re(c(Emnp)exp(-j\omega(Emnp)t)\psi_{Hmnp}(\mathbf{r}))$$

$$+\sum_{mnp} Re(c(Hmnp)exp(-j\omega(Hmnp)t)\chi_{Hmnp}(\mathbf{r}))$$

Note that

$$u_{mnp}(\mathbf{r}) = cos(p\pi z/d).exp(jm\phi)J_m(\alpha_m[n]\rho/a)$$

$$v_{mnp}(\mathbf{r}) = sin(p\pi z/d).exp(jm\phi).J_m(\beta_m[n]\rho/a)$$

Thus,

$$\partial_z u_{mnp}(\mathbf{r}) = -(p\pi/d)sin(p\pi z/d)J_m(\alpha_m[n]\rho/a)exp(jm\phi)$$

The vector valued functions $\psi_{Emnp}, \chi_{Emnp}, \psi_H(mnp), \chi_H(mnp)$ possess the usual orthogonality properties. After appropriate normalization of these functions, using the orthonormality of the vector valued functions, we can express the total energy in the electromagnetic field within the cavity $C$ as

$$H_F = (\epsilon/2)\int_C |E(t,\mathbf{r})|^2 d^3r + (\mu/2)\int_{cavity} |H(t,\mathbf{r})|^2 d^3r$$

$$= \sum_{mnp}[\omega(Emnp)c(Emnp)^* c(Emnp) + \omega(Hmnp)c(Hmnp)^* c(Hmnp)]$$

and hence it is clear that after quantization, in order to obtain the correct time dependence of the coefficients, ie $c(Emnp,t)$ should vary with time as $exp(-i\omega(Emnp)t)$ while $c(Hmnp,t)$ should vary as $exp(-i\omega(Hmnp)t)$, we must enforce the Bosonic commutation relations

$$[c(Emnp), c(Em'n'p')^*] = \delta[m-m']\delta[n-n']\delta[p-p'],$$

$$[c(Hmnp), c(Hm'n'p')^*] = \delta[m-m']\delta[n-n']\delta[p-p'],$$

with all the other commutators vanishing. We then easily obtain using these commutation relations that

$$dc(Emnp,t)/dt = i[H_F, c(Emnp,t)] = -i\omega(Emnp)c(Emnp,t),$$

$$dc(Hmnp,t) = i[H_F, c(Hmnp,t)] = -i\omega(Hmnp,t)c(Hmnp,t)$$

on solving which the desired time dependence of these coefficients is obtained. These commutation relations can alternatively be obtained by requiring that the Maxwell equations follow from the Heisenberg equations for the electromagnetic field operators. We shall now briefly discuss the computation of the far field radiation pattern. The surface current density induced on the sidewalls of the cavity is given by

$$\mathbf{J}_s^s(t,\phi,z) = -\hat{\rho} \times \mathbf{H}(t,a,\phi,z)$$

$$= -H_\phi(t,a,\phi,z)\hat{z} + H_z(t,a,\phi,z)\hat{\phi}$$

that induced on the bottom surface is

$$\mathbf{J}_s^b(t,\rho,\phi) = \hat{z} \times \mathbf{H}(t,\rho,\phi,0) =$$

$$= H_\rho(t,\rho,\phi,0)\hat{\phi} - H_\phi(t,\rho,\phi,0)\hat{\rho}$$

and that on the top surface is

$$\mathbf{J}_s^t(t,\rho,\phi,d) = -H_\rho(t,\rho,\phi,d)\hat{\phi} + H_\phi(t,\rho,\phi,d)\hat{\rho}$$

By using the formulas

$$\hat{\rho} = \hat{x}.cos(\phi) + \hat{y}.sin(\phi), \hat{\phi} = -\hat{x}.sin(\phi) + \hat{y}.cos(\phi)$$

we can evaluate the far field radiated magnetic vector potential as

$$\mathbf{A}(\omega,\mathbf{r}) = (\mu/4\pi)(exp(-jkr)/r)\int_{SC} \mathbf{J}_s(\omega,\mathbf{r}')exp(jk\hat{r}.\mathbf{r}')dS(\mathbf{r}')$$

where $SC$ denotes the boundary surface of the CDRA. The far field quantum electromagnetic field radiated by the CDRA is easily seen to be a linear function of $c(Emnp), c(Emnp)^*, c(Hmnp), c(Hmnp)^*, m,n,p \in \mathbb{Z}$. This is because, the magnetic field is linear in these observables and hence the surface current density on the CDRA is also linear in these observables. Thus, we can express the radiated electromagnetic field in the form

$$\mathbf{E}(t,\mathbf{r}) = \sum_{mnp}[c(Emnp)\mathbf{F}_1(mnp,\mathbf{r})exp(-i\omega(Emnp)t)$$

$$+c(Emnp)^*\bar{\mathbf{F}}_1(mnp,\mathbf{r})exp(i\omega(Emnp)t)]$$

$$+\sum_{mnp}[c(Hmnp)\mathbf{F}_2(mnp,\mathbf{r}).exp(-i\omega(Hmnp)t)+$$

$$c(Hmnp)^*\bar{\mathbf{F}}_2(mnp,\mathbf{r}).exp(i\omega(Hmnp)t)]$$

and likewise for the radiated magnetic field:

$$\mathbf{H}(t,\mathbf{r}) = \sum_{mnp}[c(Emnp)\mathbf{G}_1(mnp,\mathbf{r})exp(-i\omega(Emnp)t)$$

$$+c(Emnp)^*\bar{\mathbf{G}}_1(mnp,\mathbf{r})exp(i\omega(Emnp)t)]$$

$$+\sum_{mnp}[c(Hmnp)\mathbf{G}_2(mnp,\mathbf{r}).exp(-i\omega(Hmnp)t)+$$
$$c(Hmnp)^*\bar{\mathbf{G}}_2(mnp,\mathbf{r}).exp(i\omega(Hmnp)t)]$$

where $\mathbf{F}_k, \mathbf{G}_k, k = 1,2$ are complex $3 \times 1$ vector valued functions of position only. The far field electric field pattern has only the $1/r$ dependence and hence we can express it in the form

$$\mathbf{E}(t,\mathbf{r}) = r^{-1}\sum_{mnp}[c(Emnp)exp(-i\omega(Emnp)(t-r/c))\mathbf{Q}_1(mnp,\hat{r}))+$$
$$c(Emnp)^*.exp(i\omega(Emnp)(t-r/c))\bar{\mathbf{Q}}_1(mnp,\hat{r})]$$
$$+r^{-1}\sum_{mnp}[c(Hmnp)exp(-i\omega(Hmnp)(t-r/c))\mathbf{Q}_2(mnp,\hat{r})+$$
$$c(Hmnp)^*.exp(i\omega(Emnp)(t-r/c))\bar{\mathbf{Q}}_2(mnp,\hat{r})]$$

with a similar expression for the far field magnetic field pattern. In fact, using the Maxwell equation

$$curlE = -j\omega\mu H$$

it easily follows that the far field magnetic field pattern is given by

$$\mathbf{H}(t,\mathbf{r}) = r^{-1}\eta^{-1}.\sum_{mnp}[-c(Emnp)exp(-i\omega(Emnp)(t-r/c))\hat{r}\times\mathbf{Q}_1(mnp,\hat{r}))+$$
$$c(Emnp)^*.exp(i\omega(Emnp)(t-r/c))\hat{r}\times\bar{\mathbf{Q}}_1(mnp,\hat{r})]$$
$$+r^{-1}\eta^{-1}\sum_{mnp}[-c(Hmnp)exp(-i\omega(Hmnp)(t-r/c))\hat{r}\times\mathbf{Q}_2(mnp,\hat{r})+$$
$$c(Hmnp)^*.exp(i\omega(Emnp)(t-r/c))\hat{r}\times\bar{\mathbf{Q}}_2(mnp,\hat{r})]$$

where

$$\eta = \sqrt{\mu/\epsilon}$$

From these expressions, it is immediate that the far field time averaged quantum Poynting vector field pattern is given by

$$\mathbf{P}(\mathbf{r}) = (2\hat{r}/\eta)r^{-2}[\sum_{mnp}c(Emnp)^*c(Emnp)|\mathbf{Q}_1(mnp,\hat{r})|^2$$
$$+c(Hmnp)^*c(Hmnp)|\mathbf{Q}_2(mnp,\hat{r})|^2]$$

We can calculate the quantum average of the fields and Poynting vector in any state of the photons, for example in a coherent state of the CDRA. In such a state $|\phi(u)>$, we have

$$<\phi(u)|c(Emnp)|\phi(u)>= u(Emnp),$$
$$<\phi(u)|c(Emnp)^*|\phi(u)>= \bar{u}(Emnp)$$

where $u = ((u(Emnp), u)(Hmnp)))$ is an infinite dimensional complex vector that parametrizes the state of the photons in the quantum electromagnetic field within the CDRA. In order to calculate the covariance of quantum fluctuations in the electromagnetic field in a coherent state, or the average value the Poynting vector in a coherent state or more generally, the higher order moments of the field in a coherent state, we require identities like

$$< \phi(u)|c(Emnp)^* c(Emnp)|\phi(u) >= |u(Emnp)|^2$$

$$< \phi(u)|\Pi_{k=1}^{r} c(Em_k n_k p_k)^{*q_k} \Pi_{l=1}^{s} c(Em'_l n'_l p'_l)^{t_l}|\phi(u) >=$$

$$\Pi_{k=1}^{r} \bar{u}(Em_k n_k p_k)^{q_k} \Pi_{l=1}^{s} u(Em'_l n'_l p'_l)^{t_l}$$

These identities should be combined with the commutation rules to ensure that the moments of these field creation and annihilation operators taken in any order can be expressed as linear combinations of the moments of these operators in the normal order, ie, in each term, all the creation operators appear to the left of all the annihilation operators. It should be noted that these moments can also be computed in a state of the field corresponding to a finite number of photons in prescribed number states. For example the state

$$|\psi >= |N_1(m_1 n_1 p_1), ..., N_k(m_k n_k p_k) >= \bigotimes_{j=1}^{k} |N_j(m_j n_j p_j) >$$

represents a state in which the $(m_j n_j p_j)^{th}$-mode photon is in the $N_j^{th}$ number state. Thus, we have

$$c(mnp)|N_j(m_j n_j p_j) >= \delta[m-m_j]\delta[n-n_j]\delta[p-p_j]\sqrt{N_j(m_j n_j p_j)}|N_j(m_j n_j p_j)-1 >$$

and

$$c(mnp)^*|N_j(m_j n_j p_j) >= |1(mnp), N_j(m_j n_j p_j) >$$

provided that $(mnp) \neq (m_j n_j p_j)$ and otherwise

$$c(m_j n_j p_j)^*|N_j(m_j n_j p_j) >= \sqrt{N_j(m_j n_j p_j) + 1}|N_j(m_j n_j p_j) + 1 >$$

Exercise: Using the above formulae, calculate the expected value of

$$\Pi_{j=1}^{r} c(Ep_j q_j r_j)^{*s_j} \Pi_{k=1}^{m} c(Ep'_k q'_k r'_k)^{t_k}$$

in the above state $|\psi >$.

## 10.3 Interaction between electrons, positrons and photons within a cylindrical dielectric resonator antenna

The free Dirac field is

$$i\partial_t \psi(t,r) = ((\alpha, -i\nabla) + \beta m)\psi(t,r)$$

where $\alpha_1, \alpha_2, \alpha_3, \beta$ are the Dirac $4 \times 4$ matrices mutually anticommuting with each having the identity as its square. $\psi(t, r)$ is a 4-component operator wave field. We must solve this equation within the cavity with the boundary condition that $\psi$ vanishes on the boundary surface of the cavity. Thus, we take our basis functions as

$$\chi_{mnp}(\mathbf{r}) = N(mnp)J_m(\alpha_m[n]\rho/a)exp(im\phi)sin(p\pi z/d), (mnp) \in \mathbb{Z}^3$$

where $N(mnp)$ are appropriate normalizing constants. We expand the wave field as

$$\psi(t, r) = \sum_{mnp\sigma} [a(mnp\sigma)\chi_{mnp}(\mathbf{r})\mathbf{u}(mnp\sigma)exp(-i\omega(1mnp)t)+$$

$$b(mnp\sigma)^*\bar{\chi}_{mnp}(\mathbf{r})\mathbf{v}(mnp\sigma)exp(i\omega(2mnp)t)]$$

Using the orthonormality of the functions $\chi_{mnp}$ and their gradients, we find that the condition that $\psi$ satisfy the Dirac equation is that

$$\omega(1mnp)\mathbf{u}(mnp\sigma) = [\int_C \bar{\chi}_{mnp}(\mathbf{r})(\alpha, -i\nabla\chi_{mnp}(\mathbf{r}))d^3r + \beta m]\mathbf{u}(mnp\sigma)$$

and

$$-\omega(2mnp)\mathbf{v}(mnp\sigma) = [\int_C \chi_{mnp}(\mathbf{r})(\alpha, -i\nabla\bar{\chi}_{mnp}(\mathbf{r}))d^3r + \beta m]\mathbf{v}(mnp\sigma)$$

This implies that on defining

$$\mathbf{P}(mnp) = \int_C \bar{\chi}_{mnp}(\mathbf{r})(-i\nabla\chi_{mnp}(\mathbf{r})d^3r$$

which implies by integration by parts that $\mathbf{P}(mnp)$ is a real vector that

$$\omega(1mnp)\mathbf{u}(mnp\sigma) = [(\alpha, \mathbf{P}(mnp)) + \beta m]\mathbf{u}(mnp\sigma),$$

$$-\omega(2mnp)\mathbf{v}(mnp\sigma) = [-(\alpha, \mathbf{P}(mnp)) + \beta m]\mathbf{v}(mnp\sigma),$$

from which it follows that $\omega(2mnp) = \omega(1mnp) = \omega(mnp)$ being the positive eigenvalue of the $4 \times 4$ matrix $(\alpha, \mathbf{P}(mnp)) + \beta m$, ie,

$$\omega(mnp) = \sqrt{m^2 + \mathbf{P}(mnp)^2}$$

Then, we have that $\mathbf{u}(mnp\sigma), \sigma = 1, 2$ are two orthonormal eigenvectors of $(\alpha, \mathbf{P}(mnp)) + \beta m$ corresponding to the eigenvalue $\omega(mnp)$ while $\mathbf{v}(mnp\sigma)$ are two orthonormal eigenvectors of $(\alpha, \mathbf{P}(mnp)) - \beta m$ corresponding to the same eigenvalue.

When the quantum electromagnetic field within the resonator cavity interacts with the electron-positron field within, the total second quantized Hamiltonian is given by

$$H(t) = H_F + H_D + H_I(t)$$

where

$$H_F = \sum_{mnp} [\omega(Emnp)c(Emnp)^* c(Emnp) + \omega(Hmnp)c(Hmnp)^* c(Hmnp)]$$

is the Hamiltonian of the free constrained electromagnetic field,

$$H_D = \int \psi(\mathbf{r})^*((\alpha, -i\nabla) + \beta m)\psi(\mathbf{r})d^3r$$

$$= \sum_{mnp\sigma} \omega(mnp)(a(mnp\sigma)^* a(mnp\sigma) + b(mnp\sigma)^* b(mnp\sigma))$$

is the Hamiltonian of the free constrained Dirac field and

$$H_I = e\int \psi(\mathbf{r})^*(\alpha, \mathbf{A}(t,\mathbf{r}))\psi(\mathbf{r})d^3r$$

is the interaction Hamiltonian between the Dirac field of electrons and positrons. We can express it in terms of the photon and electron-positron creation and annihilation operators as follows. First note that since $divE = 0$ within the resonator and we are adopting the Coulomb gauge for which $divA = 0$, it follows from the Maxwell theory that $\nabla^2\Phi = 0$ and hence $\Phi = 0$ where $\Phi$ is the electric scalar potential. This is because the equation $divE = 0$ implies that there is zero charge density inside. Hence the electric field and magnetic vector potential are related by

$$E(t, \mathbf{r}0 = -\partial_t A(t, \mathbf{r})$$

whence

$$\mathbf{A}(t,\mathbf{r}) = \sum_{mnp} Re((c(Emnp)/j\omega(Emnp))exp(-j\omega(Emnp)t)\psi_{Emnp}(\mathbf{r}))$$

$$+\sum_{mnp} Re((c(Hmnp)/j\omega(Hmnp))exp(-j\omega(Hmnp)t)\chi_{Emnp}(\mathbf{r}))$$

or more precisely, in operator theoretic notation,

$$\mathbf{A}(t,\mathbf{r}) =$$

$$\sum_{mnp}[(c(Emnp)/2j\omega(Emnp))exp(-j\omega(Emnp)t)\psi_{Emnp}(\mathbf{r})$$

$$-(c(Emnp)^*/2j\omega(Hmnp))exp(j\omega(Emnp)t)\bar{\psi}_{Emnp}(\mathbf{r})]$$

$$+\sum_{mnp}[(c(Hmnp)/2j\omega(Hmnp))exp(-j\omega(Hmnp)t)\chi_{Emnp}(\mathbf{r})$$

$$-(c(Hmnp)^*/2j\omega(Hmnp))exp(j\omega(Hmnp)t)\bar{\chi}_{Emnp}(\mathbf{r})]$$

The interaction Hamiltonian can therefore be expressed in the form

$$H_I(t) =$$

$$e\int[a(mnp\sigma)\chi_{mnp}(\mathbf{r})\mathbf{u}(mnp\sigma)exp(-i\omega(1mnp)t)+b(mnp\sigma)^*\bar{\chi}_{mnp}(\mathbf{r})$$
$$\mathbf{v}(mnp\sigma)exp(i\omega(2mnp)t)]^*$$
$$.[(\alpha,\sum_{mnp}[(c(Emnp)/2j\omega(Emnp))exp(-j\omega(Emnp)t)\psi_{Emnp}(\mathbf{r})$$
$$-(c(Emnp)^*/2j\omega(Hmnp))exp(j\omega(Emnp)t)\bar{\psi}_{Emnp}(\mathbf{r})]$$
$$+\sum_{mnp}[(c(Hmnp)/2j\omega(Hmnp))exp(-j\omega(Hmnp)t)\chi_{Emnp}(\mathbf{r})$$
$$-(c(Hmnp)^*/2j\omega(Hmnp))exp(j\omega(Hmnp)t)\bar{\chi}_{Emnp}(\mathbf{r})])]$$
$$.[a(mnp\sigma)\chi_{mnp}(\mathbf{r})\mathbf{u}(mnp\sigma)exp(-i\omega(1mnp)t)+b(mnp\sigma)^*\bar{\chi}_{mnp}(\mathbf{r})$$
$$\mathbf{v}(mnp\sigma)exp(i\omega(2mnp)t)]d^3r$$

This interaction Hamiltonian is therefore expressible in the form

$$H_I(t)=\sum_{m_1n_1p_1\sigma_1m_2n_2p_2m_3n_3p_3\sigma_3}a(m_1n_1p_1\sigma_1)^*c(Em_2n_2p_2)$$
$$a(m_3n_3p_3\sigma_3)f_1(m_1n_1p_1\sigma_1m_2n_2p_2m_3n_3p_3\sigma_3,t)+$$
$$a(m_1n_1p_1\sigma_1)^*c(Em_2n_2p_2)b(m_3n_3p_3\sigma_3)^*f_2(m_1n_1p_1\sigma_1m_2n_2p_2m_3n_3p_3\sigma_3,t)+$$
$$+b(m_1n_1p_1\sigma_1)c(Em_2n_2p_2)a(m_3n_3p_3\sigma_3)f_3(m_1n_1p_1\sigma_1m_2n_2p_2m_3n_3p_3\sigma_3,t)+$$
$$b(m_1n_1p_1\sigma_1)c(Em_2n_2p_2)b(m_3n_3p_3\sigma_3)^*f_4(m_1n_1p_1\sigma_1m_2n_2p_2m_3n_3p_3\sigma_3,t)+$$
$$a(m_1n_1p_1\sigma_1)^*c(Hm_2n_2p_2)a(m_3n_3p_3\sigma_3)f_5(m_1n_1p_1\sigma_1m_2n_2p_2m_3n_3p_3\sigma_3,t)+$$
$$a(m_1n_1p_1\sigma_1)^*c(Hm_2n_2p_2)b(m_3n_3p_3\sigma_3)^*f_6(m_1n_1p_1\sigma_1m_2n_2p_2m_3n_3p_3\sigma_3,t)+$$
$$+b(m_1n_1p_1\sigma_1)c(Hm_2n_2p_2)a(m_3n_3p_3\sigma_3)f_7(m_1n_1p_1\sigma_1m_2n_2p_2m_3n_3p_3\sigma_3,t)+$$
$$b(m_1n_1p_1\sigma_1)c(Hm_2n_2p_2)b(m_3n_3p_3\sigma_3)^*f_8(m_1n_1p_1\sigma_1m_2n_2p_2m_3n_3p_3\sigma_3,t)+$$
$$+H.c$$

where $H.c$ denotes Hermitian conjugate of the previous eight terms. It should be noted that the operator $a(mnp\sigma)(b(mnp\sigma)$ annihilates an electron (positron) of spin $\sigma$ at the $(mnp)^{th}$ node of the Dirac field while $a(mnp\sigma)^*(b(mnp\sigma)^*$ creates an electron (positron) of spin $\sigma$ at the $(mnp)^{th}$ mode. These operators commute with the photon annihilation and creation operators $c(Em'n'p')$,

$$c(Hm'n'p'),c(Em'n'p')^*,c(Hm'n'p')^*$$

and they satisfy the canonical anticommutation relations amongst themselve:

$$\{a(mnp\sigma),a(m'n'p'\sigma')^*\}=\delta[m-m']\delta[n-n']\delta[p-p']\delta(\sigma,\sigma')$$
$$\{b(mnp\sigma),b(m'n'p'\sigma')^*\}=\delta[m-m']\delta[n-n']\delta[p-p']\delta(\sigma,\sigma')$$
$$\{a(mnp\sigma),b(m'n'p'\sigma')\}=0,$$
$$\{b(mnp\sigma),b(m'n'p'\sigma')\}=0,$$

This interaction Hamiltonian between the electrons-positrons and photons can be used to calculate scattering, absorption and emission probabilities. However, if we are interested in the state of the photon field on the retinal screen, we must proceed as follows: First, at time $t=0$, let $\rho(0)=\rho_{ph}(0)\otimes\rho_{ep}(0)$ denote the state of the photons and electron-positrons. After time $t$, this state evolves under the Hamiltonian $H(t)$ to the state

$$\rho(t) = U(t)(\rho_{ph}(0) \otimes \rho_{ep}(0))U(t)^*$$

in the interaction picture where

$$U(t) = T\{exp(-i \int_0^t \tilde{H}_I(s)ds)\}$$

Let $D$ denote the two dimensional region spanned by the retinal screen. The moments of the electric field on the screen at time $t$ are then computed using the formula

$$Tr(\rho(t). \bigotimes_{k=1}^{n} E_{i_k}(t, \mathbf{r}_k))$$

Evaluating this is a hard task but we can obtain approximate formulas for this by expressing $U(t)$ as a Dyson series and then truncating this series to an appropriate number of terms.

## 10.4 Other models for the interaction between light and matter

Let $B^a_\mu(x)$ be non-Abelian gauge fields and $A_\mu(x)$ be the photon field. The matter wave function $\psi(x)$ satisfies the matter wave equation

$$[i\gamma^\mu \nabla_\mu - m]\psi(x) = 0$$

where

$$\nabla_\mu = \partial_\mu - ieA_\mu - ieB^a_\mu \tau_a$$

with $\tau_a$ being the Hermitian generators of the gauge group $G \subset U(n)$. In turn, the matter currents associated with the photon field and gauge field are

$$J^\mu(x) = -e\psi(x)^* \gamma^0 \gamma^\mu \psi(x),$$

$$J^\mu_a(x) = -e\psi(x)^*(\gamma^0\gamma^\mu \otimes \tau_a)\psi(x)$$

and the matter field equations are

$$\partial_\nu F^{\mu\nu} = J^\mu,$$

$$(D_\nu F^{\mu\nu})_a = J^\mu_a$$

where

$$D_\nu F^{\mu\nu}_a = \partial_\nu F^{\mu\nu}_a + eC(abc)B^b_\nu F^{\mu\nu c}$$

with $C(abc)$ denoting the structure constants of the gauge group and

$$F^{\mu\nu a} = B^a_{\nu,\mu} - B^a_{\mu,\nu} + eC(abc)B^b_\mu B^c_\nu$$

We write

$$B^a_\mu = B^{0a}_\mu + \delta B^a_\mu$$

where $B^{0a}_\mu$ is the classical background Yang-Mills gauge field assumed to be classical and $\delta B^a_\mu$ is the small quantum fluctuation in this field. We then express the Lagrangian density of the Yang-Mills gauge field as

$$F^a_{\mu\nu}F^{\mu\nu a} = (F^{0a}_{\mu\nu} + \delta F^a_{\mu\nu}).(F^{0\mu\nu a} + \delta F^{\mu\nu a})$$

where

$$F^{0a}_{\mu\nu} = B^{0a}_{\nu,\mu} - B^{0a}_{\mu,\nu} + eC(abc)B^{0b}_\mu B^{0c}_\nu$$

is the classical component, ie, background Yang-Mills field tensor and

$$\delta F^a_{\mu\nu} = \delta B^a_{\nu,\mu} - \delta B^a_{\mu,\nu} + eC(abc)(B^{0b}_\mu \delta B^c_\nu + \delta B^b_\mu.B^{0c}_\nu) + eC(abc)\delta B^b_\mu \delta B^c_\nu$$

is the purely quantum component of the Yang-Mills field tensor. The linear-quadratic part of the Lagrangian density of the Yang-Mills gauge field is given by

$$L_{quadYM} = (\delta B^a_{\nu,\mu} - \delta B^a_{\mu,\nu}).(\delta B^{\nu,\mu a} - \delta B^{\mu,\nu a})$$

$$+2F^{0a}_{\mu\nu}(\delta B^a_{\nu,\mu} - \delta B^a_{\mu,\nu})$$

$$+2C(abc)F^{0a}_{\mu\nu}\delta B^b_\mu \delta B^c_\nu$$

We can easily compute the propagator of this quantum component of the Lagrangian density. In fact, if we denote the fields $\delta B^a_\mu(x)$ by $\phi_k(x), k = 1, 2, ..., N$, then this quadratic part of the Lagrangian density has the general form

$$L_q = \int K_{ij}(x,y)\phi_i(x)\phi_j(y)d^4xd^4y + \int M_i(x)\phi_i(x)d^4x$$

and we can compute easily the propagator of $\phi$ as a Gaussian field:

$$D_\phi(x,y) = \int exp(iS_q)D\phi =$$

$$(det(iK/2))^{1/2}exp((i/4)\int M_i(x)K_{ij}(x,y)M_j(y)d^4xd^4y)$$

We can now analyze the effects of cubic and four degree terms on the gauge field propagator using perturbation theory for path integrals according to the method laid out by Richard Feynman: We write the cubic and quadratic perturbations to the above Lagrangian density as

$$\delta L_q = \int P_{ijk}(x,y,z)\phi_i(x)\phi_j(y)\phi_k(z)d^4xd^4yd^4z$$

$$+\int Q_{ijkm}(x,y,z,v)\phi_i(x)\phi_j(y)\phi_k(z)\phi_m(v)^4xd^4yd^4zd^4v$$

Note that these cubic and fourth degree terms represent the terms

$$eC(abc)(\delta B^a_{\nu,\mu} - \delta B^a_{\mu,\nu})(\delta B^{\mu b}\delta B^{\nu c})$$

$$+e^2C(abc)C(apq)\delta B^b_\mu \delta B^c_\nu \delta B^{\mu p}\delta B^{\nu q}$$

Suppose that in principle, we have derived an expression for this gauge propagator. We can then as the question, what are the moments of the gauge field on the retinal screen for an initial state $\rho(0)$ of the matter plus gauge field ? As before, $\psi$ denotes the matter field and $\phi$ the gauge field. The total Lagrangian of the matter plus gauge field has the form

$$L = \int M(x)^T \phi(x) d^4x + \int \phi(x)^T K(x,y)\phi(y) d^4x d^4y$$

$$+ \int P(x,y,z)^T (\phi(x) \otimes \phi(y) \otimes \phi(z)) d^4x d^4y d^4z$$

$$+ \int Q(x,y,z,v)^T (\phi(x) \otimes \phi(y) \otimes \phi(z) \otimes \phi(v)) d^4x d^4y d^4z d^4v$$

$$+ \int \psi(x)^* R(x,y) \psi(y) d^4x d^4y + \int \psi(x)^* S_k(x) \psi(x) \phi_k(x) d^4x$$

The path integral with time ranging over $[0,T]$ results in an evolution kernel $U_T(\phi_f, \psi_f | \phi_i, \psi_i)$ where $\phi_i, \psi_i$ are gauge and matter fields over space at time $t = 0$ while $\phi_f, \psi_f$ are gauge and matter fields over space at time $t = T$. By space, we mean the spatial volume region of the CDRA. Now suppose that at time $t = 0$, the state of the matter and gauge field is represented by the density matrix kernel $\rho_0(\phi_1, \psi_1; \phi_2, \psi_2)$. Then the state of these fields at time $t = T$ will be given by

$$\rho_T(\phi_1, \psi_1; \phi_2, \psi_2) =$$

$$\int U_T(\phi_1, \psi_1 | \phi_3, \psi_3) \rho_0(\phi_3, \psi_3; \phi_4, \psi_4) \bar{U}_T(\phi_2, \psi_2; \phi_4, \psi_4) D\phi_3 D\psi_3 D\phi_4 D\psi_4$$

Once we know this state at time $T$, we are in a position to calculate the statistical moments of the gauge field on the screen at time $T$.

## 10.5 Quantum mechanics of the fluorescent effect

Suppose photons are incident upon atoms located on the screen. They cause the electrons in the atoms to jump into excited states and when the source of photons is removed, the electrons make a transition from these excited states to lower energy states thereby emitting quanta of radiation. We wish to describe this effect quantum mechanically. Let

$$\mathbf{A}(t,\mathbf{r}) = 2\sum_{mnp} [Re(c(Emnp)exp(-j\omega(Emnp)t)\chi(Emnp,\mathbf{r})) + Re(c(Hmnp)exp(j\omega(Hmnp)t)\chi(Hmnp,\mathbf{r}))]$$

be the magnetic vector potential of the quantum electromagnetic field of photons in the CDRA that is incident upon an atom at the screen. For simplicity of notation, we write this quantum vector potential as

$$\mathbf{A}(t,\mathbf{r}) = 2\sum_{mnp} Re(c(mnp).exp(-j\omega(mnp)t).\chi(mnp,\mathbf{r}))$$

The interaction Hamiltonian between the electron bound to the nucleus of the atom located at $\mathbf{r}$ and this quantum electromagnetic field is given by (in the Dirac picture)

$$H_I(t) = e(\alpha, \mathbf{A}(t,\mathbf{r})) =$$

$$e\sum_{mnp}[(\alpha, \chi(mnp,\mathbf{r}))exp(-j\omega(mnp)t)c(mnp) + e(\alpha, \bar{\chi}(mnp,\mathbf{r})exp(j\omega(mnp)t)c(mnp)^*]$$

The transition probability under this interaction for the electron between two stationary states $|r>$ and $|s>$ in time $[0,T]$ assuming that the photons of the electromagnetic field are in the coherent state $|\phi(u)>$ with $u = ((u(mnp)))$ is given upto first order perturbation theory by

$$P_T(|r> \to |s>) =$$

$$e^2|\int_0^T <s\otimes\phi(u)|H_I(t)|r,\phi(u)> exp(i(E(mn)t))dt|^2$$

$$= e^2|\sum_{mnp}\int_0^T [<r|(\alpha, \chi(mnp,\mathbf{r}))u(mnp)exp(-j\omega(mnp)t) + (\alpha, \bar{\chi}(mnp,\mathbf{r}))\bar{u}(mnp)$$
$$exp(j\omega(mnp)t)|s>]exp(iE(mn)t)dt|^2$$

where

$$E(mn) = E(m) - E(n)$$

is the difference between the two energy levels of the electron's stationary states in units in which $h/2\pi = 1$. For obtaining the Fluorescence pattern, we assume that initially, we have a certain number distribution of electrons $N(n)$ in the different stationary states $|n>$ and after the excitation by the quantum electromagnetic field, this number distribution gets altered to

$$\tilde{N}(m) = \sum_n N(n)P_T(|n> \to |m>)$$

Then, assuming that all these excited electrons after the excitation is removed jump back to the ground state, it follows that the resulting emitted radiation will have an intensity distribution proportional to $\tilde{N}(m)$ at the frequency $\omega(m) = (E(m) - E(0))/h$. This distribution can be computed for different coherent states $|\phi(u)>$. We can also consider a situation in which the photons are in superpositions of coherent states. For example, suppose the photon state is specified by giving the number of photons in each mode $|\psi> = |N(m_k n_k p_k), k = 1,2,...,L>$. Then to evaluate the above transition probability, we require the evaluation of the matrix elements

$$<\psi|c(mnp)|\psi>$$

This is directly done using the formula

$$c(mnp)|\psi>=\sum_{k=1}^{L}\delta[m-m_k,n-n_k,p-p_k]\sqrt{N(m_kn_kp_k)}|N(m_jn_jp_j),j\neq k, N(m_kn_kp_k)-1>$$

Another way to do this is to express $|\psi>$ as a superposition of coherent states: $|phi(u)>=exp(-|u|^2/2)\sum_{mnp}(\Pi_{mnp}u(mnp)^{N(mnp)}/\sqrt{N(mnp)!})|\{N(mnp):m,n,p\in\mathbb{Z}_+\}>$ and hence

$$<\phi(u)|\psi>=exp(-|u|^2/2)\Pi_{k=1}^{L}u(m_kn_kpp_k)^{N(m_kn_kp_k)}/\sqrt{N(m_kn_kp_k)!}$$

and then using the Glauber-Sudarshan non-orthogonal decomposition of the identity in terms of coherent states,

$$|\psi>=C\int|\phi(u)><\phi(u)|\psi>du$$

$$=C\int|\phi(u)>exp(-|u|^2/2)[\Pi_{k=1}^{L}u(m_kn_kp_k)^{N(m_kn_kp_k)}/\sqrt{N(m_kn_kp_k)!}]du$$

Thus, we can formulate the general problem of computing the matrix element

$$<\phi(u)|c(mnp)|phi(v)>$$

between two coherent states. Indeed, this would enable us to compute

$$<\psi|c(mnp)|\psi>$$

and hence the transition probability for the electron when the photons are in the state $|\psi>$ where $|\psi>$ is expressed as a superposition of coherent states:

$$|\psi>=\int\chi(u)|u>du$$

We get

$$<\phi(u)|c(mnp)|\phi(v)>=exp(-|u|^2/2-|v|^2/2+<u|v>)v(mnp)$$

Now consider light interacting with an array of $M$ atoms located on the screen at the spatial points $\mathbf{r}_k, k=1,2,...,M$ with the photon field in the coherent state $|\phi(u)>$. The transition probability of the electrons in these atoms for the transitions $|r_k>\to|s_k>, k=1,2,...,M$ is given by

$$P_T(|r_1,...,r_M>\to|s_1,...,s_M>)$$

$$=\Pi_{k=1}^{M}e^2|\int_0^T<s_k,\phi(u)|H_I(t)|r_k,\phi(u)>exp(iE(s_kr_k)t)dt|^2$$

$$e^{2M}\Pi_{k=1}^{M}|\int_0^T \sum_{mnp}[< s_k|u(mnp)(\alpha,\chi(mnp,\mathbf{r}_k))exp(-j\omega(mnp)t)+$$

$$\bar{u}(mnp)(\alpha,\bar{\chi}(mnp,\mathbf{r}_k))exp(j\omega(mnp)t)|r_k > exp(iE(s_k r_k)t)]dt|^2$$

provided that we assume that there is no interaction between the electrons in the different atoms. A more accurate way to model fluorescence on the screen would be to assume that the electrons and positrons on the screen are described by a second quantized wave operator field $\psi(t,r), r \in D$ and to consider the interaction between this field and the photon field as being described by the second quantized interaction Hamiltonian

$$H_I(t) = e\int_D \psi(t,r)^*(\alpha, A(t,r))\psi(t,r)d^3r$$

In this expression, both $A, \psi$ are field operators. We then use this interaction Hamiltonian to compute the probability of transition of the electron-positron field on the screen between two states, the initial one being one in which there are $M$ electrons and positrons with specified four momenta and spins and the final one being one in which there are $M'$ electrons and positrons with specified momenta and spins. From this data, it is easy to see how fluorescence can be computed by looking at the probability distribution of the number of electrons, positrons and their energies after they have interacted with the photons in the coherent state.

## 10.6 On the computation of the directivity of a quantum antenna

Let us write the Fermionic Dirac wave operator field as

$$\psi(t,r) = \sum_{mnp\sigma}[a(mnp\sigma)exp(-i\omega(mnp)t)\mathbf{u}(mnp\sigma)\chi(mnp\sigma,\mathbf{r})+b(mnp\sigma)^*exp(i\omega(mnp)t)\bar{\chi}(mnp\sigma,\mathbf{r})\mathbf{v}(mnp\sigma)]$$

The Dirac four current density corresponding to this Fermionic wave field is given by

$$J^\mu(t,\mathbf{r}) = -e\psi(t,\mathbf{r})^*\alpha^\mu\psi(t,\mathbf{r}), \alpha^\mu = \gamma^0\gamma^\mu$$

and the electromagnetic four quantum potential of radiation generated by this current is given by the standard retarded potential formula

$$A^\mu(t,\mathbf{r}) = K.\int_C J^\mu(t-|r-r'|/c,r')d^3r'/|r-r'|, K = \mu/4\pi$$

The far field magnetic vector potential is then given by

$$(A^r(t,r) : r = 1,2,3) = \mathbf{A}(t,\mathbf{r}) = (K/r).\int_C \mathbf{J}(t-r/c+\hat{r}.r'/c,r')d^3r'$$

and in the frequency domain, this can be expressed as

$$\mathbf{A}(\omega,\mathbf{r}) = (K/r)\int \mathbf{J}(\omega,\mathbf{r}')exp(j(\omega/c)\hat{r}.\mathbf{r}')d^3r'$$

We evaluate the Fourier transform of the current density operator field:

$$\mathbf{J}(\omega,\mathbf{r}) = -e\int \psi(\omega'-\omega,\mathbf{r})^*\alpha.\psi(\omega',\mathbf{r})d\omega'$$

where

$$\alpha = (\alpha^r, r = 1,2,3)$$

and we have made use of the convolution theorem for Fourier transforms. We have that

$$\psi(\omega,\mathbf{r}) = \sum_{mnp\sigma}[a(mnp\sigma)\mathbf{u}(mnp\sigma)\delta(\omega-\omega(mnp))\chi(mnp,\mathbf{r})+b(mnp\sigma)\mathbf{v}(mnp\sigma)\delta(\omega+\omega(mnp)\bar{\chi}(mnp,\mathbf{r})]$$

and hence we get

$$\mathbf{J}(\omega,\mathbf{r}) = -e\int d\omega'[\sum_{mnp\sigma}[a(mnp\sigma)^*\bar{u}(mnp\sigma)\bar{\chi}_{mnp}(r)\delta(\omega'-\omega-\omega(mnp))+$$

$$b(mnp\sigma)\bar{v}(mnp\sigma)\chi_{mnp}(r)\delta(\omega'-\omega+\omega(mnp)]$$

$$\times[\sum_{mnp\sigma}[a(mnp\sigma)u(mnp\sigma)\chi_{mnp}(r)\delta(\omega'-\omega(mnp))+$$

$$b(mnp\sigma)^*v(mnp\sigma)\bar{\chi}_{mnp}(r)\delta(\omega'+\omega(mnp)]$$

$$= -e\sum_{mnp\sigma m'n'p'\sigma'}[a(mnp\sigma)^*a(m'n'p'\sigma')\bar{u}(mnp\sigma)u(m'n'p'\sigma')$$

$$\bar{\chi}_{mnp}(r)\chi_{m'n'p'}(r)\delta(\omega(m'n'p')-\omega(mnp)-\omega)$$

$$+a(mnp\sigma)^*b(m'n'p'\sigma')^*\bar{u}(mnp\sigma)v(m'n'p'\sigma')\bar{\chi}_{mnp}(r)$$

$$\bar{\chi}_{m'n'p'}(r)\delta(\omega(m'n'p')+\omega(mnp)+\omega)$$

$$+b(mnp\sigma)a(m'n'p'\sigma')\bar{v}(mnp\sigma)u(m'n'p'\sigma')$$

$$\chi_{mnp}(r)\bar{\chi}_{m'n'p'}(r)\delta(\omega(m'n'p')+\omega(mnp)-\omega)$$

$$+b(mnp\sigma)b(m'n'p'\sigma')^*\bar{v}(mnp\sigma)v(m'n'p'\sigma')$$

$$\bar{\chi}_{mnp}(r)\bar{\chi}_{m'n'p'}(r)\delta(\omega(mnp)-\omega(m'nnp')-\omega)]$$

It is now a simple matter to evaluate the far field vector potential and hence the far field electromagnetic field and hence also the far field Poynting vector and determine a state $|\eta>$ of the electron-positron field such that if $S(\omega,\mathbf{r}) = P(\omega,\hat{r})\hat{r}/r^2$ denotes the far field Poynting vector, then for a given frequency (ie, one among the set $\omega(mnp)\pm\omega(m'n'p')$) such that $<\eta|P(\omega,\hat{r})|\eta>$ is a maximum for a given direction $\hat{e}ta$. To compute this quantum average, all we require are the matrix elements

$$< \eta|a(mnp\sigma)^*a(m'n'p'\sigma')|\eta >, < \eta|a(mnp\sigma)^*b(m'n'p'\sigma')|\eta >,$$

$$< \eta|b(mnp\sigma)a(m'n'p'\sigma')|\eta >, < \eta|b(mnp\sigma)b(m'n'p'\sigma')^*|\eta >,$$

## 10.7 Average quality factor of a cylindrical dielectric resonator antenna

We consider a quantum electromagnetic field within a CDRA described by an electromagnetic field given by

$$\mathbf{E}(t,\mathbf{r}) = \sum_{mnp} Re(c(Emnp)\psi(Emnp,\mathbf{r})exp(-i\omega(Emnp)t))$$

$$+\sum_{mnp} Re(c(Hmnp)\chi(Emnp),\mathbf{r})exp(-i\omega(Hmnp)t))$$

and likewise,

$$\mathbf{H}(t,\mathbf{r}) = \sum_{mnp} Re(c(Emnp)\psi(Hmnp,\mathbf{r})exp(-i\omega(Emnp)t))$$

$$+\sum_{mnp} Re(c(Hmnp)\chi(Hmnp),\mathbf{r})exp(-i\omega(Hmnp)t))$$

where $c(Emnp)$ and $c(Hmnp)$ are respectively the modal amplitudes of the $z$-components of the electric and magnetic field within the CDRA. We wish to compute the $Q$-factor of such an resonator when the field is in the coherent state $|\phi(u)>$ where $u = ((u(Emnp), u(Hmnp)))$. The total energy of the field within the cavity can be expressed as shown earlier as

$$U = \sum_{mnp}[\omega(Emnp)c(Emnp)^*c(Emnp) + \omega(Hmnp)c(Hmnp)^*c(Hmnp)]$$

and its average in the coherent state $|\phi(u)>$ easily seen to be given by

$$< \phi(u)|U|\phi(u) >= \sum_{mnp}[\omega(Emnp)|u(Emnp)|^2 + \omega(Hmnp)|u(Hmnp)|^2]$$

We now evaluate the average power dissipated in the side walls of the cavity assuming that these sidewalls have a conductivity $\sigma$. The tangential component of the side surface electric field is given by

$$E_T(t,\mathbf{r}) = E_\phi(t,a,\phi,z)\hat{\phi} + E_z(t,a,\phi,z)\hat{z}$$

and this field propagates within the conducting wall region decaying proportional to $exp(-\xi/\delta(\omega))$ where $\xi$ is the distance of the point from the surface and $\delta(\omega)$, the skin depth at frequency $\omega$ is approximately equal to the $1/\alpha$ where $\alpha = \alpha(\omega)$ is the real part of $\sqrt{j\omega\mu(\sigma + j\omega\epsilon)}$ which is approximately equal to $(\omega\mu\sigma/2)^{1/2}$ for $\sigma >> \omega\epsilon$. The total average power dissipated at frequency $\omega$ in the sidewalls is then

$$P_{diss} = \int_0^\infty \int_0^{2\pi} \int_0^d (\sigma/2)|E_T(\omega, a, \phi, z)|^2 exp(-2\xi/\delta(\omega))d\xi.ad\phi dz$$

$$= (\sigma\delta(\omega)/4) \int_0^{2\pi} \int_0^d |E_T(\omega, a, \phi, z)|^2 ad\phi.dz$$

When all the frequencies of the resonator are taken into account, the total power dissipated becomes a sum of the above terms over all $mnp$ with $\omega = \omega(Emnp), \omega(Hmnp)$ and with $E_T(\omega(Emnp), a, \phi, z)$ being obtained by

$$\mathbf{E}_T(\omega(Emnp), a, \phi, z) = \sum_{mnp} c(Emnp)\psi_T(Emnp, a, \phi, z)$$

and

$$\mathbf{E}_T(\omega(Hmnp), a, \phi, z) = \sum_{mnp} c(Hmnp)\chi_T(Emnp, a, \phi, z)$$

It follows therefore that the operator that represents the average power dissipated in the conducting walls of the CDRA is given by

$$P_{diss} = \sum_{mnp} [(\sigma\delta(\omega(Emnp))/4)c(Emnp)^* c(Emnp) \int_0^{2\pi} \int_0^d |\psi_T(Emnp, a, \phi, z)|^2 ad\phi dz +$$

$$\sigma\delta(\omega(Hmnp)/4)c(Hmnp)^* c(Hmnp) \int_0^{2\pi} \int_0^d |\chi_T(Emnp, a, \phi, z)|^2 ad\phi dz$$

In short, the average power dissipated when the bath in in the coherent state $|\phi(u)>$ is given by

$$<\phi(u)|P_{diss}|\phi(u)>=$$

$$\sum_{mnp} [(\sigma\delta(\omega(Emnp))/4)|u(Emnp)|^2 \int_0^{2\pi} \int_0^d |\psi_T(Emnp, a, \phi, z)|^2 ad\phi dz +$$

$$\sigma\delta(\omega(Hmnp)/4)|u(Hmnp)|^2 \int_0^{2\pi} \int_0^d |\chi_T(Emnp, a, \phi, z)|^2 ad\phi dz$$

The Q-factor of the CDRA can now be defined for the $(Emnp)^{th}$ or the $(Hmnp)^{th}$ mode as

$$Q(Emnp) = \frac{\omega(Emnp)}{(2\pi/\omega(Emnp))\sigma\delta(\omega(Emnp))/4) \int_0^{2\pi} \int_0^d |\psi_T(Emnp, a, \phi, z)|^2 ad\phi dz}$$

$$Q(Hmnp) = \frac{\omega(Hmnp)}{(2\pi/\omega(Hmnp))\sigma\delta(\omega(Hmnp))/4)\int_0^{2\pi}\int_0^d |\chi_T(Hmnp, a, \phi, z)|^2 a d\phi dz}$$

namely the ratio of the average energy stored in the CDRA per cycle to the average energy dissipated in the walls per cycle for each mode. The $(Emnp)$-mode represents the $TM_{mnp}$ mode while the $(Hmnp)$-mode represents the $TE_{mnp}$-mode.

## 10.8 Basics of quantum electrodynamics based on the Maxwell-Dirac theory, Bosonic and Fermionic coherent states, Feynman diagrams for calculating amplitudes of processes

[1] Quantization of the electromagnetic field: Our understanding of light since Sir Isaac Newton started with first assuming light to be corpuscles, namely small particles of different colour corresponding to the different frequencies. This idea was first propounded by Newton and came to be known as the corposcular theory of light. Several years later, based on experiments involving interference of light, Huygens came to understand that light is to be regarded as being composed not of particles but rather is a wave and that any light field can be regarded as a superposition of plane waves of different frequencies and wave numbers. This meant that during the process of interference involving computation of the total intensity of light coming from different sources, one should not add the respective intensities, but rather the amplitudes, then square the resultant amplitude and then form its time average. This meant that in the resultant total intensity, there would appear cross terms which are oscillatory and which lead us to the the phenomenon of interference. Later on, Fermat propounded the ray theory of light in order to explain Snell's theorems on reflection and refraction of light, according to which a light ray always follows the path of minimum time. The wave nature of light as propounded by Christiaan Huygens gained more support when Maxwell unified electromagnetism with light by showing that the basic equations of electromagnetism imply that that the electric and magnetic fields in space satisfy the wave equation with velocity of propagation being equal to that of light. This fundamental discovery gave firm support to the wave theory of light. However, with the birth of the quantum theory in the early part of the nineteenth century, it became clear that light comes in discrete packets called quanta and each quantum of radiation carries an energy proportional to its frequency and that the intensity of light is proportional to the number of quanta. These quanta came to be known as photons and it appeared therefore that with the advent of the quantum theory, physicists had reverted back to

Newton's corpuscular theory of light ! However, while Planck was busy creating his quantum theory of light, simultaneously Einstein wrote some beautiful papers proving that light can behave sometimes as a particle and sometimes as a wave, specifically showing that the variance of fluctuations in the energy of black-body radiation has two components, a particle component in agreement with Planck's quantum hypothesis and a wave component in agreement with Rayleigh's theory on the number of "modes" of a wave field within a given energy shell. The particle theory of light got further support from Einstein's special theory of relativity giving the relationship between the energy and momentum of a particle of any given mass and in particular for light particles having zero rest mass. However, the crucial breakthrough into the wave-particle duality matter and in particular of light came with De-Broglie's discovery of the fundamental relationship between the wavelength of a wave and the momentum of the associated particle. The De-Broglie theory was applicable to all of matter not just light. It gave a conclusive evidence that even particles like electrons exhibit wave-particle duality, ie, they sometimes behaved like particles and sometimes like waves. The final crunch describing the wave-particle duality came with the wave equation for particles discovered by Erwyn Schrodinger using which he was able to calculate the energy levels of an electron bound to a nucleus in terms of the eigenvalues of a partial differential equation. The corresponding eigenfunctions were complex wave fields which Schrodinger could not immediately interpret. These eigenfunctions approximately corresponded to the kind of "matter-waves" predicted by Louis-De-Broglie having frequency determined by Planck's quantum hypothesis relating frequency of waves to the energy of the quanta and having wavelength determined by De-Broglie's theory relating the wavelength of waves to the momenta of the particles. Schrodinger's wave equation was a non-relativistic wave equation in that it could be "derived" by assuming the non-relativistic relationship between energy and momentum and then assigning operators to energy and momentum. The correct interpretation of the eigenfunctions of Schrodinger's equations was provided about a year later by Max Born with the suggestion that the modulus square of the wave function gave the probability density of the particle to be present at a given point in space at a given time. Equivalently, this modulus square could be interpreted as the number density of particles or the intensity of the wave at that point in space at that time. In fact, Schrodinger's equation could also be generalized to determine the wave functions and energy levels for any particle with a prescribed energy-momentum relationship by replacing the energy and momentum with appropriate operators and applying both sides of this relationship to a wave function. Thus, Schrodinger's formalism could be applied to determine the kind of De-Broglie waves associated with a particle having any given energy-momentum relation. The De-Broglie wave solutions then have the interpretation that their modulus square represents the intensity/probability density of particles in space at a given time. In special relativity, the energy-momentum relationship for a relativistic particle is a quadratic form in the energy and momenta unlike the Newtonian-non-relativistic case where it is linear in the energy and quadratic in the momenta. Consequently the Schrodinger equation is a linear pde that is of the first order in time and second order in spatial derivatives.

The non-relativistic Schrodinger equation as a consequence then leads to a unitary evolution of the wave function with time and with the additional pleasing property that evolution from time zero to time $t_1 + t_2$ can be expressed as the composition of evolution from time zero to time $t_1$ followed by evolution from time $t_1$ to time $t_1 + t_2$. In other words, Schrodinger evolution follows the semigroup property. This is in contrast to the case of special relativity wherein the wave equation is quadratic in time leading to the wave function at a given time being a superposition of a forward propagating solution and a backward propagating solution and the loss of unitarity of the evolution. Thus, the total probability will be conserved in Schrodinger's non-relativistic theory but not so in the relativistic theory. This difficulty was finally resolved by Paul Dirac who factorized the energy-momentum relationship of special relativity into linear factors in energy and momentum using $4 \times 4$ complex anticommuting matrices and it led to Dirac's relativistic theory of the electron which along with Maxwell's equations is at the heart of quantum electrodynamics describing the interaction of light with matter. Dirac's equation being linear in the space and time derivatives is a truly relativistic wave equation for quantum theory because firstly it respects Lorentz invariance in that space and time are treated on an equal footing and that we can find a spinor representation of the Lorentz group under which Dirac's equation remains invariant and secondly it is first order in time and therefore preserves the unitary semigroup property of the evolution guaranteeing therefore conservation of total probability.

Light as we understand today is a second quantized electromagnetic field in which the vector potential components in the spatial frequency domain represent creation and annihilation operators of photons with different momenta and helicities. This is owing to the fact that the total energy of the electromagnetic field can in the Coulomb gauge be expressed as a quadratic functional of the spatial frequency components of the vector potential and each spatial frequency component of the vector potential, according to Maxwell's equations, evolves harmonically with time with a frequency being $\omega = \pm|\mathbf{k}|c$ where $\mathbf{k}$ is the wave vector. Such a picture is in perfect accord with the Hamiltonian theory of a collection of independent quantum harmonic oscillators. However, this model of light can actually be traced back to the work of Satyendranath Bose who proposed a statistical method for deriving Planck's law of black-body radiation wherein we are interested in distributing a total amount of energy $E$ having $N$ quanta at frequency $\nu$ so that $E = Nh\nu$, amongst $p$ oscillators with the quanta being regarded as indistinguishable particles. When an oscillator has $n$ such quanta, we say in the modern language of quantum mechanics that it has been excited to the $n^{th}$ energy level by an application of $n$ creation operators to the vacuum. In this way, the entire field of black-body radiation is simply a collection of quantum harmonic oscillators and if we maximize the entropy, ie, the total number of ways of distributing this energy then we end up with the famous Bose-Einstein statistics which gives us the relative number of photons at each frequency. This in turn enables us to determine the intensity of black-body radiation as a function of frequency. Summarizing the modern point of view, the entire photon field described as an operator electromagnetic four potential field is just a superposition of plane waves of different wave vectors whose coefficients are creation and annihilation operators at the different wave vectors and a given state of the photon field is actually a linear combination of number states wherein a number state is specified by specifying the number of photons that occupy each state of definite wave-vector/momentum and definite helicity/spin. A particular kind of photon state called a coherent state is a state constructed by an appropriate linear combination of this kind that turns out to be an eigenstate for all the photon annihilation operators. This model for the quantum electromagnetic field implicitly contains both the particle nature and the wave nature of light. The particle nature is contained in the presence of the creation and annihilation operators while the wave nature is contained in the plane waves that act as carrier signal fields for the creation and annihilation operators.

Second quantization means a quantization of a classical field theory. It is called second quantization for the following reason. The first quantization is simply a wave equation like the three dimensional wave equation, the Schrodinger wave equation for a single or a finite number of quantum particles, the Klein-Gordon equation or the Dirac equation. If such a classical wave equation is quantized, then it describes an infinite number of quantum particles. This can be seen clearly from the following example. Take the Klein-Gordon equation which is the wave equation described above corresponding to Einstein's energy-momentum relation with the energy and momenta replaced by appropriate operators. We expand the solution wave field as a three dimensional Fourier series within a cube of side length $L$. The coefficients of this Fourier series then become quantum operators each one and its adjoint, ie, Hermitian conjugate describe a single quantum particle. Thus, a second quantized field can equivalently be described by a countably infinite number of quantum particles. For Bosonic quantum fields like the Maxwell electromagnetic photon field, the state of the field can be described for example by specifying how many particles are occupying each state of definite momentum and helicity or equivalently by specifying the momentum and helicity of each of the particles. In this case, there can be zero, one or more than one particle having a specified momentum and helicity. In the case of Fermionic fields like the second quantized Dirac field, we cannot have more than one particle having a definite value of momentum and spin. If $a(k)^*, a(k), k = 1, 2, ...$ are canonical Bosonic creation and annihilation operators, they satisfy the canonical commutation relations (CCR) $[a(k), a(m)^*] = \delta[k-m], [a(k), a(m)] = [a(k)^*, a(m)^*] = 0$ while if $c(k)^*, c(k), k = 1, 2, ...$ are canonical Fermionic creation and annihilation operators, they satisfy the canonical anti-commutation relations (CAR) $[c(k), c(m)^*]_+ = \delta[k-m], [c(k), c(m)]_+ = [c(k)^*, c(m)^*]_+ = 0$. Here, $[A, B] = AB - BA, [A, B]_+ = AB + BA$. It is clear therefore that $c(k)^2 = 0 = c(k)^{*2}$. Here, $k$ represents a definite value of momentum and spin/helicity. If $|0>$ is a vacuum Boson state, then $a(k_1)^{*n_1}...a(k_r)^{*n_r}/\sqrt{n_1!...n_r!}|0> = |(k_1, n_1), ..., (k_r, n_r)>$ represents a normalized state of the Bosonic field in which there are $n_j$ Bosons having momentum and helicity index $k_j$ for each $j = 1, 2, ..., r$. It is clear then from the CCR that

$$a(k_j)|(k_1,n_1),...,(k_r,n_r)>=\sqrt{n_j}|(k_1,n_1),...,(j_j,n_j-1),...,(k_r,n_r)>, 1\leq j\leq r$$

while if $k\notin\{k_1,...,k_r\}$, then

$$a(k)|(k_1,n_1),...,(k_r,n_r)>=0$$

In short, $a(k)$ annihilates a Boson having momentum-helicity $k$ and if there is no such Boson with momentum-helicity $k$ in the state, then $a(k)$ annihilates the entire state giving zero. Likewise $a(k)^*$ creates a Boson having momentum-helicity $k$:

$$a(k_j)^*|(k_1,n_1),...,(k_r,n_r)>=\sqrt{n_j}|(k_1,n_1),...,(k_j,n_j+1),...,(k_r,n_r)>, 1\leq k\leq r,$$

$$a(k)^*|(k_1,n_1),...,(k_r,n_r)>=|(k_1,n_1),...,(k_r,n_r),(k,1)>, k\notin\{k_1,...,k_r\}$$

It can be verified that these two rules are in agreement with the CCR. From these CCR, it is evident that $N(k)=a(k)^*a(k)$ is the number operator, ie, when acting on a state it gives the number of Bosons in that state:

$$N(k)|(k_1,n_1),...,(k_r,n_r)>=(\sum_{j=1}^{r}\delta[k-k_j]n_j)|(k_1,n_1),...,(k_r,n_r)>$$

The state described above are called the occupation number states or simply the number states. They are eigenstates of the number operators but not of the creation and annihilation operators. The second quantized Bosonic field is a superposition of the annihilation and creation operators:

$$\phi(x)=\sum_k[a(k)\chi_k(x)+a(k)^*\bar{\chi}_k(x)]$$

where $\chi_k(x)$ satisfy the classical Bosonic field equation

$$L\chi_k(x)=0, x=(t,\mathbf{r})$$

with for example

$$L=(1/2)\partial_\mu\partial^\mu+m^2$$

in the Klein-Gordon case, or in the electromagnetic field case where the Bosons are photons having zero mass, $L$ is the wave operator

$$L=(1/2)\partial_\mu\partial^{|mu}$$

(Klein-Gordon particles have zero spin while the photon has spin one with helicities $\pm 1$, the zero helicity not being allowed. This corresponds to the fact that any state of photon polarization can be expressed as a superposition of left and right circularly polarized states). It follows that to get a state of definite field amplitude for positive frequencies, the state should be an eigenstate of the annihilation operators $a(k)$ and such a state called a coherent state can be obtained (in the discrete momentum-helicity setting) as

$$\phi(u) >= exp(-(1/2)\sum_{\{}\{n_k, k \in I\}\}|u(k)|^2)\sum \Pi_{k\in I}[u(k)^{n_k}a(k)^{*n_k}/n_k!]|0>$$

$$= exp(-(1/2)\sum_{k\in I}|u(k)|^2)\sum_k \Pi_k u(k)^{n_k}/\sqrt{n_k!}|\{n_j, j \in I\}>$$

The second quantized Dirac Fermionic field can be expressed on the other hand in terms of Fermionic creation and annihilation operators. In the photon case, the antiparticle of a photon is again a photon and so its creation operator at a given momentum-helicity is the adjoint of the corresponding annihilation operator. On the the other hand, the antiparticle of the electron is another particle the positron and hence the Dirac field should be expressed as a superposition of electron annihilation operators and positron creation operators. The creation of a positron of positive energy is according to Dirac equivalent to annihilating an electron of negative energy. Thus, the Dirac field is expressed as the superposition

$$\psi(x) = \sum_k [b(k)\chi_k(x) + c(k)^*\eta_k(x)]$$

where $b(k)$ annihilates an electron with momentum-spin $k$ and $c(k)$ annihilates a positron with momentum-spin $k$. Equivalently, $c(k)^*$ creates a positron with momentum-spin $k$. $\chi_k(x)$ and $\eta_k(x)$ are solutions to the Free Dirac equation:

$$(i\gamma^\mu\partial_\mu - m)\chi_k(x) = 0,$$

$$(i\gamma^\mu\partial_\mu - m)\eta_k(x) = 0$$

Given a quantum matter field $\eta(x)$, either Bosonic or Fermionic, it satisfies a classical wave equation of the Klein-Gordon type, or the 3-D wave type or the Dirac type with certain boundary conditions. This second quantized field can therefore be expressed as a superposition of eigenfunctions corresponding to the boundary conditions with coefficients being particle creation and annihilation operators. If the field is Bosonic, these creation and annihilation operators will satisfy CCR's whle if the field is Fermionis, they will satisfy CAR's. If $H$ corresponds to the Hamiltonian of the first quantized theory, then we can write the second quantized Hamiltonian of the field as

$$\int \eta(x)^* H\eta(x)d^3x$$

This is a function of an infinite number of particle operators acting in a Bosonic or Fermionic Fock space.

## 10.9 Corrections to radiation by Fermionic current caused by interaction with the photon radiation field

When the electron-positron field interacts with the Bosonic radiation field, the Dirac current density acquires extra terms involving coupling between the Fermionic and Bosonic components. We analyze this interaction in what follows. The Dirac equation in the presence of the radiation field is given by

$$[\gamma^\mu(i\partial_\mu + eA_\mu) - m]\psi = 0$$

or equivalently,

$$[i\gamma^\mu\partial_\mu - m]\psi = -e\gamma^\mu A_\mu\psi$$

If $\psi^{(0)}$ denotes the free Dirac field, then we can write down the approximate solution to the above equation based on first order perturbation theory as

$$\psi(x) = \psi^{(0}(x) - e\int S_e(x-y)\gamma^\mu A_\mu(y)\psi^{(0)}(y)d^4y = \psi^{(0)}(x) + \psi^{(1)}(x)$$

say, where $S_e(x-y)$ is the electron propagator defined by

$$S_e(p) = \int S_e(x)exp(-ip.x)d^4x = i\gamma^\mu p_{mu} - m = i\gamma.p - m$$

Now writing the cavity constrained free Dirac field as

$$\psi^{(0)}(x) = \sum_k [b(k)\chi_k(x) + c(k)^*\eta_k(x)]$$

and the free cavity constrained electromagnetic four potential as

$$A_\mu(x) = \sum_k [a(k)\theta_k(x) + a(k)^*\theta_k(x)^*]$$

where $a(k), a(k)^*$ are the photon annihilation and creation operators while $b(k), b(k)^*$ are the electron annihilation and creation operators and $c(k), c(k)^*$

are the positron annihilation and creation operators, we get for the approximate value of the Dirac current operator,

$$J^\mu = (\psi^{(0)} + \psi^{(1)})^* \alpha^\mu (\psi^{(0)} + \psi^{(1)}) =$$

$$J^{\mu(0)} + \delta J^\mu$$

where

$$J^{\mu(0)}(x) = \psi^{(0)*} \alpha^\mu \psi^{(0)},$$

$$\delta J^\mu = \psi^{(0)*} \alpha^\mu \psi^{(1)} + \psi^{(1)*} \alpha^\mu \psi^{(0)}$$

Now,

$$J^{\mu(0)} = \sum_{km} (b(k)\chi_k(x) + c(k)^* \eta_k(x))^* \alpha^\mu (b(m)\chi_m(x) + c(m)^* \eta_m(x))$$

$$= \sum_{k,m} [b(k)^* b(m) \chi_k(x)^* \alpha^\mu \chi_m(x) + c(k)c(m)^* \eta_k(x)^* \alpha^\mu \eta_m(x)$$

$$+ b(k)^* c(m)^* \chi_k(x)^* \alpha^\mu \eta_m(x) + c(k)c(m)^* \eta_k(x)^* \alpha^\mu \eta_m(x)]$$

is the free Dirac current ie, in the absence of interactions with the photon field. We have already indicated how to compute the far field radiation pattern produced by this field and how to evaluate the moments of this field. Specifically, if $G(x-y)$ denotes the causal Green's function for the wave operator, then the electromangetic four potential produced by the Dirac current is given by

$$A_\mu(x) = \int G(x-y) J^\mu(y) d^4y = \int G(x-y) J^{\mu(0)}(y) d^4y + \int G(x-y) \delta J^\mu(y) d^4y$$

$$= A_\mu^{(0)}(x) + \delta A_\mu(x)$$

Remark: We can consider a Fermionic coherent state rather than a Fermionic number state. Such a state is parametrized by a Grassmannian vector variable $\gamma = (\gamma_b(k), \gamma_c(k))_k$ and is denoted by $|\phi(\gamma)>$. The action of the electron and positron annihilation operators on this state is

$$b(k)|\phi(\gamma)> = \gamma_b(k)|\phi(\gamma)>,$$

$$c(k)|\phi(\gamma)> = \gamma_c(k)|\phi(\gamma)>$$

In order that the CAR

$$[b(k), b(m)]_+ = [c[k], c(m)]_+ = [b(k), c(m)]_+ = 0$$

hold good, we require that the Grassmannian parameters satisfy the anticommutation rules

$$\gamma_b(k)\gamma_b(m) + \gamma_b(m)\gamma_b(k) = 0,$$

$$\gamma_c(k)\gamma_c(m) + \gamma_c(m)\gamma_c(k) = 0,$$

$$\gamma_b(k)\gamma_c(m) + \gamma_c(m)\gamma_b(k) = 0,$$

Further, by our analogy with Bosonic coherent states, we impose the requirement that

$$[\partial/\partial\gamma_r(k), \gamma_s(m)]_+ = \delta(r,s)\delta(k,m), r, s = b, c$$

and that

$$b(k)^*|\phi(\gamma)>= (\partial/\partial\gamma_b(k))|\phi(\gamma)>,$$

$$c(k)^*|\phi(\gamma)>= (\partial/\partial\gamma_c(k))|\phi(\gamma)>$$

We then get

$$<\phi(\gamma)|b(k)^*b(m)|\phi(\gamma)>=<\phi(\gamma)|b(k)^*\gamma_b(m)|\phi(\gamma)>$$

$$=<b(k)\phi(\gamma)|\gamma_b(m)|\phi(\gamma)>= \gamma_b(k)^*\gamma_b(m)$$

on the one hand, while on the other,

$$<\phi(\gamma)|b(m)b(k)^*|\phi(\gamma)>=$$

$$<\phi(\gamma)|b(m)(\partial/\partial\gamma_b(k))|\phi(\gamma)>=$$

$$<\phi(\gamma)|(\partial/\partial\gamma_b(k))b(m)|\phi(\gamma)>$$

$$=<\phi(\gamma)|(\partial/\partial\gamma_b(k))\gamma_b(m)|\phi(\gamma)>$$

$$= \delta(k,m)-<\phi(\gamma)|\gamma_b(m)(\partial/\partial\gamma_b(k))|\phi(\gamma)>$$

$$= \delta(k,m) - \gamma_b(m) <\phi(\gamma)|b(k)^*|\phi(\gamma)>$$

$$= \delta(k,m) - \gamma_b(m) <b(k)\phi(\gamma)|\phi(\gamma)>=$$

$$\delta(k,m) - \gamma_b(m)\gamma_b(k)^*$$

This is in agreement with the CAR

$$[b(m), b(k)^*]_+ = \delta(m,k)$$

provided that we assume the CAR

$$[\gamma_b(m), \gamma_b(k)^*]_+ = \delta(m,k)$$

and likewise,

$$[\gamma_c(m), \gamma_c(k)^*]_+ = \delta(m,k)$$

By imposing such restrictions, we can calculate easily the moments of the current density field and hence of the radiated field is a state that is jointly coherent for the Bosons (ie photons) and for the Fermions. We observe that the perturbation to the current density of the Dirac field caused by the interactions between the electron-positron field and the photon field is given upto first order in the photon field and second order in the Fermion field by an expression of the form

$$\delta J^\mu = \psi^{(0)*}\alpha^\mu\psi^{(1)} + \psi^{(1)*}\alpha^\mu\psi^{(0)}$$

$$= -e\psi^{(0)*}(x)\alpha^\mu \int S_e(x-y)\gamma^\mu A_\mu(y)\psi^{(0)}(y)d^4y$$

$$+h.c$$

This expression is manifestly trilinear in the operators. Specifically, it is quadratic in the electron-positron field and linear in the photon field, totally yielding a trilinear term. It can be expressed as

$$\delta J^\mu(x) =$$

$$\sum_{k,m,q}(F_1^\mu(x|k,m,q)b(k)^*b(m) + F_2^\mu(x|k,m,q)b(k)^*c(m)^*+$$

$$F_3^\mu(x|k,m,q)c(k)b(m) + F_4(x|k,m,q))c(k)c(m)^*)a(q) + h.c)$$

It should be noted that the photon operatrors $a(q), a(q)^*$ commute with all the electron-positron operators $b(k), b(k)^*, c(k), c(k)^*$. From this expression, it is clear that the photon operators tend to couple the other modes of the electron-positron field and hence produce additional terms in the far field radiation pattern. If we have a state $|\psi>$ of the electron-positron-photon field in which there are $n_e(k) = 0,1$ electrons with momentum-spin index $k$, $n_p(k) = 0,1$ positrons with momentum-spin index $k$ and $n_{ph}(k) = 0,1,2,...$ photons with momentum-helicity index $k$ for $k = 1,2,...$, then we can calculate easily the moments of the current fluctuation field $\delta J^\mu(x)$ in this state by simply applying the rules

$$b(k)|n_e(k) = 0>= 0, b(k)|n_e(k) = 1>= |n_e(k) = 0>,$$

$$c(k)|n_p(k) = 0>= 0, c(k)|n_p(k) = 1>= |n_p(k) = 0>,$$

$$b(k)^*|n_e(k) = 0>= |n_e(k) = 1>, b(k)^*|n_e(k) = 1>= 0,$$

$$c(k)^*|n_p(k) = 0>= |n_p(k) = 1>, c(k)^*|n_p(k) = 1>= 0,$$

in view of the Pauli-exclusion principle. and likewise for the photon number states

$$a(k)|n_{ph}(k)>= \sqrt{n_{ph}(k)}|n_{ph}(k) - 1>,$$

$$a(k)^*|n_{ph}(k)>= \sqrt{n_{ph}(k) + 1}|n_{ph}(k) + 1>$$

More precisely, we can evaluate the moments

$$<\psi|\delta J^{\mu_1}(x_1) \otimes ... \otimes \delta J^{\mu_m}(x_m)|\psi>$$

by noting that the quantity

$$\delta J^{\mu_1}(x_1) \otimes ... \otimes \delta J^{\mu_m}(x_m)$$

is a homogeneous polynomial of degree $3m$ in the electron-positron-photon operators with the photon operators appearing with a total degree of $m$ and the electron-positron operators appearing with a total degree of $2m$. We can also evaluate the above moment in a joint coherent state $|\phi_{ep}(\gamma) \otimes \phi_{ph}(u)>$.

The reference for the material on Fermionic coherent state has been taken from the Master's thesis of Greplova on "Fermionic Gaussian states".

## 10.10 Analysis of a conical resonator antenna regarded as a patch on a spherical antenna surface

Eye as a conical resonator antenna (CORA)

Let $h$ be the height of the cone and $r$ its radius. We may thus assume the conical region to be specified by the equation

$$0 \le z \le d, x^2 + y^2 \le a^2 z^2, a = tan(\alpha/2)$$

where $\alpha$ is the cone angle. This cone has its apex at the origin and its surface is a surface of revolution of a line passing through the origin making an angle $\alpha/2$ with the $z$-axis. Use the spherical polar coordinate system to expand the fields as

$$\mathbf{E}(r, \theta, \phi) = E_r(r, \theta, \phi)\hat{r} + E_\theta(r, \theta, \phi) + \hat{E}_\phi(r, \theta, \phi)\hat{\phi}$$

and likewise for the magnetic field. The boundary conditions are

$$E_r = E_\phi = 0, H_\theta = 0, \theta = \alpha/2$$

corresponding to the PEC boundary conditions, ie, the vanishing of the tangential components of the electric field and the normal component of the magnetic field on the conical surface. Further if the top of the cone at $z = h$ is again a PEC, then the boundary conditions corresponding to this is

$$E_\theta = E_\phi = 0, H_z = 0, z = h$$

Note that

$$\hat{z} = cos(\theta)\hat{r} - sin(\theta)\hat{\theta}$$

and hence the last boundary condition can also be expressed as

$$H_r.cos(\theta) - H_\theta sin(\theta) = 0, r.cos(\theta) = h$$

We now write down the Maxwell curl equations in the spherical polar coordinate system. Quantum image processing via the Belavkin filter:

$$curlE = det \begin{pmatrix} \hat{r}/r^2 sin(\theta) & \hat{\theta}/r.sin(\theta) & \hat{\phi}/r \\ \partial/\partial r & \partial/\partial\theta & \partial/\partial\phi \\ E_r & r.E_\theta & r.sin(\theta)E_\phi \end{pmatrix}$$

$$= -j\omega\mu(H_r\hat{r} + H_\theta\hat{\theta} + H_\phi\hat{\phi})$$

and likewise,

$$curlH = det \begin{pmatrix} \hat{r}/r^2 sin(\theta) & \hat{\theta}/r.sin(\theta) & \hat{\phi}/r \\ \partial/\partial r & \partial/\partial\theta & \partial/\partial\phi \\ H_r & r.H_\theta & r.sin(\theta)H_\phi \end{pmatrix}$$

$$= j\omega\epsilon(E_r\hat{r} + E_\theta\hat{\theta} + E_\phi\hat{\phi})$$

Writing out the components explicitly gives us

$$(sin(\theta)H_\phi)_{,\theta} - H_{\theta,\phi} = -j\omega\mu r.sin(\theta)H_r$$

$$E_{r,\phi} - sin(\theta)(r.E_\phi)_{,r} = -j\omega\mu r.sin(\theta)H_\theta$$

$$(rE_\theta)_{,r} - E_{r,\theta}) = -j\omega\mu r H_\phi$$

and likewise its dual with $\mathbf{E} \rightarrow \mathbf{H}, \mathbf{H} \rightarrow -\mathbf{E}$ and $\epsilon < - - - > \mu$. Also these Maxwell curl equations imply the Helmholtz equation

$$(\nabla^2 + k^2)(\mathbf{E}, \mathbf{H}) = \mathbf{0}, k^2 = \omega^2\epsilon\mu$$

We require to write down these Helmholtz equations in the spherical polar coordinates. For that we note that

$$\partial\hat{r}/\partial r = 0, \partial\hat{r}/\partial\theta = \hat{\theta}, \partial\hat{r}/\partial\phi = sin(\theta)\hat{\phi},$$

$$\partial\hat{\theta}/\partial r = 0, \partial\hat{\theta}/\partial\theta = -\hat{r}, \partial\hat{\theta}/\partial\phi = cos(\theta)\hat{\phi},$$

$$\partial\hat{\phi}/\partial r = 0, \partial\hat{\phi}/\partial\theta = 0, \partial\hat{\phi}/\partial\phi = -\hat{x}.cos(\phi) - \hat{y}.sin(\phi) = -\hat{\rho} = -sin(\theta)\hat{r} - cos(\theta)\hat{\theta}$$

Thus,

$$\nabla^2(E_r\hat{r}) = (\nabla^2 E_r)\hat{r} + 2(\nabla E_r, \nabla)\hat{r} + E_r\nabla^2\hat{r}$$

with

$$(\nabla E_r, \nabla)\hat{r} = r^{-2}E_{r,\theta}\hat{\theta} + r^{-2}.(sin\theta)^{-1}E_{r,\phi}\hat{\phi}$$

and

$$\partial^2\hat{r}/\partial\theta^2 = \partial\hat{\theta}/\partial\theta = -\hat{r}$$

$$\partial^2\hat{r}/\partial\phi^2 = sin(\theta)\partial\hat{\phi}/\partial\phi = -sin(\theta)(sin(\theta)\hat{r} + cos(\theta)\hat{\theta})$$

so that

$$\nabla^2\hat{r} = (cot(\theta)/r^2)\partial\hat{r}/\partial\theta + (1/r^2)\partial^2\hat{r}/\partial\theta^2 + (1/r^2 sin^2(\theta))\partial^2\hat{r}/\partial\phi^2$$

$$= (cot(\theta)/r^2)\hat{\theta} - (1/r^2)\hat{r} + (1/r^2)(-\hat{r} - cot(\theta)\hat{\theta})$$

$$= -2\hat{r}/r^2$$

Thus,

$$\nabla^2(E_r\hat{r}) = \hat{r}(\nabla^2 E_r - 2E_r/r^2) + 2r^{-2}E_{r,\theta}\hat{\theta} + 2r^{-2}.(sin\theta)^{-1}E_{r,\phi}\hat{\phi}$$

Next,

$$\nabla^2(E_\theta\hat{\theta}) = (\nabla^2 E_\theta)\hat{\theta} + 2(\nabla E_\theta, \nabla)\hat{\theta} + E_\theta\nabla^2\hat{\theta}$$

and

$$(\nabla E_\theta, \nabla)\hat{\theta} =$$

$$= -r^{-2}E_{\theta,\theta}\hat{r} + (r.sin(\theta))^{-2}E_{\theta,\phi}cos(\theta)\hat{\phi}$$

and further,

$$\nabla^2\hat{\theta} = (cot(\theta)/r^2)\partial\hat{\theta}/\partial\theta + (1/r^2)\partial^2\hat{\theta}/\partial\theta^2$$

$$+(r.sin(\theta))^{-2}\partial^2\hat{\theta}/\partial\phi^2$$

$$= (cot(\theta)/r^2)(-\hat{r}) + (1/r^2)(-\hat{\theta}) + (r.sin(\theta))^{-2}(-sin(\theta).cos(\theta)\hat{r} - cos^2(\theta)\hat{\theta})$$

$$= -2cot(\theta)\hat{r}/r^2 - \hat{\theta}/r^2 sin^2(\theta)$$

Thus,

$$\nabla^2(E_\theta\hat{\theta}) = (\nabla^2 E_\theta)\hat{\theta}$$

$$-2r^{-2}E_{\theta,\theta}\hat{r} + 2(r.sin(\theta))^{-2}E_{\theta,\phi}cos(\theta)\hat{\phi}$$

$$+E_\theta(-2cot(\theta)\hat{r}/r^2 - \hat{\theta}/r^2 sin^2(\theta))$$

$$= -2\hat{r}(E_{\theta,\theta}/r^2 + E_\theta cot(\theta)/r^2)$$

$$+\hat{\theta}(\nabla^2 E_\theta - E_\theta/r^2 sin^2(\theta))$$

$$+\hat{\phi}(2cot(\theta)/r^2.sin(\theta))E_{\theta,\phi}$$

Finally,

$$\nabla^2(E_\phi\hat{\phi}) = (\nabla^2 E_\phi)\hat{\phi} + E_\phi\nabla^2\hat{\phi} + 2(\nabla E_\phi, \nabla)\hat{\phi}$$

Now,

$$(\nabla E_\phi, \nabla)\hat{\phi} = -r^{-2}E_{\phi,\phi}(\hat{r}.sin(\theta) + \hat{\theta}.cos(\theta))$$

and

$$\nabla^2\hat{\phi} = -\hat{\phi}/r^2 sin^2(\theta)$$

Thus,

$$\nabla^2(E_\phi\hat{\phi}) =$$

$$\hat{r}(-r^{-2}E_{\phi,\phi}sin(\theta)) + \hat{\theta}(-r^{-2}cos(\theta)E_{\phi,\phi}) + \hat{\phi}(\nabla^2 E_\phi - E_\phi/r^2 sin^2(\theta))$$

The Helmholtz equation for **E** gives on equating the $\hat{r}, \hat{\theta}$ and the $\hat{\phi}$ components, the following three equations:

$$\nabla^2 E_r - 2E_r/r^2 - 2(E_{\theta,\theta}/r^2 + E_\theta cot(\theta)/r^2) - E_\phi/r^2 + k^2 E_r = 0,$$

$$\nabla^2 E_\theta + 2r^{-2}E_{r,\theta} - E_\theta/r^2 sin^2(\theta) - cos(\theta)E_{\phi,\phi}/r^2 + k^2 E_\phi = 0,$$

$$\nabla^2 E_\phi - E_\phi/r^2 sin^2(\theta) + 2r^{-2}.(sin\theta)^{-1}E_{r,\phi} + (2cot(\theta)/r^2.sin(\theta))E_{\theta,\phi} = 0$$

and likewise with **E** replaced by **H**. It is hard to solve this Helmholtz equation taking into account the above Maxwell curl equations. Before delving into special cases of these equations, we simplify matters assuming that the entire electromagnetic field is replaced by a single scalar field $\psi(r, \theta, \phi)$ within the cone. The Helmholtz equation is

$$(2/r)\psi_{,r} + \psi_{,rr} + r^{-2}(cos(\theta)\psi_{,\theta} + \psi_{,\theta\theta}) + r^{-2}\psi_{,\phi\phi} + k^2\psi = 0$$

with the boundary condition that $\psi$ vanishes on all the surfaces of the cone. We can equivalently express this equation as

$$(2/r)\psi_{,r} + \psi_{,rr} - L^2\psi/r^2 + k^2\psi = 0$$

where

$$L^2 = -(-\frac{1}{sin\theta}\frac{\partial}{\partial\theta}sin(\theta)\frac{\partial}{\partial\theta} + \frac{1}{sin^2(\theta)}\frac{\partial^2}{\partial\phi^2})$$

is the angular part of the Laplacian or what is known in quantum mechanics as the square of the angular momentum operator. Separation of variables gives us

$$\psi(r,\theta,\phi) = R(r)Y_{lm}(\theta,\phi)$$

where $Y_{lm}$ are the spherical harmonics. They satisfy the eigen-relations

$$L^2Y_{lm} = l(l+1)Y_{lm}, L_zY_{lm} = mY_{lm}, L_z = -i\frac{\partial}{\partial\phi}$$

Thus, the radial component of the wave function $R(r)$ satisfies

$$2rR'(r) + r^2R''(r) + (k^2r^2 - l(l+1))R(r) = 0$$

We substitute into this equation,

$$R(r) = r^aF(r)$$

to get

$$2r(ar^{a-1}F(r) + r^aF'(r)) + r^2(a(a-1)r^{-2}F(r) + 2ar^{a-1}F'(r) + r^aF''(r))$$
$$+(k^2r^2 - l(l+1))r^aF(r) = 0$$

or equivalently,

$$r^2F''(r) + F'(r)(2r + 2ar) + F(r)(a(a+1) + k^2r^2 - l(l+1)) = 0$$

Choosing $a = -1/2$ gives us

$$r^2F''(r) + rF'(r)/r + F(r)(k^2r^2 - (l+1/2)^2) = 0$$

which is the Bessel equation of order $l + 1/2$. Its solution is

$$F(r) = J_{l+1/2}(kr)$$

and hence

$$R(r) = r^{-1/2}J_{l+1/2}(kr)$$

The general solution to the three dimensional Helmholtz equation then can be expressed as a superposition

$$\psi(r,\theta,\phi) = \sum_{l\geq 0,|m|\leq l} c(l,m)r^{-1/2}J_{l+1/2}(kr)Y_{lm}(\theta,\phi)$$

The first boundary condition is

$$0 = \psi(r, \alpha/2, \phi) = \sum_{lm} c(l,m) J_{l+1/2}(kr) Y_{lm}(\alpha/2, \phi) = 0$$

or equivalently writing the spherical harmonics in terms of the modified Legendre polynomials,

$$Y_{lm}(\theta, \phi) = P_{lm}(cos(\theta)).exp(im\phi)$$

this boundary condition gives us

$$\sum_l c(l,m) P_{lm}(cos(\alpha/2)) J_{l+1/2}(kr) = 0, \forall m \in \mathbb{Z}, 0 \leq r \leq L$$

where $L = \sqrt{h^2 + a^2}$ is the length of the side of the cone. The other boundary condition is $\psi$ vanishes on the circular lid at $z = h, 0 \leq \theta \leq \alpha/2$

$$0 = \psi(h.tan(\theta), \theta, \phi)$$

so that

$$\sum_l c(l,m) P_{lm}(cos(\theta)) J_{l+1/2}(kh.tan(\theta)) = 0, \forall m \in \mathbb{Z}, 0 \leq \theta \leq \alpha/2$$

By multiplying these equations by linearly independent functions of $r$ and $\theta$ respectively and integrating over the appropriate range, we get a sequence of linear homogeneous equations for $c(l,m), l \geq 0$ for each $m$ which therefore have a nontrivial solution iff the corresponding infinite determinant vanishes. This gives as a nonlinear equation for $k$ from which one can in principle obtain the characteristic frequencies of oscillation.

We now come back to the electromagnetic field case and analyze it in a different way using the multipole expansion method based on vector valued spherical harmonics (Reference:J.D.Jackson, "Classical Electrodynamics", Wiley). Define an electric field by

$$\mathbf{E}_{lm} = f_l(r) \mathbf{L} Y_{lm}(\theta, \phi)$$

where

$$\mathbf{L} = -i\mathbf{r} \times \nabla$$

is the usual angular momentum vector operator in quantum mechanics. Then,

$$\hat{r}.\mathbf{E}_{lm} = 0$$

Also, the Helmholtz equation for $\mathbf{E}_{lm}$ and the fact that $\nabla^2$ commutes with $\mathbf{L}$ implies that

$$(2/r) f_l'(r) + f_l''(r) - (l(l+1)/r^2) f_l(r) + k^2 f_l(r) = 0$$

or equivalently,

$$r^2 f_l''(r) + 2r f_l'(r) + (k^2 r^2 - l(l+1)) f_l(r) = 0$$

which has solutions

$$f_l(r) = h_l(kr)$$

where $h_l$ are the Hankel functions. $\mathbf{E}_{lm}$ also satisfies the Gauss' equation

$$div\mathbf{E}_{lm} = 0$$

since

$$\hat{r}.\mathbf{L} = 0$$

and

$$\nabla.\mathbf{L}\psi = -i\nabla.(\mathbf{r} \times \nabla\psi) = i\nabla.(\nabla \times (\mathbf{r}\psi)) = 0$$

since $\nabla \times \mathbf{r} = \mathbf{0}$. We can repeat the same analysis for the magnetic field by defining

$$\mathbf{H}_{lm} = h_l(kr)\mathbf{L}Y_{lm}(\theta, \phi)$$

Thus, the general solution to the Helmholtz equation in spherical-polar coordinates can be expressed as

$$\mathbf{E}(\omega, \mathbf{r}) = \sum_{lm}[c(l,m)h_l(kr)\mathbf{L}Y_{lm}(\hat{r}) + d(l,m)curl(h_l(kr)\mathbf{L}Y_{lm}(\hat{r}))/j\omega\epsilon]$$

$$\mathbf{H}(\omega, \mathbf{r}) = \sum_{lm}[d(l,m)h_l(kr)\mathbf{L}Y_{lm}(\hat{r}) - c(l,m)curl(h_l(kr)\mathbf{L}Y_{lm}(\hat{r}))/j\omega\mu]$$

It is a simple matter to verify that these fields satisfy Maxwell's curl equations:

$$curl\mathbf{E} = -j\omega\mu\mathbf{H}, curl\mathbf{H} = j\omega\epsilon\mathbf{E}$$

by making use of the relations,

$$div(h_l(kr)\mathbf{L}Y_{lm}(\hat{r})) = 0, (\nabla^2 + k^2)(h_l(kr)\mathbf{L}Y_{lm}(\hat{r})) = \mathbf{0}$$

For our cone problem, the coefficients $c(l,m), d(l,m)$ and the characteristic oscillation frequencies $\omega = k/\sqrt{\epsilon\mu}$ are obtained from the boundary conditions

$$E_r(r, \alpha/2, \phi) = 0, H_\theta(r, \alpha/2, \phi) = 0, 0 \le r \le L, 0 \le \phi < 2\pi$$

$$E_z(h.tan(\theta), \theta, \phi) = 0, 0 \le \theta < \alpha/2, 0 \le \phi < 2\pi$$

Denote the characteristic frequencies by $\omega[n], n = 1, 2, ...$ and for each $n$, we have eigenvectors $(c(l,m,n))_h, (d(l,m,n))_n$. Then the general solution in the time domain can be expressed as

$$\mathbf{E}(t, r, \theta, \phi)$$

$$= Re[\sum_{lmn}[c(l,m,n)h_l(k[n]r)\mathbf{L}Y_{lm}(\hat{r})exp(-j\omega[n]t)+$$

$$d(l,m,n)curl(h_l(k[n]r)\mathbf{L}Y_{lm}(\hat{r}))exp(-j\omega[n]t)/j\omega[n]\epsilon]$$

$$\mathbf{H}(t, r, \theta, \phi) = Re[\sum_{lmn}[d(l,m,n)h_l(k[n]r)\mathbf{L}Y_{lm}(\hat{r})exp(-j\omega[n]t)$$

$$-c(l,m,n)curl(h_l(k[n]r)\mathbf{L}Y_{lm}(\hat{r}))exp(-j\omega[n]t)/j\omega[n]\mu]$$

When we compute the energy in this electromagnetic field within the cone, we obtain a diagonalized quadratic form in the coefficients $c(l,m,n), d(l,m,n)$ in view of the orthogonality of the eigenfunctions as follows from the fact that these are eigenfunctions of the Helmholtz operator which is self-adjoint. We can write the total energy of the field in the cone as earlier:

$$H_F = \sum_n \omega[n]a[n]^* a[n]$$

where for each $n$, $a[n]$ is a linear combination of the coefficients $c(l,m,n), d(l,m,n), l \geq 0, |m| \leq n$ and obviously from the above formula, $a[n]$ varies with time as $exp(-i\omega[n]t)$ and $a[n]^*$ as $exp(i\omega[n]t)$. Likewise we can also expand the second quantized Dirac wave field in terms of the eigenfunctions $u_{nlm}(r,\theta,\phi) = r^{-1/2}J_{l+1/2}(k[n]r)Y_{lm}(\theta,\phi)$ of the scalar Helmholtz operator discussed above and write it as

$$\psi(t,r,\theta,\phi) = \psi(t,\mathbf{r}) =$$

$$\sum_{nlm\sigma} [b[nlm\sigma]\mathbf{f}(nlm\sigma)u_{nlm}(\mathbf{r})exp(-i\omega[n]t) + c[nlm]^*\mathbf{g}(nlm\sigma)\bar{u}_{nlm\sigma}(\mathbf{r})exp(i\omega[n]t)]$$

with $b[nlm\sigma], c[nlm]\sigma]$ denoting respectively the electron and positron annihilation operators and $\mathbf{f}(nlm\sigma), \mathbf{g}(nlm\sigma)$ being constant $4 \times 1$ complex vectors chosen to satisfy the Dirac wave equation as earlier. The entire analysis as for example, calculating the radiated field by the electrons and positrons after they have interacted with the photon field is valid in this situation also. The only difference is in the nature of the eigenfunctions used for the electromagnetic field and for the Dirac field. For example, one can in principle compute the entire power series solution in the form of a Dyson series expansion for the Dirac field in the presence of the electromagnetic field and vice versa:

$$[\gamma^\mu[i\partial_\mu - m]\psi(x) = -e\gamma^\mu\psi(x)A_\mu(x)$$

$$\partial^\alpha\partial_\alpha A^\mu(x) = e\psi(x)^*\alpha^\mu\psi(x)$$

Formally, we can write down the perturbation series solution to these equations as

$$\psi(x) = \sum_{n\geq 0} e^n\psi^{(n)}(x), A^\mu(x) = \sum_{n\geq 0} e^n A^{\mu(n)}(x)$$

where the electronic charge is taken as our perturbation parameter and equating terms of the same powers of $e$ gives us

$$[\gamma^\mu[i\partial_\mu - m]\psi^{(0)}(x) = 0$$

$$\partial^\alpha\partial_\alpha A^{\mu(0)}(x) = 0$$

and

$$[\gamma^\mu[i\partial_\mu - m]\psi^{(n+1)}(x) = -\sum_{k=0}^{n}\gamma^\mu\psi^{(k)}(x)A_\mu^{(n-k)}(x)$$

$$\partial^\alpha \partial_\alpha A^{\mu(n+1)}(x) = \sum_{k=0}^{n} \psi^{(k)}(x)^* \alpha^\mu \psi^{(n-k)}(x)$$

Denoting the $G(x-y)$ the Green's function for the photon wave operator $\partial^\alpha \partial_\alpha$ and by $S(x-y)$ the Green's function ($4 \times 4$ matrix valued) for the Dirac operator $i\gamma^\mu \partial_\mu - m$, we can formally write down the following iterative scheme for computing the various order approximations to the photon field and the Dirac field when they interact as

$$\psi^{(n+1)}(x) = -\sum_{k=0}^{n} \int S(x-y)\gamma^\mu \psi^{(k)}(y) A_\mu^{(n-k)}(y) d^4y$$

$$A^{\mu(n+1)}(x) = \sum_{k=0}^{n} \int G(x-y)\psi^{(k)}(y)^* \alpha^\mu \psi^{(n-k)}(y) d^4y$$

for $n = 0, 1, 2, ....$ $\psi^{(0)}(x), A^{\mu(0)}(x)$ describe the free fields, ie, in the absence of electron-positron and photon interactions and for $n = 1, 2, 3, ...,$ $\psi^{(n)}(x), A^{\mu(n)}(x)$ describe the higher order interaction terms. In this way, continuing upto any order $N$ of accuracy, we can express the photon radiation field

$$A^\mu(x) \approx \sum_{n=0}^{N} e^n A^{\mu(n)}(x)$$

as a polynomial in the photon and electron-positron creation and annihilation operators and hence determine their statistical moments, ie mean, variance etc. in any given state of the electron-positron photon field.

Let $X(t, r)$ be a quantum image field built out of creation and annihilation operators. We wish To denoise this quantum image field. Let $X_0(t, r)$ denote the corresponding denoised quantum image field. We pass the noisy quantum image field X(t,r) through a spatio-temporal linear filter Having an impulse response $H(t, r)$. The output of this filter is given by

$$\hat{X}_0(t, r) = \int H(t-t', r-r') X(t', r') dt' d^3r'$$

We wish to select the filter $H(t, r)$ so that $\hat{X}_0(t, r)$ is a close approximation to $X_0(t, r)$. in a given quantum state $\rho$. This means that we select the function $H(t, r)$ so that

$$\int Tr(\rho(X_0(t, r) - \hat{X}_0(t, r))^2) dt d^3r$$

Is minimal. Setting the variational derivative of this error energy function w.r.t. $H$ to zero then gives us the optimal normal equations

$$Tr(\rho \int dt d^3r (X_0(t, r) - \hat{X}_0(t, r)) X(t-t', t-r')) = 0$$

or equivalently,

$$\int Tr(\rho\{X_0(t,r), X(t-t', r-r'))\}dtd^3r$$

$$= \int dsd^3uH(s,u) \int Tr(\rho\{X(t-s, r-u), X(t-t', t'-r')\}dtd^3r$$

Thus, to calculate the filter, we must evaluate the symmetrized quantum correlations

$$Tr(\rho\{X_0(t,r), X(t_1,r_1)\}), Tr(\rho\{X(t,r), X(t_1,r_1)\})$$

Assuming that $\rho$ is a quantum Gaussian state so that it is expressible as an exponential of a linear-quadratic form in the creation and annihilation operators $a(k), a(k)^*, k = 1, 2, \ldots,$ we express

the quantum fields $X_0, X$ as polynomial functionals in the $a(k), a(k)^*$ and computing the quantum correlations then amounts to calculating the multiple moments of the creation and annihilation operators in a Gaussian state, ie evaluating moments of the form

$$Tr(\rho.\Pi_k(a(k)^{m_k})\Pi_k(a(k)^* n_k))$$

Where

$$\rho = C.exp(-\sum_k \alpha(k)a(k) + \bar{\alpha}(k)a(k)^* - \sum_{k,m} \beta_1(k,m)a(k)a(m) + \beta_2(k,m)a(k)^*a(m)^*$$
$$-\beta_3(k,m)a(k)^*a(m))$$

The easiest way to evaluate these moments is to use the Glauber-Sudarshan resolution of the identity in terms of coherent states.

## 10.11 Average directional properties of radiation by a quantum antenna

Suppose that the electron-positron field with wave operator field $\psi(t,r)$. The four current density is then given by

$$J^\mu(x) = -e\psi(x)^*\alpha^\mu\psi(x), \alpha^\mu = \gamma^0\gamma^\mu$$

And the four vector potential generated by this four current density is then given by

$$A^\mu(x) = \int J^\mu(x')G(x-x')d^4x'$$

Where

$$G(x) = (\mu/4\pi)\delta(x^2) = (\mu/4\pi r)\delta(t - |r|/c)$$

Is the causal Green's function for the wave operator. Now using the above formula for the vector potential, the far field four potential has the form

$$A^\mu(t,r) = (\mu/4\pi r) \int J^\mu(t - r/c + \hat{r}.r'/c, r')d^3r'$$

It follows that as a function of frequency, the far field four potential has the angular amplitude pattern

$$B^\mu(\omega, r) = \int J^\mu(\omega, r') exp(jk\hat{r}.r')d^3r'$$

To evaluate the directional properties of the corresponding power pattern, we must first choose a state $|\eta>$ for the electron-positron system and compute

$$S^{\mu\nu}(\omega, r) =< \eta|B^\mu(\omega, r).B^\nu(\omega, r)^*|\eta >=$$

$$\int < \eta|J^\mu(\omega, r_1)J^\nu(\omega, r_2)^*|\eta > exp(jk\hat{r}.(r_1 - r_2))d^3r_1d^3r_2$$

In order to obtain superdirectional properties of the radiated field, we must prepare the state $|\eta>$ so that the above quantum average is large when $\mu = \nu$. First observe that in terms of the creation and annihilation operators of the electron-positron field, the Dirac wave operator field is given by

$$\psi(t, r) = \int [u(P, \sigma)a(P, \sigma)exp(-i(E(P)t - P.r)) + v(P, \sigma)b(P, \sigma)^* exp(i(E(P)t - P.r))]d^3P$$

Where $E(P) = \sqrt{m^2 + P^2}$. We then find that the temporal Fourier transform of The four current density $J^\mu(t, r) = -e\psi(t, r)^*\alpha^\mu\psi(t, r)$ is given by the convolution

$$J^\mu(\omega, r) = (-e/2\pi) \int_{\mathbb{R}} \psi(\omega' - \omega, r)\alpha^\mu\psi(\omega', r)d\omega'$$

Where $\psi(\omega, r)$, the temporal Fourier transform of $\psi(t, r)$ is given by

$$\psi(\omega, r) = (2\pi)^{-1} \int [u(P, \sigma)a(P, \sigma)exp(iP.r)\delta(\omega - E(P)) + v(P, \sigma)b(P, \sigma)^* exp(-iP.r)\delta(\omega + E(P))]d^3P$$

In our CDRA case, we have to modify this formula slightly. The possible frequencies of the Dirac field are not a continuum $E(P), P \in \mathbb{R}^3$ but rather a discrete set $\omega(mnp) = E(P(mnp))$ and at a given oscillation frequency $\omega(mnp)$, the Dirac field contributes an amount

$$\psi_{mnp}(\omega, r) = \chi_1(mnp, \mathbf{r})\delta(\omega - \omega(mnp))a(mnp)$$

If we consider the corresponding negative frequency terms also (ie, radiation from both electrons and positrons), then the result is

$$\psi_{mnp}(\omega, r) = \chi_1(mnp, \mathbf{r})\delta(\omega - \omega(mnp))a(mnp) + \chi_2(mnp, \mathbf{r})\delta(\omega + \omega(mnp))b(mnp)^*$$

The result of performing the above convolution is then

$$J^\mu(\omega, r) = \chi_1(mnp, r)$$

## 10.12 A brief description of some practical problems in quantum antenna theory

An antenna on the quantum/nano scale can be used to generate a quantum electromagnetic field whose effect on atoms and molecules can be easily studied by means of sensitive quantum measurement apparatus. For example, suppose we are given an ion or a set of $N$ ions enclosed within a cavity. We generate a quantum electromagnetic field within this cavity by inserting a quantum probe carrying current at the quantum scale and excite the quantum electromagnetic field modes within the cavity. This quantum electromagentic field can be made to interact with the ions within the cavity thereby causing transtions of the ion from one energy level to another. By means of a sensitive quantum measurement apparatus, we can determine the relative population of the ions in different energy states and hence use this information to estimate the exact current in the exciting probe. This scheme in other words enables us to estimate signals at the quantum scale by measuring its indirect effect via the quantum electromagnetic field generated by it on ions. Another application of a quantum antenna is to determine the location of atoms in space by transmitting an electromagnetic field generated by electrons and positrons with the quantum antenna to a distant (on the quantum scale) region where it will interact with atoms and ions. The quantum electromagnetic field radiated by such an antenna will be generated using the retarded potentials acting on the surface quantum current density on the antenna surface which is produced by the boundary conditions of the quantum electromagnetic field within the antenna cavity on its surface. By studying the quantum fluctuations of the electromagnetic field scattered by these distant atoms and ions, we can easily design an algorithm for locating the range and bearing of these atoms and ions. In other words, we can solve the quantum direction of arrival estimation problem. The third major application as far as the eye cavity is concerned is to determine the structure of the quantum electromagnetic field generated within this cavity from optical signals coming from outside this cavity and and hence by looking at interactions of this quantum electromagnetic field with the matter within the eye including the eye lens, calculate the mean square quantum fluctuations of the image field formed on the retinal plane. All these applications involve sensitive measurement apparatus and these apparatus are currently in existence.

## 10.13 Fundamental problems in quantum antennas

### 10.13.1 Numerical estimates of field, frequency, energy and power in quantum antenna theory

Consider a cavity resonator of one Angstrom size, ie, a cube with each side of length $a = 10^{-10}m$. The Maxwell equations in such a cube have solutions of the from

$$A_r(t,x,y,z) = \sum_{mnp} c(mnp,t)u_{r,mnp}(x,y,z), r = 1,2,3$$

where $u_{r,mnp}$ are spatial functions obtained by integrating the electric field w.r.t time. These functions are of the form

$$\{cos(m\pi x/a), sin(m\pi x/a)\}\otimes\{cos(n\pi y/a), sin(n\pi y/a)\}\otimes\{cos(p\pi z/a), sin(p\pi z/a)\}$$

multiplied by some constants depending on the indices $(m,n,p)$. We may, without loss of generality, assume that the functions $u_{r,mnp}$ are normalized so that

$$\int_C u_{r,mnp}(\mathbf{r})\bar{u}_{s,m'n'p'}(\mathbf{r})d^3r = \delta_{rs}\delta_{mm'}\delta_{nn'}\delta_{pp'}$$

The dependence of $c(mnp,t)$ on $t$ is $exp(i\omega(mnp)t)$ where $\omega(mnp)$ are the characteristic frequencies of oscillation:

$$\omega(mnp) = (\pi c/a)\sqrt{m^2+n^2+p^2}, m,n,p = 1,2,...$$

which are of the order of magnitude

$$\omega = \pi c/a$$

The electric field is

$$E_r = \partial_t A_r = \sum_{mnp} c(mnp,t)i\omega(mnp)u_{r,mnp}(\mathbf{r})$$

The magnetic field is

$$B = curlA$$

which is of the order of magnitude $|c(mnp,t)|/a$ where by $c(mnp,t)$ we actuall mean its average in a coherent state. The total electric field energy within the cavity $C$ is

$$U_E = (\epsilon_0/2)\int_C |E|^2 d^3r$$

which has components of the order of magnitude

$$\epsilon_0|\omega(mnp)c(mnp,t)|^2a^3 = \epsilon_0\omega(mnp)^2a^3|c(mnp,t)|^2$$

The total magnetic field energy within the cavity is

$$U_B = (2\mu_0)^{-1}\int_C |B|^2 d^3r$$

which is has components of the order of magnitude

$$|c(mnp,t)/a|^2a^3/\mu_0 = |c(mnp,t)|^2a/\mu_0$$

The ration of the orders of magnitude of the electric field energy and the magnetic field energy within the cavity therefore has the order of magnitude

$$U_E/U_B \approx \mu_0\epsilon_0\omega(mnp)^2a^2 \approx \omega^2a^2/c^2 \approx 1$$

as expected. The canonical commutation relations are

$$[A_r(t,\mathbf{r}), \partial_t A_s(t,\mathbf{r}')] = (ih/2\pi)\delta^3(\mathbf{r}-\mathbf{r}')$$

These yield,

$$[c_r(mnp,t), \omega(mnp)c_s(m'n'p',t)^*] = (h/2\pi)\delta_{rs}\delta_{mm'}\delta_{nn'}\delta_{pp'}$$

so that the eigenvalues of $c_r(mnp,t)^*c_r(mnp,t)$ are positive integer multiples of $h/2\pi\omega(mnp)$. This means that the field energy within the cavity when a finite number of modes are excited assumes eigenvalues that are of the same order of magnitude as positive integer multiples of $h\omega/2\pi$ as expected by Planck's quantum theory of radiation. This fact also yields the result that $|c(mnp,t)|$ is of the order of magnitude of $\sqrt{h/(2\pi\omega)}$.

Now we come to the question of computing the order of magnitude of the Poynting vector power flux at a given radial distance $R$ from the quantum cavity antenna caused by the surface current density induced by the magnetic field on on the antenna surface. The magnetic field on the surface and hence the corresponding induced surface current density both have the order of magnitudes of $|c(mnp,t)|/a$ which is of the order $a^{-1}\sqrt{h/\omega}$. Therefore, the far field magnetic vector potential at a distance $R$ from the cavity is of the order of magnitude (use the retarded potential formula) $(a/R)\sqrt{h/\omega}$ and hence the corresponding far field radiated magnetic field is of the order of magnitude $(\omega/c)(a/R)\sqrt{h/\omega}$ while the near field magnetic field is of the order of magnitude $(a/R^2)\sqrt{h/\omega}$. Actually, these expressions for the magnetic field must be multiplied by $\sqrt{N}$ where $N$ is a positive integer corresponding to the largest modal eigenvalue of the operators $(2\pi\omega(mnp)/h)c(mnp,t)^*c(mnp,t)$.

The far field Poynting vector has the order of magnitude of $B^2c/2\mu_0$ which is of the order

$$(c/2\mu_0)(\sqrt{N}.(\omega/c)(a/R).\sqrt{h/\omega})^2$$
$$= (h/2\mu_0)N.(\omega/c)(a^2/R^2)$$

and the total power radiated outward by this quantum antenna in the far field zone is thus of the order of magnitude

$$P = N(h/2\mu_0)(a^2\omega/c)$$

Now we look at the order of magnitude of the power radiated in the far field zone by the Dirac field of electrons and positrons within the cavity. The Dirac equation is

$$[i\gamma^\mu\partial_\mu - m]\psi(x) = 0$$

or more precisely in arbitrary units,

$$[(ih/2\pi)\partial_t - c(\alpha, (-ih/2\pi)\nabla) - \beta mc^2]\psi(x) = 0$$

Here, the appearance of the constants $h, m, c$ is explicitly shown. Now the $|\psi(x)|^2$ is the probability density of the electron which must integrate to unity over the cavity volume. Thus $\psi(x)$ is of the order of magnitude $a^{-3/2}$. The Dirac current density $J^\mu = e\psi^*\gamma^0\gamma^\mu\psi$ has the same order of magnitude as $e|\psi(x)|^2 c$ which is $ec/a^{3/2}$. Therefore the far field magnetic vector potential at a radial distance of $R$ from the cavity is, in accordance with the retarded potential theory of the order

$$(ec/a^{3/2}).(a^3/R) = eca^{3/2}/R$$

The electric field in the far field zone is then of the order

$$E \approx \omega.eca^{3/2}/R$$

where $\omega$ is the characteristic oscilation frequency of the Dirac current. The magnetic field is of the order

$$B \approx a^{-1}.eca^{3/2}/R = ec\sqrt{a}/R$$

If $P$ is the characteristic momentum of the electrons and positrons in a given state, for example $P$ may be the average momentum of an electron in a given state, then according to De-Broglie, $P$ is of the order $h/a$ since $a$ is the order of the electron wavelength. Then the electron energy is of the order

$$E_e = c\sqrt{m^2c^2 + P^2} \approx c\sqrt{m^2c^2 + h^2/a^2}$$

and the characteristic frequency of oscillation of the Dirac wave field is then

$$\omega = E_e/h$$

The Poynting vector corresponding to the power radiated by the Dirac field in the far field zone then has the order of magnitude

$$S \approx c(\epsilon_0 E^2 + B^2/\mu_0) = c^3\epsilon_0\omega^2 ea^3/R^2 + e^2c^3a/\mu_0 R^2$$

and the total power radiated in the far field zone is of the order

$$W = SR^2 = c^3\epsilon_0\omega^2 ea^3 + e^2c^3a/\mu_0$$

### 10.13.2 Controlling the quantum electromagnetic field in a Fermionic coherent state

Aim: The aim of this section is to present a calculation involving the computation of the quantum statistical moments of the electromagnetic field produced by an ensemble of electrons and positrons whose state is specified by a mixed state superposition of Fermionic coherent states. Fermionic coherent states are parameterized by Grassmannian/Fermionic numbers and in order to attach physical signficance to the final results, we must use the Berezin integral for Fermionic variables to determine the above mentioned superposition of Fermionic coherent states. We can incorporate some unknown real parameters into the Berezin linear combination of coherent states and estimate these parameters by minimizing the distance between the average value of the electromagnetic field generated by the Fermions and the desired electromagnetic field pattern. If need be, we may modify this cost function to be minimized by constraining the higher order quantum statistical moments of the generated quantum electromagnetic field to be specified. An example of an application of this circle of ideas is to use a quantum antenna to generate a set of desired spatial patterns at a given set of frequencies.

First consider just a single Fermion specified by the annihilation operator $a$ and the creation operator $a^*$. Thus,

$$a^2 = a^{*2} = 0, aa^* + a^*a = 1$$

Let $\gamma$ be a Grassmannian variable that will be used to specify the coherent state of this Fermion just as a complex number $z$ is used to specify a the coherent state of a single Boson. $\gamma$ anticommutes with itself, with $\gamma^*$ and with $a, a^*$, just as in the Bosonic situation, the complex number $z$ that specifies the coherent state commutes with itself, with $\bar{z}$ and with the Boson creation and annihilation operators:

$$\gamma^2 = 0, \gamma\gamma^* + \gamma^*\gamma - 0, \gamma.a + a.\gamma = 0,$$

$$\gamma^*a + a\gamma^* = 0, \gamma^*a^* + a^*\gamma^* = 0$$

Define now the Fermionic Weyl operator

$$D(\gamma) = exp(\gamma.a^* - a\gamma^*)$$

Clearly, $D(\gamma)$ is a unitary operator since it is the exponential of a skew Hermitian operator. Now,

$$D(\gamma) = 1 + \gamma a^* - a\gamma^* + (1/2)(\gamma a^* - a\gamma^*)^2$$

$$= 1 + \gamma a^* - a\gamma^* - (1/2)(\gamma a^* a\gamma^* + a\gamma^*\gamma a^*)$$

$$= 1 + \gamma a^* - a\gamma^* + (1/2)(\gamma^*\gamma a^* a - \gamma^*\gamma(1 - a^*a))$$

$$= 1 + \gamma a^* - a\gamma^* + \gamma^*\gamma(a^*a - 1/2)$$

Then,

$$aD(\gamma) = a - \gamma aa^* + \gamma^*\gamma a/2 = (1+\gamma^*\gamma/2)a - \gamma aa^*$$

$$D(\gamma)a = a + \gamma a^* a - \gamma^*\gamma a/2 = (1-\gamma^*\gamma/2)a + \gamma a^* a$$

Thus,

$$aD(\gamma) - D(\gamma)a = \gamma^*\gamma a - \gamma$$

However,

$$D(\gamma)\gamma = \gamma - \gamma^*\gamma a$$

$$\gamma.D(\gamma) = \gamma - \gamma^*\gamma a$$

ie,

$$[\gamma, D(\gamma)] = 0$$

Thus,

$$aD(\gamma) - D(\gamma)a = -D(\gamma)\gamma = -\gamma.D(\gamma)$$

These equations can be rearranged as

$$D(\gamma)aD(\gamma)^{-1} = a + \gamma,$$

$$D(\gamma)^{-1}.aD(\gamma) = a - \gamma$$

We define the Fermionic single particle coherent state as

$$|\gamma >= D(-\gamma)|0 >= D(\gamma)^{-1}|0 >$$

where $|0>$, is the vacuum, ie, zero particle state. Then

$$a|\gamma >= aD(\gamma)|0 >= D(\gamma)^{-1}.D(\gamma)a.D(\gamma)^{-1}|0 >= D(\gamma)^{-1}(a+\gamma)|0 >$$

$$= \gamma.D(\gamma)^{-1}|0 >= \gamma|\gamma >$$

This proves the desired property of a coherent state, namely that it should be an eigenvector of the annihilation operator. We observe that

$$D(\gamma)^{-1} = 1 - \gamma a^* + a\gamma^* + \gamma^*\gamma(a^*a - 1/2)$$

and hence

$$|\gamma >= |0 > -\gamma|1 > -(1/2)\gamma^*\gamma|0 >= (1-\gamma^*\gamma/2)|0 > -\gamma|1 > ---(1)$$

From this expression, we can directly verify the coherent state property:

$$a|\gamma >= \gamma.a|1 >= \gamma|0 >,$$

while

$$\gamma|\gamma >= \gamma|0 >$$

since

$$\gamma^2 = 0, \gamma\gamma^*\gamma = -\gamma^*\gamma^2 = 0$$

proving thereby the coherent state property for the state (1).

Now we are in a position to discuss physical implications for Fermionic coherent states. The first observation is that a coherent state is not parametrized by a complex number, it is parametrized by a Fermionic/Grassmannian parameter or a set of anticommuting Grassmannian parameters. Then, if we compute average values of quantities like for example the Dirac four current density in such a state, we will get a Grassmannian number. What physical significance does this have when our averages are not real or complex numbers ? The answer to this question is provided by the Berezin integral: Let $\phi(\gamma, \gamma^*)$ be a function of the Grassmannian parameters $\gamma, \gamma^*$ so that the Berezin integral

$$\rho = \int \phi(\gamma, \gamma^*)|\gamma >< \gamma|d\gamma.d\gamma^*$$

defines a mixed state. Then, the average value of a function $F(a, a^*)$ of the Fermionic operators $a, a^*$ in the state $\rho$ becomes a complex number to which we can attach physical meaning:

$$Tr(\rho.F(a, a^*)) = \int \phi(\gamma, \gamma^*) < \gamma|F(a, a^*)|\gamma > d\gamma.d\gamma^*$$

Another example involving computing average values of the electromagnetic field emitted by a field of electrons and positrons in a given coherent state of the electron-positron field. Let $a_k, k = 1, 2, ...$ denote the annihilation operators of the electrons and positrons after discretizing in momentum space. They satisfy the CAR

$$[a_k, a_m]_+ = 0, [a_k, a_m^*] = \delta_{km}$$

The current density field generated by this field is according to Dirac's theory, a quadratic function of these operators and hence the electromagnetic field generated by this current density according to the retarded potential formula, is also a quadratic function of these operators. We can express this electromagnetic field as

$$F_{\mu\nu}(x) = \sum_{k,m=1}^{N} [G_{\mu\nu}(x, k, m, 1)a_k a_m + \bar{G}_{\mu\nu}(x, k, m, 1)a_m^* a_k^*$$
$$+ G_{\mu\nu}(x, k, m, 2)a_k^* a_m], x \in \mathbb{R}^4$$

This should be a Hermitian operator field and hence

$$\bar{G}_{\mu\nu}(x, k, m, 2) = G_{\mu\nu}(x, m, k, 2)$$

The coherent state of the electrons and positrons is given by

$$|\gamma >= D(\gamma)|0 >, D(\gamma) = \Pi_{k=1}^N exp(\gamma(k)a_k^* - a_k\gamma(k)^*)$$

where $\gamma = ((\gamma(k)))_{k=1}^N$ are Fermionic/Grassmannian parameters and $\gamma(k)$ and $\gamma(k)^*$ anticommute with $\gamma(l), \gamma(l)^*, a_l, a_l^*$ for all $l$. We can write

$$D(\gamma) = \Pi_{k=1}^N (1 + \gamma(k)a_k^* - a_k\gamma(k)^* + \gamma(k)^*\gamma(k)(a_k^* a_k - 1/2))$$

The state of the electrons and positrons is assumed to be given by a Berezin integral based superposition of the coherent states:

$$\rho(\theta) = \int \phi(\gamma, \gamma^*|\theta)|\gamma >< \gamma|d\gamma.d\gamma^*$$

and hence the average electromagnetic field in this state is

$$< F_{\mu\nu}(x) > (\theta) = Tr(\rho(\theta)F_{\mu\nu}(x)) =$$

$$\int \phi(\gamma, \gamma^*|\theta) < \gamma|F_{\mu\nu}(x)|\gamma > d\gamma.d\gamma^*$$

where

$$< \gamma|F_{\mu\nu}(x)|\gamma >=$$

$$\sum_{k,m=1}^{N} [G_{\mu\nu}(x,k,m,1) < \gamma|a_k a_m|\gamma > +\bar{G}_{\mu\nu}(x,k,m,1) < \gamma|a_m^* a_k^*|\gamma > \\ +G_{\mu\nu}(x,k,m,2) < \gamma|a_k^* a_m|\gamma >], x \in \mathbb{R}^4$$

$$= \sum_{k,m=1}^{N} [G_{\mu\nu}(x,k,m,1)\gamma(k)\gamma(m)+\bar{G}_{\mu\nu}(x,k,m,1)\gamma(m)^*\gamma(k)^* \\ +G_{\mu\nu}(x,k,m,2)\gamma(k)^*\gamma(m)], x \in \mathbb{R}^4$$

We can now control the parameter vector $\theta$ so that this average electromagnetic field is as close as possible to a desired electromagnetic field $F_{d\mu\nu}(x)$ over a given space-time region $x \in D$ by minimizing

$$E(\theta) = \int_D | < F_{\mu\nu}(x) > (\theta) - F_{\mu\nu}(x)|^2 d\mu(x)$$

where $\mu(.)$ is a measure on $D$.

Remark 1: More generally, we can compute all the statistical moments of the radiation field

$$Tr(\rho(\theta)F_{\mu_1\nu_1}(x_1)...F_{\mu_k\ni_k}(x_k) >$$

in the superposed coherent state $\rho(\theta)$. This computation will involve determining coherent state expectations such as

$$< \gamma|a_{k_1}^*...a_{k_r}^* a_{s_1}...a_{s_m}|\gamma >$$

and noting that this evaluates to

$$\gamma(k_r)^*...\gamma(k_1)^*\gamma(s_1)...\gamma(s_m)$$

The reference for Fermionic coherent state for us has been the master's thesis by Greplova, title "Fermionic Gaussian States".

Remark 2: From Steven Weinberg's book, "The quantum theory of fields, vol.1", it is known that the free Dirac field can be expanded in terms of momentum-spin space electron annihilation operators $a(P,\sigma)$ and positron creation operators $b(P,\sigma)^*$ which satisfy the CAR (canonical anticommutation relations)

$$[a(P,\sigma),a(P',\sigma')^*]_+ = \delta^3(P-P')\delta_{\sigma,\sigma'},$$

$$[b(P,\sigma),b(P',\sigma')^*]_+ = \delta^3(P-P')\delta_{\sigma,\sigma'}$$

and all the other anticommutators evaluating to zero. The second quantized Dirac wave field is then the solution to Dirac's relativistic wave equation and is given by

$$\psi(x) = \psi(t,r) = \int [a(P,\sigma)u(P,\sigma)exp(-ip.x) + b(P,\sigma)^*v(P,\sigma)exp(ip.x)]d^3P$$

where

$$p^0 = E(P) = \sqrt{m^2+P^2}$$

The Dirac current density operator field is then

$$J^\mu(x) = -e\psi(x)^*\gamma^0\gamma^\mu\psi(x)$$

and it is evident that this can be expressed as a linear combination of the quadratic operators

$$a(P,\sigma)^*a(P,\sigma'), a(P,\sigma)^*b(P,\sigma')^*, b(P,\sigma)a(P,\sigma')^*, b(P,\sigma)b(P',\sigma')^*$$

Thus, using the retarded potential formula for the Maxwell equations in the form

$$A^\mu(x) = \int G(x-x')J^\mu(x')d^4x'$$

it is evident that once again $A^\mu(x)$ is expressible as a linear combination of the above quadratic operators. After discretizing the integrals in 3-momentum space, we then club all the electron and positron annihilation operators into one set $\{a_k\}$ and their adjoints into $\{a_k^*\}$ and then use the above coherent state formalism of Greplova to determine the quantum averages of the electromagnetic field.

The ultimate aim of all these computations can be formulated in very simple terms as an optimization problem: Design the control parameters $\theta$ or the control classical fields to a quantum antenna so that the error energy between the average value of the quantum electromagnetic field produced by the quantum antenna and the desired classical electromangetic field pattern is a minimum subject to the constraint that the second order central moments of the quantum electromagnetic field (ie variance of fluctuations) is smaller than a given threshold.

Remark:More generally, we can control the wave function operator of the Dirac field of electrons and positrons as well as the Maxwell photon field operators within the cavity resonator antenna by introducing classical control current and electromagnetic field sources into the cavity. The quantum cavity photon and electron-positron fields will then be expressible in terms of the free quantum fields plus additional perturbation terms involving the classical current and field sources. Once this is done, we can in principle calculate the far field antenna pattern produced by the cavity surface currents induced by the tangential components of the quantum magnetic field operators as well as that produced by the Dirac field of electrons and positrons and then design these classical control fields so that the far field quantum Poynting radiation pattern has a mean value and correlations in a given quantum coherent state of the photons and electrons-positrons within the cavity as close as possible to specified values.

## 10.14 Modal expansions of the Maxwell and Dirac field within a rectangular quantum antenna

Approximate analysis of a rectangular quantum antenna

The quantum antenna is assumed to be the cuboid region $[0, a] \times [0, b] \times [0, d]$. This rectangular cavity is assumed to comprise of photons, electrons and positrons. The exact equations governing the quantum fields corresponding to these particles are (a) The Maxwell equations for the four vector potential driven by the Dirac field current and (b) The Dirac field equations driven by an interaction between the Dirac field and the Maxwell field four vector potential. These exact field equations are:

$$\Box A_\mu(x) = \mu_0 e\psi^*(x)\alpha^\mu\psi(x), x = (t, r) - -(1)$$

$$((\alpha, -i\nabla) + \beta m_0)\psi(t, r) + eA_\mu(x)\alpha^\mu\psi(x) = i\partial_t\psi(t, r) - - - (2)$$

where

$$\alpha^\mu = \gamma^0\gamma^\mu, \beta = \gamma^0$$

and $\gamma^\mu$ are the Dirac Gamma matrices. Note that $(\gamma^0)^2 = I_4$ and hence $\alpha^0 = I_4$. The boundary conditions under which we need to solve these Maxwell-Dirac equations are that the Dirac operator wave field $\psi(x)$, the tangential components of the electric field $F_{0r} = A_{r,0} - A_{),r}, r = 1, 2, 3$ and the normal components of the magnetic field $F_{rs} = A_{s,r} - A_{r,s}, 1 \leq r < s \leq 3$ must vanish on the boundaries of the cavity. In particular, the freed Dirac field must have an expansion

$$\psi^{(0)}(t, r) = \sum_{mnp} \mathbf{c}(mnp, t)u_{mnp}(r)$$

where $m, n, p$ run over positive integers and

$$u_{mnp}(r) = (2\sqrt{2}/\sqrt{abd})sin(m\pi x/a)sin(n\pi y/b)sin(p\pi z/d)$$

Substituting this into the free Dirac equation, ie, without any electromagnetic interactions, we get

$$\sum_{mnp}(i\partial_t c(mnp,t))u_{mnp}(r) =$$

$$((\alpha,-i\nabla)+\beta m_0).\sum_{mnp}\mathbf{c}(mnp,t)u_{mnp}(r)$$

from which we derive on taking the inner products on both sides with $u_{kls}(r)$ and using the orthonormality of this set of functions over the cavity volume, ie,

$$< u_{kls}, u_{mnp} >= \int_B u_{kls}(r)u_{mnp}(r)d^3r = \delta_{km}\delta_{ln}\delta_{sp}$$

where $B$ is the cavity volume

$$B = [0,a]\times[0,b]\times[0,d],$$

the following sequence of differential equations

$$i\partial_t c(kls,t) = \sum_{mnp}[< u_{kls}, -i\partial_x u_{mnp} > \alpha_1 c(mnp,t)$$

$$+ < u_{kls}, -i\partial_y u_{mnp} > \alpha_2 c(mnp,t) + < u_{kls}, -i\partial_z u_{mnp} > \alpha_3 c(mnp,t)] + m_0\beta c(kls,t)$$

Now we evaluate

$$< u_{kls}, -i\partial_x u_{mnp} >= -i\delta_{ln}\delta_{sp}(m\pi/a)(\int_0^a (2/a)sin(k\pi x/a)cos(m\pi x/a)dx$$
$$= a_1(k,m)\delta_{ln}\delta_{sp}$$

where

$$a_1(k,m) = (-2im\pi/a^2)\int_0^a sin(k\pi x/a).cos(m\pi x/a)dx$$

Likewise,

$$< u_{kls}, -i\partial_y u_{mnp} >= a_2(l,n)\delta_{km}\delta_{sp},$$

and

$$< u_{kls}, -i\partial_z u_{mnp} >= a_3(s,p)\delta_{km}\delta_{ln}$$

Combining all these equations gives us finally,

$$i\partial_t c(kls,t) = \sum_m a_1(k,m)\alpha_1 c(mls,t)$$

$$+\sum_n a_2(l,n)\alpha_2 c(kns,t) + \sum_p a_3(s,p)\alpha_3 c(klp,t)] + m_0\beta c(kls,t)$$

Note that $\alpha_1, \alpha_2, \alpha_3, \beta$ are $4\times4$ Hermitian matrices while $c(mnp,t)$ is a $4\times1$ complex vector. Arranging the $4\times1$ vectors $c(mnp,t), m,n,p \geq 1$ in lexicographic order to give an infinite vector $\mathbf{c}(t)$ and likewise defining a block structured infinite dimensional Dirac Hamiltonian matrix $\mathbf{H}_0$ by

$$\mathbf{H}_0 = \sum_{klsm} a_1(k,m)(\mathbf{I}_4 \otimes \mathbf{e}(kls))\alpha_1(\mathbf{I}_4 \otimes \mathbf{e}(mls)^T)$$

$$+\sum_{lksn} a_2(l,n)(\mathbf{I}_4 \otimes \mathbf{e}(kls))\alpha_2(\mathbf{I}_4 \otimes \mathbf{e}(kns)^T)$$

$$+\sum_{klsp} a_3(s,p)(\mathbf{I}_4 \otimes \mathbf{e}(kls))\alpha_3(\mathbf{I}_4 \otimes \mathbf{e}(klp)^T)$$

$$+m_0\beta \otimes \mathbf{I}$$

where we may choose $\mathbf{e}(mnp), m, n, p \geq 1$ as any orthonormal basis for $l^2(\mathbb{Z}_+)$, the Hilbert space of all one sided square summable infinite sequences and define

$$\mathbf{c}(t) = \sum_{mnp} c(mnp,t)\mathbf{e}(mnp)$$

By orthonormal, we mean that

$$\mathbf{e}(kls)^T\mathbf{e}(mnp) = \delta_{km}\delta_{ln}\delta_{sp}$$

Thus the free Dirac equation in the RDRA has been put in "Standard" block matrix form:

$$i\frac{d\mathbf{c}(t)}{dt} = \mathbf{H}_0\mathbf{c}(t)$$

the general solution to which can be expressed as

$$\mathbf{c}(t) = \sum_n d(n).\mathbf{c}_n exp(-iE(n)t)$$

where $\mathbf{c}_n, n \geq 1$ form an orthonormal basis for $l^2(\mathbb{Z}_+)$ and the $d(n)'s$ are arbitrary complex numbers such that

$$\sum_n |d(n)|^2 = 1$$

$E(n)'s$ are the (energy) eigenvalues of the infinite dimensional Hermitian $H_0$:

$$det(\mathbf{H}_0 - E(n)\mathbf{I}) = 0$$

The average energy of the free Dirac field of electrons and positrons within the cavity is then

$$< \mathbf{c}(t), \mathbf{H}_0\mathbf{c}(t) >= \sum_n E(n)d(n)^*d(n)$$

It is easy to see as in the case of the Dirac equation in free space that if $E(n)$ is an eigenvalue of $\mathbf{H}_0$ then so is $-E(n)$ where the $E(n)'s$ may be taken as positive, Hence if $\mathbf{c}_{en}$ is an eigenvector of $\mathbf{H}_0$ corresponding to the eigenvalue

$E(n)$ and $\mathbf{c}_{pn}$ is an eigenvector corresponding to the eigenvalue $-E(n)$, then the solution can be expressed as

$$\mathbf{c}(t) = \sum_n [d_e(n)\mathbf{c}_{en} exp(-iE(n)t) + d_p(n)^* \mathbf{c}_{pn} exp(iE(n)t)]$$

Therefore, it is plausible in the second quantized theory, to look upon the $d_e(n)'s$ as annihilation operators of the electrons and the $d_p(n)^*$'s as the creation operators of the positrons. The actual Dirac wave function $\psi(t,r)$ in the absence of electromagnetic interactions is then

$$\psi(t,r) = \sum_{kmnp} [d_e(k)c_{ek}(mnp)u_{mnp}(r)exp(-iE(k)t) +$$
$$d_p(k)^* c_{pk}(mnp)u_{mnp}(r)exp(iE(k)t)] --- (3)$$

A simple calculation then shows that the second quantized Hamiltonian of the free Dirac field of electrons and positrons within the cavity is given by

$$H_{D0} = \int_B \psi(t,r)^*((\alpha, -i\nabla) + \beta m)\psi(t,r)d^3r$$

$$= \int_B \psi(t,r)^* i\partial_t \psi(t,r)d^3r$$

$$= \sum_k E(k)(d_e(k)^* d_e(k) - d_p(k)d_p(k)^*)$$

Now from the basic anticommuation relations for the Dirac field, we have

$$\{\psi(t,r), \psi(t,r')^*\} = \delta^3(r - r')I$$

and this immediately implies the following anticommutation relations for the electron and positron creation and annihilation operators:

$$\{d_e(k), d_e(m)^*\} = \delta_{km}, \{d_p(k), d_p(m)^*\} = \delta_{km}$$

with all the other anticommutators vanishing. This completes our description of the free Dirac field of electrons and positrons within the RDRA. Using these anticommutation relations, we immediately get that the total second quantized Hamiltonian of the free Dirac field in the cavity can equivalently be expressed as

$$H_{D0} = \sum_k E(k)(d_e(k)^* d_e(k) + d_p(k)^* d_p(k))$$

namely, the sum of the total electron and positron energies. Likewise, when we solve the free Maxwell equations within the cavity after incorporating the appropriate boundary conditions, we get that the scalar potential is zero since there are no charges while the magnetic vector potential admits an expansion obtained from

$$\mathbf{E} = -\partial_t \mathbf{A}$$

as

$$\mathbf{A}(t,r) = \sum_k [b(k)\mathbf{w}_k(r)exp(-i\omega(k)t) + b(k)^*\mathbf{w}_k(r)^*exp(i\omega(k)t)] - - - (4)$$

where now $\mathbf{w}_k(r)$ has three components that are calculated from the expansion of $E_z$ and the relationship between the transverse and longitudinal components of the electric field within the cavity. We note that the third, ie, $z$ component of $\mathbf{w}_k(r)$ where the index $k$ is identified with the modal triplet $(mnp)$ is proportional to $sin(m\pi x/a)sin(n\pi y/b)cos(p\pi z/d)$ in view of the boundary conditions on the electric field an the fact that each mode of the magnetic vector potential is proportional to the electric field ($-j\omega\mathbf{A} = \mathbf{E}$). $b(k) = b(mnp)$ is identified with a photon annihilation operator while $b(k)^*$ with a photon creation operator. They satisfy the canonical commutation relations

$$[b(k), b(m)^*] = \delta_{km}$$

Formally, we can compute both the free Dirac current density $\psi(t,r)^*\alpha^\mu\psi(t,r)$ of electrons and positrons within the cavity as well as the surface current density on the RDRA walls induced by the tangential components of the quantum magnetic field $\mathbf{B} = curl\mathbf{A}$ and obtain the far field radiation pattern generated by both of these cavity current components. Obviously, this far field radiation pattern will have its first component being a quadratic form in the electron-positron creation and annihilation operators $d_e(k), d_p(k), d_e(k)^*, d_p(k)^*$ while the second component will be linear in the photon creation-annihilation operators $b(k), b(k)^*$ and therefore, in principle, we can compute all the statistical moments of the radiation field in a joint coherent state of the photons, electrons and positrons. However, this picutre of the far field quantum radiation pattern is incomplete because it does not take into account the cavity current density terms caused by perturbation in the Dirac wave field due to interaction with the photons and it does not also take into account the cavity surface current density terms caused by perturbation in the Maxwell field caused by its interaction with the Dirac field. We shall now indicate an approximate first order calculation by which these extra correction terms may be obtained due interactions between the Maxwell field and the Dirac field.

We denote the free Dirac field within the cavity derived above by $\psi^{(0)}(t,r)$ and the corresponding momentum space wave function $c(mnp,t)$ by $c^{(0)}(mnp,t)$. Likewise, we denote the free Maxwell field within the cavity by $\mathbf{A}^{(0)}$. Let $\delta\mathbf{A}$ denote the perturbation to the Maxwell field caused by the Dirac current and $\delta\psi, \delta c(mnp,t)$ the perturbation to the Dirac field caused by the Maxwell current. Then, clearly if $S(x-y)$ denotes the electron propagator and $D(x-y)$ the photon propagator, we have using (1) and (2), approximately,

$$\delta A^\mu(t,r) = \mu_0 e \int D(t-t', r-r')\psi^{(0)*}(t',r')\alpha^\mu\psi^{(0)}(t',r')dt'd^3r'$$

$$\delta\psi(t,r) = e \int S(t-t', r-r')A^{(0)}_\mu(t',r')\alpha^\mu\psi^{(0)}(t',r')dt'd^3r'$$

where we substitute for $\psi^{(0)}$ and $A_\mu^{(0)}$ the expressions given in (3) and (4). Then,

$$\psi^{(0)*}\alpha^\mu\psi^{(0)}(t,r) =$$

$$\sum_{kmnpk'm'n'p'} [d_e(k)^*\bar{c}_{ek}(mnp)exp(iE(k)t) + d_p(k)\bar{c}_{pk}(mnp)exp(-iE(k)t)].\alpha^\mu.$$

$$.[d_e(k')c_{ek'}(m'n'p')exp(-iE(k')t)+d_p(k')^*c_{pk'}(m'n'p')exp(iE(k')t)]u_{mnp}(r)u_{m'n'p'}(r)$$

$$=\sum[d_e(k)^*d_e(k')\bar{c}_{ek}(mnp)\alpha^\mu c_{ek'}(m'n'p')exp(i(E(k)-E(k'))t)u_{mnp}(r)u_{m'n'p'}(r)]$$

$$+\sum[d_e(k)^*d_p(k')^*\bar{c}_{ek}(mnp)\alpha^\mu c_{pk'}(m'n'p')exp(i(E(k)+E(k'))t)u_{mnp}(r)u_{m'n'p'}(r)]$$

$$+\sum[d_p(k)d_e(k')\bar{c}_{pk}(mnp)\alpha^\mu c_{ek'}(m'n'p')exp(-i(E(k)+E(k'))t)u_{mnp}(r)u_{m'n'p'}(r)]$$

$$+\sum[d_p(k)d_p(k')^*\bar{c}_{pk}(mnp)\alpha^\mu c_{ek'}(m'n'p')exp(-i(E(k)-E(k'))t)u_{mnp}(r)u_{m'n'p'}(r)]$$

We see that the frequencies of the Dirac current that generate the perturbation to the quantum electromagnetic field are $E(k) \pm E(k'), k, k' = 1, 2, ...$ or more precisely, these divided by Planck's constant. Here $E(k)$ was obtained by solving the free Dirac eigenvalue equation inside the rectangular cavity with zero boundary conditions. The $E(k)'s$ were obtained as the eigenvalues of the Dirac Hamiltonian. From basic principles of special relativity, it is easy to see that these $E(k)'s$ are of the order

$$c\sqrt{m_0^2c^2 + P^2}$$

where

$$P^2 = (h/2\pi)^2((m\pi/a)^2 + (n\pi/b)^2 + (p\pi/d)^2)$$

with $m, n, p$ being positive integers determined by the mode of oscillation of the field within the cavity. Now this current is of the general form

$$\psi^{(0)*}\alpha^\mu\psi^{(0)}(t,r)$$

$$\sum_{k,k'}[d(k)^*d(k')f_{kk'}(t,r) + d(k)d(k')g_{kk'}(t,r) + d(k)^*d(k')^*\bar{g}_{k'}(t,r)]$$

where the $d(k)'s$ are the annihilation operators of the electrons and positrons and their adjoints $d(k)^*$ are the corresponding creation operators. The functions $f_{kk'}, g_{kk'}$ are constructed by superposing $exp(\pm i(E(k)\pm E(k'))u_{mnp}(r)u_{m'n'p'}(r)$ and these components are easily seen to be expressible as superpositions of space-time sinusoids with the temporal frequencies being $E(k) \pm E(k')$ or their negatives and the spatial frequencies, ie, wave-numbers being
$m\pi/a, n\pi/b, p\pi/d, m'\pi/a, n'\pi/b, p'\pi/d.$
Let

$$J^\mu(k) = e\int \psi^{(0)*}(x)\alpha^\mu\psi^{(0)}(x)exp(-ik.x)d^4x$$

$$= \int \psi^{(0)*}(t,r)\alpha^\mu\psi^{(0)}(t,r)exp(-i(k^0t - K.r))dtd^3r$$

where

$$k = (k^\mu) = (k^0, K)$$

denote the space-time four dimensional Fourier transform of the unperturbed Dirac four current density. Then, we can write down the space-time Fourier transform of the correction $\delta A^\mu(x), x = (t, r)$ to the electromagnetic four potential caused by this Dirac current as

$$\delta A^\mu(k) = \int \delta A^\mu(x).exp(-ik.x)d^4x$$

$$= \mu_0 J^\mu(k)/k^2, k^2 = k_\mu k^\mu = (k^0)^2 - |K|^2$$

in units where $c = 1$. It should be noted that by the convolution theorem for Fourier transforms, if $\psi^{(0)}(k)$ denotes the space-time Fourier transform of $\psi^{(0)}(x)$, then

$$J^\mu(k) = (2\pi)^{-4} \int \psi^{(0)*}(k' - k)\alpha^\mu \psi^{(0)}(k')d^4k'$$

and hence, the perturbation to the electromagnetic four potential in the space-time Fourier domain, ie, in four momentum space of the photon can be expressed as

$$\delta A^\mu(k) = (\mu_0/(2\pi)^4 k^2) \int \psi^{(0)*}(k' - k)\alpha^\mu \psi^{(0)}(k')d^4k'$$

Remark: The unperturbed electromagnetic field is in the Coulomb gauge, ie, $div\mathbf{A}^{(0)} = 0$ and also since there is no charge/current for the unperturbed field, the unperturbed electric scalar potential is a matter field which is identically zero, ie, $A^{(0)0} = 0$. Hence, we are guaranteed that the unperturbed electromagnetic potentials also satisfy the Lorentz gauge conditions, ie, $div\mathbf{A}^{(0)} + \partial_t A^{(0)0} = 0$. This means that while computing the perturbations to the electromagnetic potentials caused by currents coming from the Dirac field, we can safely work in the Loretnz gauge.

Likewise, the change in the Dirac field caused by interaction with the electromagnetic field within the cavity is given upto first order perturbation theory by

$$\delta\psi(x) = \delta\psi(x) = e \int S(x - x')A^{(0)}_\mu(x')\alpha^\mu \psi^{(0)}(x')d^4x'$$

$$= e \int S(x - x')A^{(0)}_r(x')\alpha^r \psi^{(0)}(x')d^4x'$$

$$= -e \int S(x - x')(\alpha, \mathbf{A}^{(0)}(x'))\psi^{(0)}(x')d^4x'$$

$$= -e \int S(t-t', r-r') \sum_k [b(k)(\alpha, \mathbf{w}_k(r'))exp(-i\omega(k)t') + b(k)^*(\alpha, \mathbf{w}_k(r')^*)exp(i\omega(k)t')].$$

$$.[\sum_{kmnp}[d_e(k)c_{ek}(mnp)u_{mnp}(r')exp(-iE(k)t')+d_p(k)^*c_{pk}(mnp)u_{mnp}(r')exp(iE(k)t')]dt'd^3r'$$

$$= -e\sum_{kk'mnp} b(k)d_e(k')\int S(t-t',r-r')(\alpha,\mathbf{w}_k(r'))c_{ek'}(mnp)u_{mnp}(r')exp(-i(\omega(k)+E(k'))t')dt'd^3r'$$

$$-e\sum_{kk'mnp} b(k)d_p(k')^*\int S(t-t',r-r')(\alpha,\mathbf{w}_k(r')^*)c_{pk'}(mnp)u_{mnp}(r')exp(-i(\omega(k)-E(k'))t')dt'd^3r'$$

$$-e\sum_{kk'mnp} b(k)^*d_e(k')\int S(t-t',r-r')(\alpha,\mathbf{w}_k(r'))c_{ek'}(mnp)u_{mnp}(r')exp(i(\omega(k)-E(k'))t')dt'd^3r'$$

$$-e\sum_{kk'mnp} b(k)^*d_p(k')^*\int S(t-t',r-r')(\alpha,\mathbf{w}_k(r')^*)c_{pk'}(mnp)u_{mnp}(r')exp(i(\omega(k)+E(k'))t')dt'd^3r'$$

From this expression, it is clear that the characteristic frequencies of the interaction term between the electromagnetic potentials and the Dirac field and hence the characteristic frequencies of the perturbation in the Dirac field caused by electromagnetic interaction are $\pm\omega(k)\pm E(k')$. In terms of the compact notation introduced above, namely using the same symbol $d(k)$ for both electron and positron annihilation operators and likewise $d(k)^*$ for both electron and positron creation operators, we can write

$$\delta\psi(x) = \int S(t-t',r-r')[\sum b(k)d(k')h_{1kk'}(t',r') + b(k)d(k')^*h_{2kk'}(t',r')+$$

$$+b(k)^*d(k')h_{3kk'}(t',r') + b(k)^*d(k')^*h_{4kk'}(t',r')]dt'd^3r'$$

where the functions $h_{mkk'}(t,r)$ are built by superposing the functions $exp(i\pm(\omega(k)\pm E(k'))t)(\alpha,w_k(r))c_{k'}(mnp)u_{mnp}(r)$ and the same expression with $w_k(r)$ replaced by its complex conjugate $w_k(r)^*$. Here, the symbol $c_{k'}(mnp)$ stands for either $c_{ek'}(mnp)$ or $c_{pk'}(mnp)$.

In particular this expression shows that the perturbation to the Dirac field caused by electromagnetic interactions have frequencies $\pm\omega(k)\pm E(k')$, namely linear combinations of the unperturbed electromagnetic characteristic frequencies and the unperturbed Dirac characteristic frequencies. This represents a new feature of our model.

Before proceeding further, observe that we can write in the four dimensional momentum/space-time frequency domain,

$$\delta\psi(k) = S(k)\mathcal{F}(eA_\mu^{(0)}\alpha^\mu\psi^{(0)})(k)$$

where

$$S(k) = (k^0 - (\alpha,\mathbf{K}) - \beta m_0 + i0)^{-1}$$

is the electron propagator in the four momentum domain $k = (k^\mu) = (k^0,\mathbf{K})$ and

Control of the quantum electromagnetic field and the Dirac field of electrons and positrons within the rectangular cavity by means of a classical electromagetic field coming from a laser source connected to the cavity plus a classical current source coming from a probe inserted into the cavity: Let $A_\mu^c(x)$ denote the classical electromagnetic four potential from the laser and $J_\mu^c(x)$ the classical current density coming from the probe insertion. The relevant equations are

$$\Box \mathbf{A}_\mu = -e\mu_0\psi^*\alpha_\mu\psi + \mu_0\mathbf{J}_\mu^c,$$

$$((\alpha, -i\nabla) + \beta m)\psi = [-e(\alpha, \mathbf{A}) - e(\alpha, \mathbf{A}^c)]\psi$$

The first order perturbative solution to these equations is with $x = (t, \mathbf{r})$,

$$\psi(x) = \psi^{(0)}(x) + \delta\psi(x),$$

$$A^r(x) = A^{r(0)}(x) + \delta A^r(x), r = 1, 2, 3$$

where

$$\psi^{(0)}(x) =$$

$$\sum_{kmnp}[d_e(k)c_{ek}(mnp)u_{mnp}(r)exp(-iE(k)t) + d_p(k)^* c_{pk}(mnp)u_{mnp}(r)exp(iE(k)t)]$$

$$A^{r(0)}(x) = \sum_k [b(k)w_k^r(\mathbf{r})exp(-i\omega(k)t) + b(k)^*\bar{w}_k^r(\mathbf{r})exp(i\omega(k)t)]$$

$$\delta\psi(x) =$$

$$-e\int S_e(x-y)[(\alpha, \mathbf{A}^{(0)}(y)) + (\alpha, \mathbf{A}^c(y))]\psi^{(0)}(y)d^4y$$

$$= -e\int S_e(x-y)[\alpha^r A_r^{(0)}(y) + \alpha^r A_r^c(y)]\psi^{(0)}(y)d^4y$$

$$= \delta\psi_1(x) + \delta\psi_{ctr}(x),$$

$$\delta A_r(x) = -e\mu_0\int D(x-y)(\psi^*\alpha_r\psi)(y)d^4y + \mu_0\int D(x-y)J_r^c(y)d^4y$$

$$\approx -e\mu_0\int D(x-y)(\psi^{(0)*}\alpha_r\psi^{(0)})(y)d^4y + \mu_0\int D(x-y)J_r^c(y)d^4y$$

where the classically controllable part of the Dirac field is

$$\delta\psi_{ctr}(x) = -e\int S_e(x-y)\alpha^r A_r^c(y)\psi^{(0)}(y)d^4y$$

and this component contains a classical field component $A_r^c$ and a quantum field component $\psi^{(0)}$, while the part of the Dirac field perturbation that is not controllable is

$$\delta\psi_1(x) = -e\int S_e(x-y)\alpha^r A_r^{(0)}(y)\psi^{(0)}(y)d^4y$$

On the other hand, the controllable part of the electromagnetic field is purely classical:

$$\delta A_{r,ctr}(x) = \mu_0 \int D(x-y) J_r^c(y) d^4y$$

If we go one step further in the perturbation series, then we get an additional term in the controllable part of the electromagnetic field so that the above equation gets modified to:

$$\delta A_{r,ctr}(x) = \mu_0 \int D(x-y) J_r^c(y) d^4y$$

$$-e\mu_0 \int D(x-y)(\delta\psi_{ctr}(y)^* \alpha_r(\psi^{(0)} + \delta\psi_1)(y) d^4y$$

$$-e\mu_0 \int D(x-y)(\psi^{(0)*} + \delta\psi_1^*)(y)\alpha_r \delta\psi_{ctr}(y) d^4y$$

Note that in this analysis, the perturbation parameter is the electron charge $e$ and if we neglect $O(e^2)$ terms, then the above expression for the controllable part of the electromagnetic field simplifies to

$$\delta A_{r,ctr}(x) = \mu_0 \int D(x-y) J_r^c(y) d^4y +$$

$$-e\mu_0 \int D(x-y) \delta\psi_{ctr}(y)^* \alpha_r \psi^{(0)}(y) d^4y$$

$$-e\mu_0 \int D(x-y) \psi^{(0)*}(y) \alpha_r \delta\psi_{ctr}(y) d^4y$$

In the particular case of the rdra considered here, we find that the controllable part of the Dirac field has the expansion

$$\delta\psi_{ctr}(x) = -e \int S_e(x-y) \alpha^r A_r^c(y) \psi^{(0)}(y) d^4y$$

$$= -e \int S_e(t-t', r-r') \alpha^r A_r^c(t', r') [\sum_{kmnp} [d_e(k) c_{ek}(mnp) u_{mnp}(r') exp(-iE(k)t') +$$

$$d_p(k)^* c_{pk}(mnp) u_{mnp}(r') exp(iE(k)t')] dt' d^3r'$$

Now define the following Fourier components of the control classical laser generated electromagnetic field w.r.t the cavity boundary conditions and the energy spectrum of the free Dirac field in the cavity after:

$$\int A_r^c(t', r') u_{mnp}(r') exp(-iK.r') exp(i\omega t') d^3r' dt' = C_{A,r}(\omega, K|m, n, p)$$

Then, we can express the above controllabe part of the Dirac field in the following form in the spatio-temporal Fourier domain:

$$\int \delta\psi_{ctr}(t, r).exp(i(\omega t - K.r) dt d^3r =$$

$$\sum_{mnpk}[-ed_e(k)S(\omega,K)c_{ek}(mnp)\alpha^r[C_r(\omega-E(k),K|mnp)$$

$$-ed_p(k)^*S(\omega,K)c_{pk}(mnp)\alpha^r C_r(\omega+E(k),K|mnp)]$$

$$=-eS(\omega,K)\sum_{mnpk}[d_e(k)c_{ek}(mnp)\alpha^r C_r(\omega-E(k),K|mnp)+$$

$$d_p(k)^*c_{pk}(mnp)\alpha^r C_r(\omega+E(k)|mnp)]$$

The controllable part of the Dirac four current density is then given upto first order perturbation terms by $(x=(t,r))$

$$\delta J^\mu(t,r)=-e\psi^{(0)*}(x)\alpha^\mu\delta\psi_{ctr}(x)-e\delta\psi_{ctr}(x)^*\alpha^\mu\psi^{(0)}(x)$$

and it is immediately clear from the above expression that the far field radiated electromagnetic potential generated by this controllable current field can be expressed in the form

$$\delta A^\mu_R(t,r)=\int D(t-t',r-r')\delta J^\mu(t',r')dt'd^3r'$$

$$=\sum_{mnpkrm'n'k's} d_e(k)^*d_e(k')\int C_r(\omega-E(k),K|mnp)\bar{C}_s(-\omega'-E(k'),K'|m'n'p')$$

$$\times F^\mu rs(t,r|\omega,K,\omega',K',mnpk,m'n'p'k')d\omega d^3Kd\omega'd^3K'$$

plus three other similar terms involving $d_e(k)^*d_p(k')^*, d_p(k)d_e(k'), d_p(k)d_p(k')^*$. In compact notation, the expected value of this controllable far field pattern can be expressed as a Hermitian quadratic form in the complex numbers $C_r(\omega,K|mnp)$, $\omega\in\mathbb{R}, K\in\mathbb{R}^3, m,n,p\in\mathbb{Z}_+$. These complex numbers are controllable since they represent in some sense the spatio-temporal components of the Fourier components of the classical control electromagnetic field $A^c_\mu$.

For Product Safety Concerns and Information please contact our EU
representative GPSR@taylorandfrancis.com
Taylor & Francis Verlag GmbH, Kaufingerstraße 24, 80331 München, Germany

www.ingramcontent.com/pod-product-compliance
Lightning Source LLC
Chambersburg PA
CBHW052117230326
41598CB00080B/3785
*9781032384177*